Olfaction and Taste V

INTERNATIONAL COMMISSION

Dr. Lloyd Beidler,
Department of Biological Sciences,
The Florida State University,
Tallahassee, Florida 32306

Dr. Derek Denton,
Howard Florey Institute of
Experimental Physiology & Medicine,
University of Melbourne,
Parkville, Vic. 3052, Australia.

Professor Jacques Le Magnen,
College de France,
11 Place Marcelin Berthelot,
75231 Paris Cedex 05, France.

Dr. Carl Pfaffmann, (Chairman)
Vice President,
The Rockefeller University,
New York, N.Y. 10021

Dr. Masayasu Sato,
Institute of Physiology,
College of Medicine,
Kumamoto University,
Kumamoto, Japan.

Dr. Dietrich Schneider,
Max-Planck-Institut fur
Verhaltensphysiologie,
8131 Seewiesen,
Germany.

Professor Yngve Zotterman,
Office of President,
International Union of
Physiological Sciences,
Wenner-Gren Center,
Sveavagen 166,
Stockholm S0113 46,
Sweden.

LOCAL ORGANIZING COMMITTEE

Dr. Derek Denton (Chairman)
Dr. John Coghlan
Professor Ian Darian-Smith
Mr. James Carroll
Dr. Richard Weisinger
Miss Suzanne Abraham
Mrs. Kaye Thorpe

Olfaction
and Taste

V

PROCEEDINGS OF THE FIFTH INTERNATIONAL SYMPOSIUM
held at the Howard Florey Institute of
Experimental Physiology & Medicine
University of Melbourne, Australia
OCTOBER, 1974

EDITED BY

Derek A. Denton

John P. Coghlan

Howard Florey Institute
University of Melbourne
Melbourne, Australia

ACADEMIC PRESS, INC.   New York   San Francisco   London   1975
A Subsidiary of Harcourt Brace Jovanovich, Publishers

ACADEMIC PRESS, INC.
111 Fifth Avenue, New York, New York 10003

United Kingdom Edition published by
ACADEMIC PRESS, INC. (LONDON) LTD.
24/28 Oval Road, London NW1

Library of Congress Cataloging in Publication Data

International Symposium on Olfaction and Taste, 5th,
    Howard Florey Institute of Experimental Physiology
    & Medicine, 1974.
    Olfaction and taste V.
    Includes index.
    1.   Smell—Congresses.   2.   Taste—Congresses.
I.   Denton, Derek A.   II.   Coghlan, John P.   III.   Howard
Florey Institute of Experimental Physiology and Medicine.
IV.   Title.   [DNLM: 1.   Smell—Congresses.   2.   Taste—
Congresses.   W3 OL45]
QP455.I57 1974   612'.86   75-14069
ISBN 0—12—209750—5

CONTENTS

CONTENTS

INVITED LECTURE

MARKET PLACE SYMPOSIUM

CONTENTS

GENERAL OLFACTION
Chairman: W. Whitten

CHEMORECEPTION IN INSECTS
Chairman: J. Boeckh

CONTENTS

OLFACTION IN BIRDS
Chairmen: J. Le Magnen and F. Bell
Moderator: B. Wenzel

OLFACTION AND PHEROMONES IN ANIMAL BEHAVIOUR
Chairmen: R. Bradley and J. Coghlan

CONTENTS

PARTICIPANTS

Abraham, S. Howard Florey Institute of Experimental Physiology and Medicine, University of Melbourne, Parkville, Vic. 3052, Australia.

Akaike, N. Department of Physiology, Kumamoto University Medical School, Kumamoto, Japan.

Amoore, J. Western Regional Research Laboratory, U.S. Department of Agriculture, Berkeley, California 94710, U.S.A.

Aron, C. Université Louis Pasteur, Faculté de Médecine, Institut d'Histologie, 67 – Strasbourg, France.

Baldaccini, N. Institute of General Biology, University of Pisa, Pisa, Italy.

Bardach, J. Hawaii Institute of Marine Biology, University of Hawaii, P.O. Box 1346, Kaneohe, Hawaii 96744.

Beidler, L. Department of Biological Sciences, Florida State University, Tallahassee, Florida, 32306, U.S.A.

Bell, F. Department of Medicine, Royal Veterinary College, London N.W.1, England.

Benvenuti, S. Institute of General Biology, University of Pisa, Pisa, Italy.

Bernard, R. Department of Physiology, Michigan State University, E. Lansing, Michigan, U.S.A.

Bernays, E. Centre for Overseas Pest Research, College House, Wrights Lane, London W8, England.

Bertmar, G. Department of Biology, Section of Ecological Zoology, Umea University, 901 87 Umea, Sweden.

Blaine, E. Department of Physiology, University of Pittsburgh School of Medicine, Pittsburgh, Pennsylvania, 15261, U.S.A.

Blaney, W. Zoology Department, Birkbeck College, Malet Street, London WC1, England.

Boeckh, J. Department of Biology, University of Regensburg, Regensburg, Germany.

Bonsall, R. Department of Psychiatry, Emory University School of Medicine, Atlanta, Georgia, 30322, U.S.A.

Bradley, R. Department of Oral Biology, School of Dentistry, University of Michigan, Ann Arbor, Michigan, U.S.A.

Brandon, A. Laboratory of Psychophysiology, Claude Bernard University, Lyon, France.

Breipohl, W. Institut fur Anatomie, Geb, MA 5/52, D-4630 Bochum, Germany.

Brower, K. Department of Chemistry, NMIMT, Socorro, NM, U.S.A.

Caprio, J. Department of Biological Sciences, Florida State University, Tallahassee, Florida 32306, U.S.A.

Cattarelli, M. Laboratory of Psychophysiology, Claude Bernard University, Lyon, France.

Chanel, J. Laboratory of Psychophysiology, Claude Bernard University, Lyon, France.

Chapman, R. Centre for Overseas Pest Research, College House, Wrights Lane, London W8, England.

Cheal, M. Zoology Department, University of Michigan, Ann Arbor, Michigan, 48104, U.S.A.

Clarke, B. Carlton & United Breweries Ltd., P.O. Box 753 F, Melbourne, Vic. 3001. Australia.

Coghlan, J. Howard Florey Institute, University of Melbourne, Parkville, Vic. 3052. Australia.

Culvenor, C. Division of Animal Health, C.S.I.R.O. Private Bag No. 1, Parkville, Vic. 3052. Australia.

Darian-Smith, I. Department of Physiology, University of Melbourne, Parkville, Vic. 3052. Australia.

Daval, G. Laboratoire de Neurophysiologie Comparee, Universite Paris, Paris VI, France.

Denton, D. Howard Florey Institute, University of Melbourne, Vic. 3052. Australia.

Dennis, B. Department of Human Physiology and Pharmacology, University of Adelaide, North Terrace, Adelaide, 5000. Australia.

Den Otter, C. Department of Zoology, Groningen State University, Haren, The Netherlands.

xiii

Diesendorf, M. Department of Applied Mathematics, Australian National University, A.C.T., Australia.
Edgar, J. Division of Animal Health, C.S.I.R.O. Private Bag No. 1, Parkville, Vic. 3052. Australia.
Ernst, K. Department of Biology, University of Regensburg, Regensburg, Germany.
Farbman, A. Department of Anatomy, Northwestern University, Chicago, Ill., U.S.A.
Ferkovich, S. Insect Attractants, Behaviour and Basic Research Laboratory, Agricultural Research Service, U.S.D.A., Gainesville, Florida 32604, U.S.A.
Fiaschi, V. Institute of General Biology, University of Pisa, Pisa, Italy.
Fitzgerald, J. Apt. 194, 4800 North Stanton Street, El Paso, Texas, 79902, U.S.A.
Frank, M. The Rockefeller University, New York, N.Y., U.S.A.
Garcia, J. Department of Psychology, University of California, Los Angeles, California 90024, U.S.A.
Gesteland, R. Department of Biological Sciences, Northwestern University, Chicago, Ill., U.S.A.
Giachetti, I. Laboratoire de Physiologie Sensorielle, Collège de France, Paris, France.
Giannakakis, A. Department of Zoology, University of Sydney, Sydney, N.S.W., Australia.
Glaser, D. Anthropologisches Institut, Universitat Zurich, 8001 Zurich, Kunstlergasse 15, Schweiz.
Goodrich, B. McMaster Laboratory, Division of Animal Health, C.S.I.R.O. Private Bag No. 1, P.O. Glebe, N.S.W. 2037, Australia.
Graziadei, P. Department of Biological Science, Florida State University, Tallahassee, Florida, 32306, U.S.A.
Halpern, B. Department of Psychology, Cornell University, Ithaca, New York 14835, U.S.A.
Hankins, W. Department of Psychology, University of California, Los Angeles, California, U.S.A.
Hara, T. Department of the Environment, Freshwater Institute, Winnipeg, Manitoba, Canada R3T 2N6.
Haug, M. Laboratoire de Psycholphysiologie, Strasbourg, France.
Hellekant, G. Department of Physiology, Kungl. Veterinarhogskolan, S-104 05 Stockholm 50, Sweden.
Hidaka, I. Faculty of Fisheries, Mie Prefectural University, Tsu, Mie Prefecture, Japan.
Iino, M. Department of Physiology, Gunma University School of Medicine, Maebashi-city, Japan.
Jahan-Parwar, B. Worcester Foundation for Experimental Biology, Shrewsbury, Massachusetts 01545, U.S.A.
Kagi, D. Carlton & United Breweries Ltd., P.O. Box 753 F, Melbourne, Vic. 3001. Australia.
Kamo, N. Faculty of Pharmaceutical Sciences, Hokkaido University, Sapporo, Japan.
Kerr, D. Department of Human Physiology and Pharmacology, University of Adelaide, North Terrace, Adelaide, S.A. 5001. Australia.
Kikuchi, T. Department of Biological Science, Tohoku University, Kawauchi, Sendai 980, Japan.
Kissileff, H. Department of Physical Therapy, School of Allied Medical Professions, University of Pennsylvania, Philadelphia, Pennsylvannia 19104. U.S.A.
Kiyohara, S. Fisheries Laboratory, Faculty of Agriculture, Nagoya University, Nagoya, Japan.
Kobatake, Y. Faculty of Pharmaceutical Sciences, Hokkaido University, Sapporo, Japan.
Kramer, E. Max-Planck-Institut, Abteilung Schneider, 8131 Seewiesen, Germany.
Kratzing, J. Department of Veterinary Anatomy, University of Queensland, St. Lucia, Qld. 4067, Australia.
Kurihara, K. Faculty of Pharmaceutical Sciences, Hokkaido University, Sapporo, Japan.
Laing, D. C.S.I.R.O. Division of Food Research, North Ryde, N.S.W. 2113, Australia.
Laverack, M. Gatty Marine Laboratory, St. Andrews, Fife, Scotland.
Le Magnen, J. Laboratoire de Neurophysiologie Sensorielle et Comportementale, College de France, 11 Place Marcelin Berthelot, 75231 Paris, Cedex 05, France.
Leveteau, J. Laboratoire de Neurophysiologie Comparée, Université Paris VI, France.

MacDonald, J. Department of Zoology, University of Auckland, Private Bag, Auckland, New Zealand.

Mackay-Sim, A. School of Biological Sciences, Macquarie University, North Ryde, N.S.W. 2113, Australia.

Macleod, P. Laboratoire de Physiologie Sensorielle, College de France, 11 Place Marcelin Berthelot, 75231 Paris Cedex 05, France.

Maes, F. Department of Zoology, Groningen State University, Haren, The Netherlands.

Mayer, M. Insect Attractants, Behaviour and Basic Biology Research Laboratory, Agric. Res. Serv., USDA, Gainesville, Florida, 32604, U.S.A.

McBride, R. C.S.I.R.O. Division of Food Research, P.O. Box 52, North Ryde, N.S.W. 2113, Australia.

McKinley, M. Howard Florey Institute, University of Melbourne, Parkville, Vic. 3052. Australia.

Michell, A. Department of Medicine, Royal Veterinary College, London NW1, England.

Michael, R. Department of Psychiatry, Emory University School of Medicine, Atlanta, Georgia 30322, U.S.A.

Miller, I. Jnr. Department of Anatomy, Bowman Gray School of Medicine, Wake Forest University, Winston-Salem, North Carolina 27103, U.S.A.

Mistretta, C. Department of Oral Biology, School of Dentistry, University of Michigan, Ann Arbor, Michigan, U.S.A.

Morrison, A. Department of Veterinary Medicine, University of Pennsylvania, Philadelphia, Pa. 19174, U.S.A.

Moulton, D. Monell Chemical Senses Center, University of Pennsylvania, Philadelphia, Pennsylvania 19104, U.S.A.

Mykytowycz, R. C.S.I.R.O. Division of Wildlife Research, P.O. Box 84, Lyneham, A.C.T. 2602, Australia.

Nelson, J. Howard Florey Institute, University of Melbourne, Parkville, Vic. 3052. Australia.

Nicolaidis, S. Laboratoire de Neurophysiologie Sensorielle et Comportementale, College de France, 11 Place Marcelin Berthelot, 75231 Paris, Cedex 05, France.

Oakley, B. Department of Zoology, University of Michigan, Ann Arbor, Michigan, 48104, U.S.A.

Oshima, Y. Department of Physiology, Gunma University School of Medicine, Maebashi-City, Japan.

Pager, J. Laboratoire d'Electrophysiologie, Universite Claude Bernard, 69100 Lyon-Villeurbanne, France.

Papi, F. Institute of General Biology, University of Pisa, Pisa, Italy.

Perrotto, R. Department of Psychology, University of Delaware, Newark, Delaware, U.S.A.

Pfaffmann, C. The Rockefeller University, New York, N.Y. 10021, U.S.A.

Pietras, R. Department of Physiology, University of California School of Medicine, Los Angeles, California 90024, U.S.A.

Rausch, R. University of California, Brain Research Institute, Los Angeles, U.S.A.

Rieke, G. Department of Anatomy, Hahnemann Medical College, 230 N. Broad St., Philadelphia, Penna. 19102. U.S.A.

Rogeon, A. 17-19 Rue de Marguettes, 75 012 Paris, France.

Ropartz, P. Laboratoire de Psychophysiologie, Strasbourg, France.

Sass, H. Department of Biology, University of Regensburg, Regensburg, Germany.

Sato, M. Department of Physiology, Kumamoto University Medical School, Kumamoto, Japan.

Serafetinides, E. University of California, Brain Research Institute, Los Angeles, U.S.A.

Schafer, R. Department of Zoology, University of Michigan, Ann Arbor, Michigan, U.S.A.

Schneider, D. Max-Planck-Institut, 8131 Seewiesen, Germany.

Scott, T. Department of Psychology, University of Delaware, Newark, Delaware, U.S.A.

Shallenberger, R. Department of Biology, University of California, Los Angeles, California, U.S.A.

Shulkes, A. Howard Florey Institute, University of Melbourne, Parkville, Vic. 3052. Australia.

Sly, J. Department of Medicine, Royal Veterinary College, London N.W.1, England.

Smith, D. Department of Psychology, University of Wyoming, Laramie, Wyoming 82071, U.S.A.

Smith, J. Department of Psychology, Florida State University, Tallahassee 32306, U.S.A.

Stange, G. Department of Neurobiology, Research School of Biological Sciences, Australian National University, Canberra, A.C.T. 2600, Australia.

Stuart, A. Department of Zoology, University of Massachusetts, Amherst, Massachusetts 01002, U.S.A.

Sutterlin, A. Environment Canada, Biological Station, St. Andrews, N.B., Canada.

Takagi, S. Department of Physiology, Gunma University School of Medicine, Maebashi-City, Japan.

Tanabe, T. Department of Physiology, Gunma University School of Medicine, Maebashi-City, Japan.

Tarnecki, R. Nencki Institute of Experimental Biology, Warsaw, Poland.

Traynier, R. C.S.I.R.O. Division of Entomology, P.O. Box 109, Canberra City, A.C.T. 2601. Australia.

Tucker, D. Department of Biological Science, Florida State University, Tallahassee, Florida 32306, U.S.A.

Ueda, T. Faculty of Pharmaceutical Sciences, Hokkaido University, Sapporo, Japan.

Vinnikov, Y. Laboratory of Evolutionary Morphology, Sechenov Institute of Evolutionary Physiology and Biochemistry, Academy of Sciences of the U.S.S.R., Leningrad, U.S.S.R.

Vernet-Maury, E. Laboratory of Psychophysiology, Claude Bernard University, Lyon, France.

Waldow, U. Department of Biology, University of Regensburg, Regensburg, Germany.

Warner, P. Department of Psychiatry, Emory University School of Medicine, Atlanta, Georgia 30322, U.S.A.

Weeks, K. Kraft Food Ltd., G.P.O. Box 1673N, Melbourne, Vic. 3001, Australia.

Weisinger, R. Howard Florey Institute, University of Melbourne, Parkville, Vic. 3052. Australia.

Wenzel, B. Department of Physiology, University of California, Los Angeles, California 90024, U.S.A.

Whipp, G. Howard Florey Institute, University of Melbourne, Parkville, Vic. 3052. Australia.

Whitten, W. The Jackson Laboratory, Bar Harbor, Maine 04609, U.S.A.

Zippel, H. Physiologisches Institut II, Humboldtallee 7, D-3400 Gottingen, Germany,

Zotterman, Y. Office of President, International Union of Physiological Sciences, Wenner-Gren Centre, Sveavagen 166, Stockholm S-113 46, Sweden.

PREFACE

The First International Symposium on Olfaction and Taste was held in Stockholm under Professor Yngve Zotterman's chairmanship in 1962 prior to the official organization of the Commission on Olfaction and Taste of the International Union of Physiological Sciences. ISOT, along with ISFF, was one of the first IUPS Congress satellites. The full schedule of ISOT meetings is as follows:

		Location	Chairman
1962	ISOT I	Stockholm, Sweden	Yngve Zotterman Wenner Gren Center
1965	ISOT II	Tokyo, Japan	Takashi Hayashi Keio University
1968	ISOT III	New York, N.Y.	Carl Pfaffmann The Rockefeller University
1971	ISOT IV	Starnberg, Germany	Dietrich Schneider Max-Planck-Institut
1974	ISOT V	Melbourne, Australia	Derek Denton Howard Florey Inst. Exp. Physiology and Medicine

The proceedings of each have been published.

The 1977 meeting will be held in France under the chairmanship of Dr. Jacques LeMagnen of College de France.

The Commission on Chemoreceptors of IUPS was organized in 1969 when the late Dr. Wallace O. Fenn was Union President, with Professor Zotterman as Chairman, L.M. Beidler, M. Sato, D. Schneider, and C. Pfaffmann as members. The purposes of the Commission were two-fold: one, to provide a channel for expediting ways in which IUPS might assist the chemoreceptor group in organizing their meetings or in other ways, and second, to obtain from the commission any advice to help the IUPS in fostering the advancement of physiology in the general area of interest of the Commission. Prof. Zotterman succeeded to the Presidency of the IUPS in 1971, Dr. C. Pfaffmann to the chairmanship of the Commission, and

Drs. Denton and LeMagnen joined the Commission. The status and nature of IUPS Commissions were formalized in 1972. By then the number of Commissions had increased markedly.

The aim of the Commissions of IUPS is to survey the entire subject field on an international scale and to endeavor to foster the activities in the respective field both by suitable suggestions to IUPS and by their own initiative.

The activities of the Commissions should therefore include:

(a) Suggestions concerning the selection of topics and the choice of speakers at Congresses, Symposia, and other scientific sessions organized on behalf of IUPS.

(b) The organization of Symposia and panel discussions, sponsored by IUPS.

(c) The coordination of the diversified scientific meetings in the specific field of their subject.

(d) The exploration of possibilities of liaison with societies or groups concerned with the specific subject field.

(e) The approval of definition of terms of the specific subject field collected by the IUPS Board for Definition of Terms.

(f) The preparation and/or the sponsoring of books related to the specific subject field.

(g) The fostering of teaching at graduate and postgraduate level.

ISOT has been most active in holding the triennial international symposium but has begun to be more generally active under all the above headings. A primary aim of the original organizers was to stimulate greater interest in research in this field. In this it feels it has been successful, witnessed by the increasing output of work from its members, their students, and other scientists who have developed an interest in the fascinating and important problems of chemoreception reflected in an ever increasing attendance at ISOT. There is fear that the symposia may grow too large for easy and informal interchange among participants. New methods of communication are being experimented with. At this meeting a so-called "market place" symposium was scheduled. A number of tables and booths were set up so that participants could move from one to the other, examine exhibits and data, and engage the presenters in face to face discussion. It is hoped this and other experiments in communication will be continued.

I came away from this meeting with a sense of optimism that certain classical problems were on the verge of solution and many new approaches and techniques were being developed with great promise for advances in this field of physiological science. Over the years since the first ISOT, we have seen significant advances. Anatomical knowledge of receptors, their ultra structure, details of innervation of receptor fields, and the processes of receptor "turnover" represent one such advance. Taste modifiers and receptor proteins have come upon the scene. The classical concern with afferent coding is approaching a resolution more so perhaps in taste than olfaction where both transduction and coding remain difficult problems. Advances in CNS recording have shown how the sensory code for taste and olfaction are processed and sharpened. Conditioned taste aversions and other taste learning effects in food and fluid intake have become clearer in the generation of specific hungers and appetites, but certain classic mysteries as to the exact "trigger" for the salt appetite in salt need still eludes us. Pheromones have become an evermore active field both with regard to receptor mechanisms and behavioral control in invertebrates and vertebrates. By and large psychophysicists have put scaling techniques to good use in taste and olfaction to correlate with electrophysiological data on man. We would wish for more precise delineation of a physico-chemical dimension or set of parameters with which to better characterize our stimuli. Further insights on the adaptive role of these senses in a phylogenetic and evolutionary context from this meeting are apparent in the volume that follows.

Carl Pfaffmann
The Rockefeller University
New York, New York 10021

ACKNOWLEDGEMENTS

The full financial support to the Howard Florey Institute of Experimental Physiology and Medicine for the organization and conduct of the Fifth International Symposium on Olfaction and Taste was generously provided by the H.J. Heinz Company Australia Pty. Ltd.

The organizing Committee of ISOT V records its appreciation to the Government of the State of Victoria for the State Reception generously tendered overseas guests.

INTRODUCTION

The successful ISOT IV at Seeweisen in 1971 under the Chairmanship of Dr. Dietrich Schneider was orientated to receptor physiology and transduction. We decided in accepting the invitation of the International Commission to hold ISOT V at the Howard Florey Institute of Experimental Physiology and Medicine in Melbourne, Australia to place the major emphasis on behaviour and the evolutionary emergence of the chemoreceptor systems. The proposition to a number of potential contributors was that the basic modalities of taste − sweet, sour, bitter, salt (and water) are represented throughout the vertebrate phylum. Presumably the four modalities are not consistently present by sheer chance. Powerful selection pressures operated to contrive phylogenetic emergence of these modalities with attendant survival advantage. Accordingly, it would be of great interest to look at each modality within the sensory organization of the species set against environmental circumstances during evolution which might be postulated as favouring its emergence and refinement, e.g. the emergence of bitter and relation to poisoning. Whereas the exercise might involve some speculation there was much solid fact which might be put together in a new way in such a synthesis. Parallel to this there should also be presentation and analysis of the ontogenesis of taste, and also consideration of some special instances such as chemoreception in aquatic animals. On the side of olfaction it was more difficult, but presumably, putative primary modalities in olfaction might be examined, with the possible presence of undifferentiated and innate responses to different classes of odour in the light of distinguishable survival advantage accruing. This, again, might be done separately at different phylogenetic levels.

The response to the idea of attempting this emphasis and line of questioning for the Symposium was generally enthusiastic. The issue has, of course, been raised often before. Lord Adrian in his introductory statement at ISOT III in New York said,

"I hope somebody will explain why taste has developed the way it has done, with endings for acid and sweet and salt and bitter, and apparently nothing more; and how are the different endings arranged in animals with different ways of feeding. Carnivorous animals that bolt their food have little time to taste it, and one would think that sheep and cattle could do with different sensory equipment in their mouths. Pigs and men and monkeys surely have varying needs according to what they eat and when they eat it".

Also in discussion during the Conference "The Chemical Senses and Nutrition" Ed. Morley Kare and Owen Maller, Johns Hopkins Press 1967, we note the Chairman of the International Commission on Olfaction and Taste, Dr. Carl Pfaffmann, remarking,

"The thing that is more impressive to me is that there are so many similarities and not so many differences between animals — especially with regard to the use of something like sugar and saccharin — as a reinforcer.

It might be an interesting exercise to see if some order or reason could be found for these organisms that do respond positively to the sugars, that have the "sweet tooth". I'm perfectly willing to accept the fact that there may not be a simple biologically derived basis".

A number of presentations to ISOT V have given substantial consideration to the phylogenetic aspects of chemoreception and behaviour, and have contributed new data in this field. The conference has equally included presentation of interesting new work in several aspects of chemoreception research traditionally represented at the ISOT symposia.

The local Organizing Committee wishes to warmly thank the participants, the large majority of whom made long voyages, for making the meeting a most stimulating scientific occasion and for entering wholeheartedly into the substantial discussion periods. The ready participation of guests in the extra-sessional practical work in olfaction and taste was also most appreciated.

Derek Denton
Chairman
ISOT V
24th January, 1975.

PHYLOGENETIC EMERGENCE OF SWEET TASTE

Chairman: Y Zotterman
Moderator: C. Pfaffmann

Phylogenetic Origins of Sweet Sensitivity

Carl Pfaffmann

The Rockefeller University

New York, New York 10021 U.S.A.

The range of organisms displaying sensitivity to and a preference for sugar is wide. At ISOT IV, Adler (1971) reviewed his studies of chemotaxis of motile bacteria. E coli is attracted to amino acids and sugar, and normally metabolizes them, but metabolism is not required nor sufficient for attraction. Bacteria detect the attractant directly by means of distinct chemoreceptor systems. Mutants were found defective in one or more of the specific chemoreceptor for galactose, glucose, fructose, ribose, maltose, mannitol, aspartate, and serine.

Among invertebrates, the evidence for sweet sensitivity is most striking. The blowfly, fleshfly, and butterfly have been studied by the electrophysiological method and reveal the presence of specific sugar receptors at the base of sensory hairs (Dethier, 1955; Morita, 1959; Takeda, 1961). The recent demonstration (Shimada, et al., 1974) of a double sugar receptor earlier suggested by Evans (1963) is an important advance. Genetic analysis of insect chemoreception opens another line of investigation (Kikuchi, this symposium).

Behavioral evidence does indicate sensitivity to some sugars in several fish species (Bardach, 1971). The bottom feeding catfish have taste buds over the external body and in barbels. The first recordings of any taste nerve impulses were made by Hoagland (1933) from the facial nerve of this species. Tateda (1964) and Konishi, et al. (1966) found that in bullhead and sea catfish, salts, acids, and to a lesser degree quinine, elicited a good response, but sugar was much less effective. More striking electrophysiological evidence of sugar sensitivity was obtained in carp (Cyprinus carpio (L)), in the IXth nerve innervating the palatal organ (Konishi & Zotterman, 1961). The responses to sugar and to acid were both strong compared with weaker response to 0.5 M NaCl. Sugar sensitivity occurred in three of the seven fiber groups, but was never highly specific, usually accompanied by sensitivity to NaCl, acetic acid, and some factor in human saliva. Type III fibers had high sensitivity to sugar, glycerol and glycol, the latter two sweet to man, but saccharin had little effect. Sugar thresholds ranged from 0.005 M to 0.01 M; quinine from 0.0002 M to 0.0005 M; NaCl less than 0.0005 M. Japanese carp had a lower sensitivity to sugar but higher to quinine.

3

Amino acids comprise another large class of substances sweet to man. D and L amino acids stimulate sugar sensitive fibers in bullhead (Bardach & Atema, 1971). In carp, Hidaka & Yokota (1967) found a synergistic interaction between sucrose and glycine. $HgCl_2$ blocked sucrose more than glycine. This plus the recently demonstrated high sensitivity in catfish to amino acids with moderate sucrose sensitivity (Caprio, this symposium) suggest that amino acids may stimulate another set of fibers in addition to those for sugar. Preference for sugar to the degree shown by mammals is not obvious.

Pumphrey (1935) reported the first recording from frog taste afferents in the IXth nerve, primarily for touch, salt, and acid. Later studies revealed sensitivity to sugar and quinine, as well as salt and acid. Kusano (1960) and Sato & Kusano (1960) classified single fibers roughly into four groups but with individual departures from strict categorization. D units responded mainly to divalent ions, also to sucrose and in most cases, water; M units mainly to monovalent salts, such as NaCl, LiCl. KCl, etc. Quinine (Q units) or acid (A units) gave little or no response to salts.

Zotterman (1950) was the first to describe water taste in frogs. Isotonic NaCl abolished this response; isotonic sucrose did not. He warned that some sugar taste responses might be only "apparent" due to solvent (H_2O). Sato and Kusano (1960) found sugar and water to be true D-fiber stimuli. Single frog receptor cells respond to sucrose and other taste solutions (T. Sato, 1972). Sugar sensitivity in frogs is often associated with sensitivity to other stimuli.

Reptiles (snakes) show good discrimination in feeding (Burghardt, 1966), but this may be largely olfactory or vomeronasal. The tongues of snakes are usually smooth and slender with a bifurcated tip but devoid of taste buds (Bradley, 1971). The occurrence of taste buds on the reptilian tongue is variable. They are said to be absent in the snapping turtle but present in the tortoise and lizard.

Birds are derived from a major reptilian radiation. Tongue morphology is well correlated with feeding habits; fish feeders have sharp stiff spines pointing backward to facilitate holding slippery prey. Seed eaters have large muscular tongues and ducks have rows of hairs for filtering small food particles from the water. Nectar feeders (humming birds) have protrusible tongues tube-like or rolled in cross section (Bradley, 1971). Birds, at first thought not to have taste buds, do have relatively small numbers on the posterior tongue unassociated with papillae. In the pigeon, they number 27-59; in the chicken 24; and bullfinch 42-50 (Fare, 1971).

Kitchell, Strom and Zotterman (1959) found electrophysiological responses to distilled water, NaCl, glycerine, glycol, quinine and acetic acid in the lingual branch of the glossopharyngeal nerve in chickens and pigeons, but there was no response to sucrose. Halpern (1962) found small but consistent responses in the chicken to 1.0 M sucrose, glucose and .5 M xylose; NaCl showed a clear concentration dependent response at .2 M up to 1.0 M. HCl gave a slight response (.002 N to .01 N) but 0.1 N acetic a large response that inactivated the preparation.

Certain experiments indicate a slight sugar preference in birds, whereas others did not (Kare, 1971; Jacobs, 1961). Conditions of the experiment, i.e., degree of food deprivation and hydration, etc., might account for discrepancies. Nectar feeders show clearer sugar preference. Their stomach contents contain many insects, but little nectar, due to the latter's high absorption rate. Hummingbirds cannot survive on nectar alone, the mixture of nectar plus insects being necessary for balanced nutrition.

Hovering flight requires a good source of energy. When fed sucrose solutions, hummingbirds are 97-99% efficient in utilization (Hainsworth, 1974). Most nectars are composed of only sucrose, glucose, and/or fructose, equivalent to 0.24-2.10 M. Hummingbirds' lick rate of 2.6 to 3.8 per second was uniform, not affected by sugar concentrations up to 1.0 M. In taste tests hummingbirds detect differences in concentrations of sugar and reject acid solutions (Scheithauer, 1967). Weymouth, et al. (1964) report seeing only few taste buds on the basal portion of the tongue with no special papillae. These workers, however, were most interested in the tongue musculature and the quasi-tubular form adapted for nectar feeding. The relation of hummingbirds to the plants upon which they feed is an interesting case of reciprocal evolution. Grant & Grant (1968) suggest that the ancestors of nectariferous hummingbirds may have first sought insects in flowers. The gradual transition to a predominantly nectar diet then followed. Linnets, orioles and other birds, they report, eagerly feed on sugar water when provided in feeders and sometimes seek nectar. "The demand for nectar is evidently more widespread among birds than the ability to get it successfully." Energetics, metabolism, even structure of muscles and gross morphology of the tongue have been emphasized, with but little regard to the gustatory sensory apparatus itself which seems uniquely attuned to a role in such ecological coadaptation.

The teleosts, as descendants of a collateral line in fish evolution, were not antecedents to any amphibian, reptile, bird or mammal. Birds represent another specialization, not ancestral to mammals. The early mammals were quite small, in the size range of living shrews, moles, mice and rats. Present-day opossums and hedgehogs are quite similar to those of the Cretaceous period nearly 100 million years ago. Other mammals have survived with little change since the Eocene period 50 million years ago,

Chorda Tympani Responses Relative to NaCl or NH$_4$Cl*

	.5 NH$_4$Cl	.05 HCl	.05 Suc.	.02 Q
Opossum (a)	1.0	0.55	0.09	0.11
		.01 HCl	.5 Suc.	
Cat (b)	1.0	0.36	0-0.20	0.31
Dog (b)	1.0	0.16	0.27	0.09
Rabbit (b)	1.0	0.56	0.52	0.48
	.1 NaCl			
Rat (b)	1.0	0.61	0.21	0.20
Hamster (b)	1.0	0.85	0.75	0.33
Guinea Pig (b)	1.0	0.44	0.62	0.24
	.5 NaCl	.2 acetic		
Calf (c)	1.0	0.66	0.16	0.33
	.5 NH$_4$Cl			
Lamb (d)	1.0	1.1	0.19	0.09
		.05 HCl	.05 Suc.	.04 Q
Bat Brown (a)	1.0	0.95	0.10	0.05
	.1 NaCl	.01 HCl	.1 Suc.	.02 Q
Frug. (e)	1.0	0.55	0.4	0.5
	.3 NH$_4$Cl			.01 Q
Squirrel Monkey (f)	1.0	0.65	0.82	0.65
	.3 NaCl		.3 Suc.	
Macaque Monkey (g)	1.0	0.72	0.76	0.70
	.2 NaCl	.02 citric	.5 Suc.	.002 Q
Man (h)	1.0	1.3	0.94	0.70

* Molar concentrations
(a) – (h). See references for data sources.

as, for example, dogs, pigs, and lemurs. Rodents and carnivores have followed independent developmental courses since the late Cretaceous and early Paleocene periods. Primates evolved as a specialized branch of the insectivore line (shrews, moles, hedgehogs, etc.). "Rats were never ancestral to cats nor were cats ancestral to primates; rather each represents a different evolutionary lineage" (Hodos & Campbell, 1969).

The sense of taste, in contrast with vision or hearing, appears to have undergone relatively little evolutionary development in morphology of the receptor and central neural connections. The taste cortical representation is not extensive (Burton & Benjamin, 1971). There appears to be less encephalization of function so that brain stem pontine and medullary areas may modulate feeding and ingestion in a basic manner in vertebrates (Herrick, 1905; Macht, 1951; Norgren & Leonard, 1973).

Mammalian taste receptors are concentrated in the oral cavity, primarily upon the tongue, the mobile member for manipulating food in gnawing, chewing, and swallowing. Food will be accepted or rejected as its properties are sensed. In mammals the mouth functions as part of the digestive system especially for the mechanical reduction of food. It serves

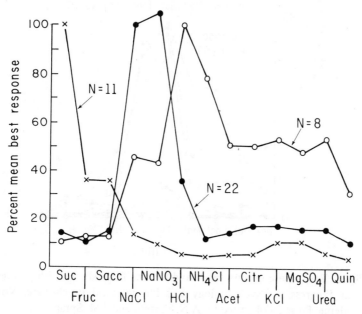

Fig. 1. Response profiles of sugar-best, salt-best and acid-best units.

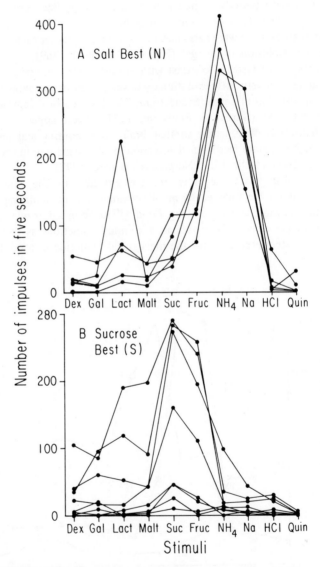

Fig. 2. Response contours for N and S units. Reproduced with permission from *Progress in Psychobiology and Physiological Psychology*, Vol. 6, Academic Press 1974 edited by A.N. Epstein and J.M. Sprague.

only secondarily as an organ of prehension, its primary function in reptiles. The mammalian tongue is a whole new evolutionary structure, the sublingua representing the remains of the reptilian tongue (Davis, 1961). The receptor field of the tongue is clearly made up of two components. The anterior tongue region is innervated by the chorda tympani of the VIIth cranial nerve; the posterior by the IXth and Xth nerves. The following table samples relative chorda tympani responses across species. In nearly all, there is a good salt and acid response, the best sugar sensitivity ($> .75$) is in hamster, squirrel monkey, macaque monkey, and man.

The multiple sensitivity of a single unit of fish and frog taste units also occurs in mammals. Frank (1973) applied her "best-stimulus" categorization to 79 hamster units, Statistical analysis of sugar-best units showed that the sweet receptors were more specific and independent than other sensitivities. Her data, rearranged in Fig. 1, shows how response to sugar and other sweeteners peak sharply (as does that for salt). Acid is less so and its individual response contours also show greater variability. Recently Dr. Nowlis in our laboratory has recorded good hamster responses to amino acids sweet to man.

We found a good sugar response in squirrel monkey chorda with two clusters of units (Pfaffmann, et al., 1974). One, the S class, responded maximally to sucrose. The N class responded well to sugars but best to salts (See Fig. 2). Sucrose was always best sugar for S units; fructose the best sugar for N units. The suc/fruc response ratio was for S units > 1.0; for N < 1.0. Sucrose was a more effective behavioral reinforcer than fructose in bar pressing, consummatory licking and preference. We conclude that S units signal sweet which motivates the behavioral "sweet tooth" and preference.

Zotterman (1967) in dog and macaque previously reported specific sugar fibers. M. Sato (this symposium) has evidence on specificity of sweet in macaque. All evidence indicates that where significant sweet sensitivity appears in mammals, it is relatively specific with the characteristics of a "labelled line" cluster signaling sweetness to the CNS.

9

Acknowledgement
Research was supported by NSF Grant No. BMS 70-00854 A04.

References
Adler, J. 1972. Olfaction & Taste IV, p.70.
Bardach, J.E. and Atema, J. 1971. Hdbk. Sens. Physiol., IV-2, p.293.
Beidler, L.M., Fishman, I.Y. and Hardiman, C.W. 1955. Amer. J. Physiol., 181, p. 235. (b).
Bradley, R.M. 1971. Hdbk. Sens. Physiol., IV-2, p.1.
Bradley, R.M. and Mistretta, C.M. 1972. Oral Physiol., p. 239. (d).
Burghardt, G.M. 1966. Psychon. Sci., 4, p. 37.
Burton, H. and Benjamin, R.M. 1971. Hdbk. Sens. Physiol., IV-2, p. 148.
Davis, D.D. 1961. Amer. Zool., 1, p. 229.
Dethier, V.G. 1955. Quart. Rev. Biol., 30, p. 348.
Evans, D.R. 1963. Olfaction & Taste I, p. 165.
Fishman, I.F. 1963. Iowa Acad. Science, 70. p. 465. (e).
Frank, M. 1973. J. gen. Physiol., 61, p. 588.
Grant, K.A. and Grant, V. 1968. Himmingbirds and their flowers, p.87.
Hainsworth, F.R. 1974, J. comp. Physiol., 88, p. 425.
Halpern, B.P. 1962. Amer. J. Physiol., 203, p. 541.
Herrick, C.J. 1905. J. comp. neurol. Psychol., 15.
Hidaka, I. and Yokota, S. 1967. Jap. J. Physiol., 17, p. 652.
Hoagland, H. 1933. J. gen. Physiol., 16, p. 685.
Hodos, W. and Campbell, C.B.G. 1969. Psychol. Rev., 76, p. 337.
Jacobs, H.L. 1961. Physiol. & Beh. Aspects of Taste, p. 31.
Kare, M. 1971. Hdbk. Sens. Physiol., IV-2, p. 278.
Kitchell, R.L. 1963. Olfaction & Taste I, p. 235. (c).
Kitchell, R.L., Strom, L., and Zotterman, Y. 1959. Acta Physiol. Scand., 46, p. 133.
Konishi, J., Uchida, M. and Mori, Y. 1966. Jap. J. Physiol., 16, p. 194.
Konishi, J., and Zotterman, Y. 1961. Acta Physiol. Scand., 52, p. 150.
Kusano, K. 1960. Jap. J. Physiol., 10, p. 620.
Macht, M.B. 1951. Fed. Proc., 10, p. 88.
Morita, H. 1959. J. cell. comp. Physiol., 54, p. 189.
Norgren, R., and Leonard, C.M. 1973. J. comp. Neurol., 150. p. 217.
Ogawa, H., Yamashita, S., Noma, A., and Sato, M. 1972 Physiol. & Beh., 9, p. 325. (g).
Pfaffmann, C., Frank, M., Bartoshuk, L., and Snell, T. 1974. Prog. Psychobiol. Physiol. Psych., In press. (f).
Pumphrey, R.J. 1935. J. cell. comp. Physiol., 6, p. 457.
Sato, M. and Kusano, K. 1960. Electrical Activity of Single Cells, p. 77.
Sato, T. 1972. J. cell. Physiol., 80, p. 207.
Shimada, I., Shiraishi, A., Kijima, H. and Morita, H. 1974. J. Insect. Physiol., 20, p. 605.
Scheithauer, W. 1967. Hummingbirds, p. 143.
Takeda, K. 1961. J. cell. comp. Physiol., 58, p. 233.
Tamar, H. 1961. Physiol. Zool., 34, p. 86. (a).
Tateda, H. 1964. Comp. biochem. Physiol., 11, p. 367.
Weymouth, R.D., Lasiewski, R.C. and Berger, A.J. 1964. Acta Anat., 58, p. 252.
Zotterman, Y. 1950. Acta Physiol. Scand., 18, p. 181.
Zotterman, Y. 1967. Sensory Mechanisms, p. 139.
Zotterman, Y., 1971. Hdbk. Sens. Physiol. IV-2, p. 102. (h).

Can Taste Neuron Specificity
Coexist with Multiple Sensitivity?

Rudy A. Bernard

Physiol. Dept., Mich. State Univ.

E. Lansing, Michigan.

Definition of Specificity

The concept of specificity does not mean that a set of neurons is sensitive only to one kind of stimulus. Rather it states that a particular set of neurons elicits the same sensation regardless of how they are stimulated, by virtue of their central connections (12). It's not how, but which neurons are stimulated that determines the sensation. The emphasis is on the output of the neurons rather than on the input. This fundamental principle of sensory physiology, although expressed in somewhat different terms from those of Johannes Müller, is alive and well today, as is being demonstrated by the exciting work on cortical visual prostheses for the blind (15).

Multiple Sensitivity

From the earliest electrophysiological experiments it became clear that individual taste neurons and receptor cells responded to more than one of the four postulated basic taste stimuli. It is now clear that a large majority of individual neural elements is multisensitive. This has led to the concept of differential sensitivity or "specificity" and the development of the pattern theory for coding taste quality. A closely related question is whether or not there are indeed a limited number of taste qualities and corresponding classes of stimuli. The arguments in favor of primary tastes for man have been well documented by McBurney (10). I doubt that many will dispute their existence at the psychophysical level. This does not settle the question of the neural mechanism underlying the psychophysical phenomena nor the related question of the applicability of the primary taste concept to animal data, from which most of the evidence for neural mechanisms is necessarily derived.

It is instructive to examine the data from the fields of vision and audition. It is apparent that the individual elements in these systems are multisensitive, yet they are classified according to their peak sensitivities, which correspond to a small number of primaries in vision but not in audition. A similar approach has been used for taste fibers (16, 7) but

11

differences from vision and audition must be recognized. The major difficulty with the classification of taste stimuli is that there is no obvious continuum along which to locate a broad range of stimuli. Both visual and auditory stimuli have a natural continuum of light wavelength and sound frequency but there is no similar continuum within salty or sweet stimuli, much less for the transition from one quality to another. In practice the range of stimuli tested is reduced and a bias is introduced by selecting stimuli known to represent the basic qualities.

Neural Patterns

There can be no doubt that any particular stimulus will activate a particular set of neurons within the network that constitutes a complete sensory system. This would seem to be required by the very fact, and to the extent, that the stimulus can be recognised and differentiated from other stimuli. Discrimination would be the more difficult the more the stimuli resembled each other. In this sense neural patterns would seem to be a requisite for sensory discrimination.

Because of the multiple sensitivity of the individual neural elements the specific fiber theory has to contend with the fact that a salty or sweet stimulus activates more than one neuron type. This is because, as Bekesy (2) pointed out, not even pure chemicals are monogustatory stimuli. As McBurney's (10) work with cross-adaptation has shown, the taste of a series of salty or other compounds can be characterized by the different degree to which each of the other basic tastes contributes to the overall sensation. Overall quality would depend on which fiber type was predominantly rather than exclusively stimulated. Correspondingly all that is required for a defensible theory of specific fiber types is that the neurons be predominantly activated by the matching stimuli.

As I see it, the question for taste then becomes whether the neural pattern is made up of a small or an indefinitely large number of neuron types. The import of this question is best conveyed by the very large difference in the number of active first order neurons when stimulation shifts from the whole tongue to a single papilla.

Single Papilla Experiments

Bekesy's experiments on electrical (2) and chemical (3) stimulation of individual human taste papillae have been both confirmed (5) and disputed (9, 10, 13). In my view, coding of taste quality by specific fibers does not require a unique response from the single papilla, because individual papillae are innervated by the terminal branches of more than one first order neuron. Although I am not aware of any human data, a single fungiform papilla is innervated by a maximum of 7-8 first order fibers in the rat (1) and an average of 2 in the frog (calculated from data reported by Rapuzzi and Casella (14)). Assuming that a similarly small number of first order neurons innervates a single human papilla, which generally contains more than one taste bud, the pattern theory has the very difficult problem of demonstrating what sort of pattern can be common to the extreme conditions of one papilla versus all papillae being stimulated. The simplest proposal is to have the pattern consist of the distribution of activity across a limited number of neuron types, which would remain unchanged regardless of the number of papillae being stimulated. If each fiber is different in the sensation it elicits, stimulation of one papilla would produce a qualitatively different sensation from the stimulation of several papillae, which does not appear to be the case.

To explain the conflicting results of the experiments described above it is sufficient to postulate that single papillae are innervated by two or more fiber types each with sufficiently different densities of innervation to allow a preferential response to one class of stimulus. When chemical stimulation is close to threshold (3) a single response would occur whereas suprathreshold (11) and highly concentrated solutions (9) would be able to elicit all four taste qualities.

For the conflicting results of electrical stimulation it is sufficient to postulate that the stimulus parameters in the one case (2, 5) were inadvertently such as to stimulate the fiber type with the lowest threshold whereas in the other (13) they were not.

Quality Coding Time

If a specific class of neurons exists for each of the basic tastes, very simple patterns would be produced by stimuli that have a preponderant quality such as NaCl, sucrose, HCl or quinine. It could be predicted then that discriminating between these stimuli would be a very simple task, perhaps an unlearned one. Such would seem to be the case in human neonates, where the pleasurable acceptance of sucrose and the grimacing response to quinine reportedly occurs from the very first experience.

Quality identification occurs very quickly in rats, in as little as 250 msec (8), and adult human reaction time to electrical taste, which is chemical in origin, can be as little as 200 msec (4). This is the time period that corresponds to the phasic component of the taste nerve response, which has not been adequately treated by the pattern theorists, as was pointed out by Faull and Halpern (6). It is this phasic neural component which is very labile and subject to adaptation (17) and provides the only animal neural evidence for the psychophysical adaptation well documented in man.

It seems to me, then, that the initial events in the neural response are sufficient for identifying the basic quality of the taste stimulus and provide further evidence for the concept that quality coding in taste depends on the activity of specific neurons corresponding to each quality.

Summary

A review of the evidence has led this observer to unify the specific fiber and neural pattern theories for taste coding into what may be called a specific neuron pattern theory. This theory proposes that there are a limited number of neuron types corresponding to the basic taste qualities, which are commonly considered to be four in man, but which may be different in number as well as kind for animals. The theory accepts the existence of a distinct neural pattern for every discriminable taste stimulus, but proposes that the patterns are not utilized for coding primary taste qualities, but for discriminating between similarly tasting stimuli such as glucose and sucrose or NaCl and LiCl, or between stimuli that elicit complex taste sensations. In this theory multiple sensitivity may be viewed as a mechanism that enables a broad range of chemical stimuli to generate a common sensory experience as well as provide very precise information about the outside world. It enables taste to be an affective as well as an information system.

Supported in part by NIH grant NS 09168.

References

1. Beidler, L.M. 1969. In C. Pfaffmann (ed.), Olfaction and Taste III, Rockefeller Univ. Press, New York, pp. 352-369.
2. Bekesy, G. von 1964. J. Appl. Physiol. 19: 1105-1113.
3. Bekesy, G. von 1966. J. Appl. Physiol. 21:1-9.
4. Bujas, Z. 1971. In L.M. Beidler (ed.), Chemical Senses 2, Taste, Vol. IV, Handbook of Sensory Physiology, Springer-Verlag, Berlin, pp. 180-199.
5. Dzendolet, E. and Murphy, C. 1974. Chemical Senses and Flavor 1:9-15.
6. Faull, J.R. and Halpern, B.P. 1972. Science 178:73-75.
7. Frank, M. 1974. Chemical Senses and Flavor 1:53-60.
8. Halpern, B.P. and Tapper, D.N. 1971. Science 171:1256-1258.
9. Harper, H.W., Jay, J.R. and Erickson, R.P. 1966. Physiology and Behavior 1:319-325.
10. McBurney, D.H. 1974. Chemical Senses and Flavor 1:17-28.
11. McCutcheon, N.B. and Saunders, J. 1972. Science 175:214-216.
12. Mountcastle, V.B. 1974. Medical Physiology, Vol. I, C.V. Mosby, St. Louis, pp. 286-287.
13. Plattig, K.H. 1972. In D. Schneider (ed.), Olfaction and Taste IV, Wissenschaftliche Verlags-gesellschaft MBH, Stuttgart, pp. 323-328.
14. Rapuzzi, G. and Casella, C. 1965. J. Neurophysiol. 28:154-165.
15. Sterling T.D., Bering, E.A. (Jnr)., Pollack, S.V. and Vaughn, H. (Jnr). (eds.) 1971. Visual Prosthesis, Academic Press, New York.
16. Wang, M.B. and Bernard, R.A. 1969. Brain Res. 15:567-570.
17. Wang, M.B. and Bernard, R.A. 1970. Brain Res. 20:277-282.

Different types of sweet receptors in mammals

By

Goran Hellekant

It seems to be the general opinion that there is only one type of sweet receptor in mammals (Schoonoven 1973) though some authors have suggested more than one type (Beidler 1971). This makes particularly interesting the isolation of two new substances, the proteins monellin and thaumatin (Morris and Cagan 1972, van der Wel 1972 a,b). They have to man an intense sweet taste and ought to elicit a similar taste sensation in other mammals if there is a general sweet receptor structure. This presentation will summarize the observations obtained in monkey, dog, guinea pig, hamster, pig, rabbit and rat during taste stimulation with these two proteins.

Methods

Recordings of activity in the chorda tympani proper nerve were made after dissection of the nerve. The nerve activity was summated and recorded on a Statos 1 recorder. Taste stimuli were applied with a device described earlier (Andersson et al. 1971). Solutions of NaCl, quinine sulphate or quinine hydrochloride, citric acid and sucrose all made up in tap water were used as standard taste stimuli. Fresh 0.01, 0.02 or 0.04% solutions of thaumatin and monellin made up in tap water were used. The tongue was rinsed with a continuous flow of tap water between each stimulation.

Results

Observations in the monkey

A series of summated responses to the standard stimuli and 0.02% thaumatin and monellin is shown in Fig. 1. The records show that both substances elicited a significant increase of activity in the chorda tympani. If one uses the magnitude of these responses and those of sucrose as an indication of perceived sweetness, it can be concluded that the perceived sweetness of these proteins have the same magnitude in man and monkey. That is, they are 30-60000 times sweeter than sucrose calculated on molar basis.

15

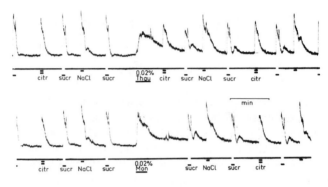

Fig. 1

The rise time of this response was slower than that to any of the standard stimuli used. Cross adaptation between sucrose and the sweet proteins can also be observed in the records. In man a similar cross adaptation between sucrose and the sweet proteins can be experienced.

Following the exposure to the sweet proteins an after-response was observed during the water rinse subsequent to stimulations with sucrose or sodium chloride. This can be seen in Fig. 1 but is better illustrated in Fig. 2, where the after-response is indicated by arrows. (The after-response which can be seen after the NaCl responses before the thaumatin and the monellin in Fig. 1 emanated from earlier application of these two substances). In Fig. 2 it should be noticed that we considered the peaks appearing 3 sec after a salt stimulus or 8 sec after a sucrose stimulation as the after-responses and not the short peak appearing at the moment the stimulus is replaced by the water rinse. Fig. 2 shows that the after-response to salt was more evident, lasted longer and came sooner after the stimulus response than that after sucrose. No after-response was observed after citric acid or when an intermittent flow of water was used. This seems to parallel an observation in man. Thus a sweet taste of water after thaumatin has been reported by van der Wel (1972). This taste can be repeatedly experienced when the mouth is rinsed with water after a previous application of thaumatin, saliva possibly serving as the salt solution in this case. In summary our observations in the monkey seem to parallel the perceptual observations in man.

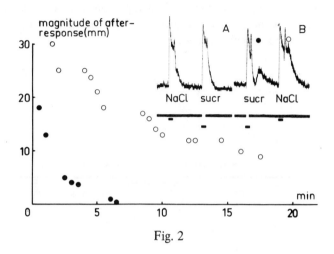

Fig. 2

Observations in the dog

Fig. 3 gives an example of recordings obtained in the dog. It shows the summated nerve activity during stimulation with the standard solutions and 0.02% thaumatin and monellin. The record shows that the standard solutions gave a good response. It is also evident that during stimulation with monellin an increase of the nerve activity was recorded, while thaumatin caused no increase. In monkey we observed cross adaptation between monellin and sucrose. Fig. 3 shows that in the dog a previous stimulation with monellin did not depress the subsequent response to sucrose. Thus cross adaptation was not found in the dog between sucrose and 0.02% monellin. Thaumatin seemed not to elicit a taste effect.

Fig. 3

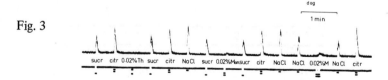

Observations in the guinea pig

The response to sucrose in the guinea pig is relatively large. A response to the sweet proteins may therefore be expected. However, neither

17

monellin nor thaumatin, which were applied at a concentration of 0.02% for almost 1 min, gave significant responses in the chorda tympani nerve. This is shown in Fig. 4. In this record the proteins were applied in cotton wool soaked in the stimulus in question. The small peaks which can be seen are probably stimulus artifacts.

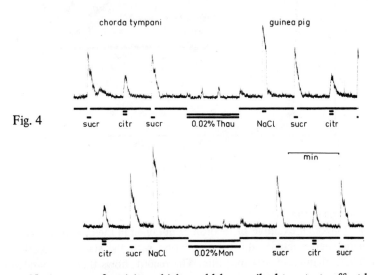

Fig. 4

No increase of activity which could be ascribed to a taste effect by the sweet proteins can be observed. The increase which can be observed after the stimulation was the result of mechanical and thermal stimulation by the subsequent water rinse. It was not observed when the method of rinsing with the stimuli was used.

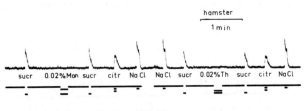

Fig. 5

Observations in the hamster

In the hamster the response to sucrose is larger than in most other species investigated. In spite of this we were not able to observe an effect by thaumatin or monellin. Fig. 5 shows such a recording during stimulation of the tongue with sucrose, 0.04% monellin and thaumatin, NaCl and citric acid. It must be concluded that this record shows no trace of any increase in nerve activity which could be related to a gustatory effect of these sweet proteins. It shows further that application of the sweet proteins on the tongue did not affect the response to sucrose. The records of the 9 animals tested gave similar negative results.

Observations in the pig

Fig. 6 shows records obtained in a pig during stimulation with similar solutions as illustrated in Fig. 5. Though the response to sucrose as well as the other standard stimuli were significant, there was no indication of an increased activity in the nerve when 0.01% monellin or thaumatin were applied as shown in the record. Further the record shows no sign of any cross adaptation between the sweet proteins and sucrose. This was not the result of a too short stimulation or too weak stimulus, because when 0.02% thaumatin was applied for a longer time period, about 20 sec, a similar absence of a response was recorded.

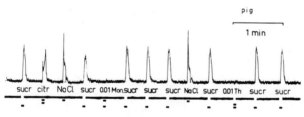

Fig. 6

In conclusion, we were not able to record a change in the chorda tympani proper activity in pig which could be ascribed to a gustatory effect of the sweet proteins.

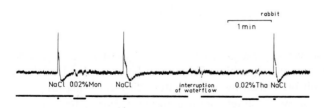

Fig. 7

19

Observations in the rabbit

Fig. 7 shows a record obtained in a rabbit. Since we knew from earlier experiments in the monkey that the response to these sweet proteins develops more slowly than that to the other stimuli we extended the period of stimulation. We could not apply the sweet proteins in an uninterrupted flow, because of our limited supply, but had to use intermittent stimulation. This allows the tongue to cool between stimulation, which affects the nerve activity. Similar changes can be observed during the interruption of the flow of water in the middle part of the record. Therefore we conclude that neither monellin nor thaumatin elicited any significant changes of the chorda tympani proper activity which could be ascribed to their gustatory effects in the rabbit.

Observations in the rat

Neither thaumatin nor monellin caused any apparent increase in the activity of the nerve. This is shown in Fig. 8 which was obtained during stimulation with 0.04% thaumatin and monellin. A slight depression is seen during stimulation.

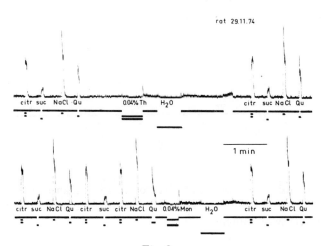

Fig. 8

A similar depression can be seen when water was applied in a similar way as the taste proteins. Thus it can be concluded that this depression was probably caused by a temperature difference. Because of our limited supply of sweet protein we had to use a pipette for application which diminished our ability to control the temperature. Fig. 8 shows no sign of cross adaptation between sucrose and the sweet proteins. After the sweet proteins a short-lasting burst during the following water rinse can be seen. This was not seen after water. Thus it is likely that rats can distinguish between water and the sweet proteins, though these proteins, judged by the neural response they evoke, do not taste sweet to the rat.

Conclusions

These results show that in the species tested monellin and thaumatin elicit a significant response only in the monkey. The parallelity of this response with the perceptual observations in man makes it probable, that to the monkey they taste sweet. Our findings in the other species did not indicate that the sweet proteins elicit a sweet sensation. We must conclude that there seems to be a fundamental difference in the gustatory effects of these proteins between man and monkey on one side and the other mammals studied on the other.

References

Andersson, I., Hellekant, G., Larsson, R., Molander, C. and Strom, L. 1971. Med. & Biol. Engin. 9, 715-717.

Beidler, L.M., 1971, Sweetness and Sweeteners, Applied Science Publishers, London, Amsterdam, New York, 1971.

Morris, J.A., and Cagan, R.H., 1972. Biochem. Biophys. Acta 261, 114-122.

Schoonhoven, L.M., 1973. Transduction mechanisms in chemoreception, pp 189-201.

van der Wel, H., 1972. Olfaction and Taste IV, p. 226-233.

van der Wel, H., 1972. FEBs letters, 21, No. 1, 88-90.

Response Characteristics of
Taste Nerve Fibers in Macaque Monkeys:
Comparison with those in Rats and Hamsters

Masayasu Sato

Department of Physiology

Kumamoto University Medical School

Kumamoto, Japan

Our recent study on the chorda tympani nerve response in macaque monkeys (1) demonstrates that they respond to a variety of chemical substances which produce sweet taste in men. Based on recordings of impulse discharges in single chorda tympani fibers in crab-eating monkeys, we reported previously (2) that single taste nerve fibers in macaque monkeys responded to two or more stimuli but that the number of fibers responding to one or two stimuli only was greater in this animal than in rats and hamsters. In this paper I will describe the results of our analyses of responses of single chorda tympani nerve fibers in crab-eating monkeys to the four basic stimuli (0.3 M NaCl, 0.3 M sucrose, 0.01 M HCl and 0.003 M quinine hydrochloride) and some sweet-tasting substances. The analyses were based on the numbers of impulses elicited during 5 sec after stimulation in 67 fibers.

Many fibers in the monkey responded to more than one kind of the four stimuli. However, a group of fibers responded almost specifically to sucrose, and another group to quinine, although many fibers yielded good responses to NaCl as well as HCl. This was statistically supported by calculating the across-fiber correlation coefficients between amounts of responses to pairs of stimuli: Correlation coefficients across 67 fibers between amounts of responses to sucrose or quinine and those to the remaining three stimuli are very low (sucrose-NaCl, -0.23; sucrose-HCl, 0.04; sucrose-quinine, -0.13; NaCl-quinine, -0.18; HCl-quinine, -0.02), while a highly significant positive correlation exists between amounts of responses to NaCl and HCl ($r = 0.45$, $p < 0.001$, t-test).

Recently Frank (3,4) classified chorda tympani fibers in hamsters and squirrel monkeys into four categories according to their highest responsiveness to one of the four basic stimuli, and compared taste sensitivities in these two animal species with each other. Therefore, I classified macaque monkey fibers into four categories depending on which stimulus excited them most strongly. Among 67 fibers 29 responded best

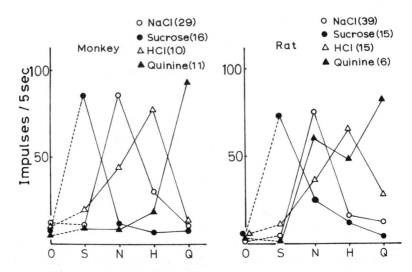

Fig. 1. Mean response profiles of macaque monkey and rat chorda tympani fibers across four taste stimuli: 0.3 M sucrose (S), 0.3 M NaCl (N), 0.01 M HCl (H) and 0.003 M quinine hydrochloride (Q) for the monkey and 0.5 M sucrose (S), 0.1 M NaCl (N), 0.01 M HCl (H) and 0.02 M quinine hydrochloride (Q) for the rat. The amount of response at point O indicates the spontaneous discharge level in each class of fibers. ○ ●△ and ▲ represent salt-best, sucrose-best, acid-best and quinine-best fibers, respectively, and numerals in parentheses indicate the number of fibers in each category.

to 0.3 M NaCl (salt-best fibers), 16 to 0.3 M sucrose (sucrose-best fibers), 10 to 0.01 M HCl (acid-best fibers) and 11 to 0.003 M quinine (quinine-best fibers). Response profiles of the four classes of fibers, represented by mean numbers of impulses for each stimulus, are shown in Figure 1 (left). Sucrose- and quinine-best fibers as a whole responded little to the three stimuli other than sucrose or quinine, respectively, although salt-best and acid-best fibers responded relatively well to HCl and NaCl, respectively.

Since we had previously recorded responses in a large number of chorda tympani fibers of rats and hamsters to 0.1 M NaCl, 0.5 M sucrose, 0.01 M HCl and 0.02 M quinine hydrochloride (5), I classified these fibers in a similar manner. As shown in figure 1 (right), each class of fibers in the rat is multiply sensitive to the stimuli, and specificity in taste sensitivity to the four basic stimuli is far less developed in the rat than in the macaque monkey. Similarly 28 hamster fibers were classified into the four

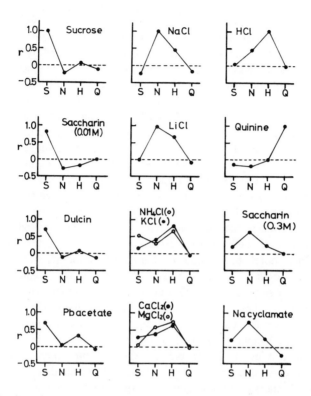

Fig. 2. Correlation profiles of a variety of taste stimuli across the four basic stimuli (0.3 M sucrose (S)), 0.3 M NaCl (N), 0.01 M HCl (H), and 0.003 M quinine hydrochloride (Q), obtained from the numbers of impulses in 5 sec in 67 monkey chorda tympani fibers, for 0.3 M sucrose, 0.3 M NaCl, 0.01 M HCl and 0.003 M quinine hydrochloride, and from those in 30-38 fibers for 0.01 M saccharin, 0.003 M dulcin, 0.1 M Pb acetate, 0.3 M NaCl, 0.3 M LiCl, 0.3 M NH_4Cl, 0.3 M KCL, 0.3 M $CaCl_2$, 0.3 M $MgCl_2$, 0.3 M saccharin and 0.1 M Na cyclamate. The r's plotted across-fiber correlation coefficients between amounts of responses to the test stimulus and to one of the four basic stimuli.

categories. Profiles of salt-best and acid-best fibers are more or less the same as those of the same categories of fibers in rats and macaque monkeys, but sucrose-best fibers are more specifically sensitive to sucrose in hamsters than in rats while sensitivity of quinine-best fibers to quinine appears to be lower in hamsters than in rats. Therefore, comparing the response profiles of chorda tympani fibers in three animal species specific

25

sensitivity of taste nerve fibers to sucrose or quinine differs from one animal to another, and it is most developed in macaque monkeys.

Saccharin produced a very good response in the macaque monkey chorda tympani (8), and impulse discharges were produced by saccharin not only in sucrose-sensitive fibers but also in quinine-sensitive and NaCl-sensitive fibers. Also 0.003 M dulcin and 0.1 M Pb acetate elicited impulse discharges in sucrose-sensitive fibers. In general, fibers responding better to sucrose tend to respond better to these sweet-tasting substances. Fig. 2 presents correlation profiles of various taste stimuli across the four basic stimuli, obtained from impulse discharges in about 30 chorda tympani fibers. As indicated in the figure, profiles for saccharin (0.01 M), dulcin and Pb acetate are similar to that for sucrose: Amounts of responses to these stimuli are highly significantly ($p < 0.001$) correlated across fibers with those for sucrose but not with those for other three basic tastes. The results suggest that in the macaque monkey these chemicals produce taste quality similar to that elicited by sucrose.

Certain behavioral studies indicate that hamsters and rats prefer sucrose and saccharin to water but do not respond to dulcin (6, 7), while squirrel monkeys prefer sucrose and dulcin but reject saccharin (6). On the other hand, rhesus monkeys show preference for saccharin (8). A recent electrophysiological study also indicates that sweet-tasting proteins, monellin and thaumatin, elicited a significant response in the green monkey but not in the guinea pig and rat (9). Our study on macaque monkey chorda tympani fiber responses showed further that, contrary to rats, hamsters and squirrel monkeys, macaque monkeys possess an ability of tasting a variety of substances which produce sweet taste in men.

References

1. Ogawa, H., Yamashita, S., Noma, A. and Sato, M. 1972. Physiol. Behav., 9, 325-331.
2. Ogawa, H., Sato, M. and Yamashita, S. Proc. Internat. Uni. Physiol. Sci., IX, 425.
3. Frank, M. 1973. J. Gen. Physiol., 61, 588-618.
4. Frank, M. 1974. Chem. Sens. Flav., 1, 53-60.
5. Ogawa, H., Sato, M. and Yamashita, S. 1968. J. Physiol. 199, 223-240.
6. Fisher, G.L. Pfaffmann, C. and Brown, E. 1965. Science, 150, 506-507.
7. Carpenter, J.A. 1965. J. Comp. Physiol. Psychol. 49, 139-144.
8. Weiskrantz, L. 1960. Nature, 187, 879-880.
9. Brouwer, J.N., Hellekant, G., Kasahara, Y., van der Wel, H. and Zottermann, Y. 1973. Acta Physiol. Scand., 89, 550-557.

Genetic Alteration of Insect Sugar Reception

Toshihide Kikuchi

Department of Biological Science, Tohoku University,

Kawauchi, Sendai 980, Japan

There has been increasing interest in the properties of sugar receptor sites or receptor substances both in vertebrates and insects. Recent studies on blockade of sugar receptor of the fleshfly with p-chloromercuribenzoate (PCMB) (1) could show the existence of two different sugar receptor sites as has been suggested by Evans (2). Besides, many experiments have been attempted in an effort to prove the hypothesis that the enzyme α-glucosidase was a sugar receptor substance (3, 4), but its relation to sugar reception has remained undefined.

In the past few years some important developments have been seen in behavior genetics: chemotaxis in a bacterium (5) and olfaction in an insect (6) have been successfully analysed, by using a mutation which indicates an alteration of one element at a time. A similar approach is expected to be fruitful in the studies of sugar receptors. I shall describe here properties of a sugar receptive mutant of the fruitfly, *Drosophila melanogaster*, with simultaneous reductions in electrophysiological responses and hydrolytic activities to compounds possessing certain glucopyranosyl moieties, which afford direct evidence that α-glucosidase plays a key role in the sugar reception of *Drosophila* (7, 8, 9).

1. Isolation of a sugar receptive mutant (7). Males of an isogenic line *(AA75-3)* of *D. melanogaster* were treated with a mutagen (EMS), and then mated singly with untreated virgin females of the marker strain *T5* having genotype $Cy/Pm;Ubx/Sb$ with an attached X-chromosome. Each F_1 progeny male of $Cy/+;Ubx/+$ was then back-crossed to virgins of the marker strain *T5* to isolate homozygotes in descendants. Of a number of isogenic lines, one line of flies (mutant strain *126B04*) revealed a marked decrease in response to D-glucose in the preference-aversion tests.

To define the regions where modifications were involved, electrophysiological responses of flies aged 0-2 days were recorded from the tips of three particular type-*L* labellar single hairs; a glass capillary filled with sugar dissolved in 100 μM NaCl solution served as a stimulator and recording electrode, while an indifferent electrode was inserted into the back of the head (10). The mutant flies (*126B04*) revealed marked reductions of impulses in response to the stimulations by D-glucose, as compared with those in the parent (*AA75-3*): 10.6 and 3.4 impulses per

TABLE 1

Responses of Mutant and Parent Flies to Sugars

Sugars	Concentrations (M)	Number of impulses/ 0.2 sec*		Differences $(Nm-Np) / Np$ %	
		Parent (Np)	Mutant (Nm)		
Trehalose	0.1	7.8	8.1	4	N***
Maltose	0.1	15.2	12.3	−19	—**
	1	25.6	22.4	−12	—
Methyl—α—D— glucoside	0.1	17.5	8.6	−51	—
D—Glucose	0.1	7.9	3.3	−58	—
	1	16.4	12.4	−24	—
Sucrose	0.1	15.6	13.0	−17	—
	1	24.0	25.6	7	N
Melezitose	0.1	15.3	11.8	−23	—
Turanose	0.1	16.5	10.7	−35	—
Palatinose	0.1	14.5	14.3	−1	N
Raffinose	0.1	13.1	13.1	0	N
D—Fructose	0.1	7.8	8.5	9	N
	1	15.7	14.9	−5	N

*Average number of impulses originating in sugar receptors of labellar single chemosensory hairs of *Drosophila* flies aged 12-36 hr during 0.2–0.4 sec after initiation of stimulations at 20°C in a relative humidity range over 90%. **Specific reductions where $(Nm-Np)/Np<$-10%. ***Normal responses.

0.2 sec to 0.1 M glucose, and 19.6 and 14.5 impulses to 0.1 M glucose for the parent and the mutant, respectively, whereas the mutant flies revealed normal responses to D-fructose. This constitutes the first direct demonstration of the existence of two different sugar receptor sites as suggested for the other dipterous insects (1, 2). The genetic analysis indicated that the mutation was autosomal recessive, since the F_1 hybrid flies revealed phenotypes similar to those of the parent in response to glucose and no significant difference in response was recognized between sexes of the same strain.

TABLE 2

Glycosidase Activities of Mutant and Parent Flies

Sugars	Concentrations (nM)	Rates of hydrolyses (nmole/min/mg protein) Parent (Rp)	Mutant (Rm)	Differences (Rm-Rp) / Rp (%)	
Trehalose	10	G* 625	1,227	96	+
Kojibiose	5	G 72	71	−1	N**
Nigerose	1	G 52	41	−21	−
Maltose	5	G 985	831	−16	−
Isomaltose	2	G 38	29	−24	−
Sucrose	10	G 2,650	2,240	−15	−
Melezitose	20	G 177	95	−46	−
Turanose	5	G 393	293	−25	−
Palatinose	5	G 49	49	0	N
Raffinose	10	Fru <85 Gal <25	<85 <25		
Methyl−α−D− glucoside	100	G <4	<4		
Phenyl−α−D− glucoside	20	Ph 1,930	1,390	−28	−
p-Nitrophenyl− α−D−glucoside	10	PNPh 720	1,014	41	+

*G, Liberated D-glucose; Fru, D-fructose; Gal, D-galactose; Ph, phenol; PNPh, p-nitrophenol. **N, Normal activities.

2. *Stimulant specificities and molecular features in nerve responses of chemosensory hairs of the mutant (8).* To know the molecular features of the stimulants, to which the mutant flies reveal abnormal responses, responses of the labellar chemosensory hairs of the mutant flies (*126B04*) to stimulations by sugars were compared with those of the parent (*AA75-3*). Table 1 shows the average number of impulses originating in the sugar receptors of single hairs of both mutant and parent flies during 0.2-0.4 second after the initiation of stimulation. It appears from this table that the mutation reduces the magnitudes of responses by 12 to 58 percent to stimulations by 0.1 M sucrose, turanose, melezitose and

methyl-α-D-glucoside, and both 1 and 0.1 M D-glucose and maltose, as compared with those of the parent (these compounds were defined as "specific"), whereas the mutant showed normal responses to 0.1 M trehalose, raffinose and palatinose, and both 1 and 0.1 M D-fructose. The labellar hairs of both mutant and parent flies were completely insensitive to 1 M D-mannose, D-xylose, 0.1 M cellobiose, melibiose, 0.3 M lactose and 0.01 M p-nitrophenyl-α-D-glucoside. From comparisons of chemical structures of these "specific compounds" with those of the "non-specific", the minimal configurational requirements in producing specificities appear (at present) to be particular α-D-glucopyranosyl moieties, which resemble the substrate specificities for the enzyme α-glucosidase (EC 3.2.1.20) isolated from other organisms.

   3. Modifications of glycosidase activities in the mutant (9). From the specificity-structure relationships discussed above, it became of particular interest to see whether or not the glycosidase activities of the mutant flies for the "specific sugars" were also genetically modified. To compare the glycosidase activities of the mutant flies with those of the parent, about 50 mg of flies of each sex aged 12 to 36 hr was homogenized for a few minutes in 5 ml of 100 mM potassium phosphate − 2% Triton X-100 buffer, pH 6.2, and gently stirred in 10 ml of the same buffer for 1 hr. The homogenate was then centrifuged at 1,500 g for 30 min, and portions of the supernatant were used for the assay (the supernatant contained about 90% of the glycosidase activities of the whole flies). 1 ml of a sugar solution dissolved in distilled water was then mixed with the same volume of the supernatant, diluted from 10- to 200-fold with the same buffer, and the reaction mixtures thus prepared were incubated at 30°C for 100 min. Liberated D-glucose (11, 12), D-galactose (13), D-fructose (14), phenol (15) and p-nitrophenol (16) were assayed routinely by the methods described.

   Table 2 shows the glycosidase activities of both mutant and parent flies represented as average rates of hydrolyses per min per mg protein for given concentrations of 13 glycosides. It is apparent from this table that the mutant flies revealed reductions in hydrolytic activities for maltose, sucrose, turanose and melezitose, to which they revealed reductions of nerve responses. For the "non-specific sugars", trehalose, raffinose and palatinose, no significant reduction of the hydrolytic activity was seen in the mutant flies, as compared with those in the parent. In the cases of kojibiose, nigerose, isomaltose, phenyl-α-D-glucoside and p-nitrophenyl-α-D-glucoside, the relation between the response and hydrolytic activity is not clear, since the responses of the labellar hairs to these stimuli have remained undefined.

For these phenomena, one explanation based on a single gene mutation is that the specific reductions of the hydrolytic activities for the seven glycosides are due to a genetic modification or defect of an isozyme of α-glucosidase, which result in compensatory increase in the activities of trehalase and other glycosidases. If it is so, the fact that the mutant flies revealed simulataneous reductions of electrophysiological responses and hydrolytic activities to sucrose, turanose, melezitose and maltose, whereas they did not reveal any reduction of hydrolytic activities to the "non-specific sugars" implies that the enzyme α-glucosidase plays a key role in sugar response of *Drosophila*. If additive mutants with abnormal responses to trehalose, raffinose, palatinose, fructose and other sugars, are isolated, the use of such mutants would offer a favorable opportunity to dissect the complex system of sugar receptors in insects.

References

1. Shimada, I., Shiraishi, A., Kijima, H. and Morita, H. 1974. J. Insect Physiol. 20, 605.
2. Evans, D.R. 1961. Science 133, 327.
3. Hansen, K. 1969. Olfaction and Taste (Ed. by Pfaffmann, C.) 3, 382. Rockefeller University Press, New York.
4. Morita, H. 1972. Olfaction and Taste (Ed. by Schneider, D.) 4, 357. Wissenschaftliche Verlagsgesellschaft, Stuttgart.
5. Hazelbauer, G.I. and Adler, J. 1971. Nature New Biol. 230, 101.
6. Kikuchi, T, 1973. Nature 243, 36.
7. Isono, K. and Kikuchi, T. 1974. Nature 248, 243.
8. Tanimura, T. and Kikuchi, T. in preparation.
9. Kikuchi, T., Nishino, T. and Tanimura, T., in preparation.
10. Isono, K. and Kikuchi, T. 1974. Jap. J. Genet. 49, 113.
11. Borel, E., Hostettler, F. and Deuel, H. 1952. Helv. Chim. Acta 35, 115.
12. Papadopoules, N.M. and Hess, W.C. 1960. Arch. Biochem. Biophys. 88, 117.
13. Avigad, G., Amaral, D., Asensio, C. and Horecker, B.L. 1962. J. Biol. Chem. 237, 2736.
14. Roe, J.H., Epstein, J.H. and Golstein, N.P. 1949. J. Biol. Chem. 178, 839.
15. Robertson, J.J. and Halvorson, H.O. 1957. J. Bacteriol. 73, 186.
16. Halvorson, H. and Ellias, L. 1958. Biochim. Biophys. Acta 30, 28.

Chairman's Summary

Phylogenetic Emergence of Sweet Taste
Yngve Zotterman
Wenner-Gren Center
S-113 46 Stockholm

Dealing with the characteristics of sweet taste must bring up the old debate about specificity of fibers as was done this morning. From the electrophysical experiments during the last few decades it has been found that the animal kingdom displays sweet receptors of many different kinds from the specific receptors of an insect which only respond to fructose to the receptors of monkeys which respond to all substances which taste sweet to man. Pigeons lack response to both quinine and sugars (Kitchell & Zotterman, 1959). Further we have found that in some dogs the fibers responding to sugars do not react on saccharine as they do in the monkey and in man. In the dogs saccharine excites the fibers responding to quinine (B. Andersson et al., 1950).

As shown in Fig. 1. sucrose, saccharine, glycerol and ethanol glycol in addition to their action on sweet receptors also excite a various number of fibers responding preferably to quinine. It was also reported by Kurihara this morning that Monellin in man gives a "heavenly" sweetness. I would thus like to suggest that sucrose and saccharine also in man excite not only the strictly specific sweet fibers but also to a less degree bitter taste fibers while the sweet protein stimulates the sweet fibers only. In this connection the experiments performed by our late friend Georg von Bekesy on single papilla are worth mentioning. When he stimulated a single papilla with weak currents, the subject sometimes responded sweet taste which the subject characterized as "heavenly" sweet. That indicates that the sweetness experienced was different from that of sucrose.

The specific sweet fibers in primates are stimulated by such a wide field of structures as sugars, saccharine, lead acetate and the sweet proteins Monellin and Thaumatin. They are also reacting on Miraculin, the glucoprotein, which has no taste in itself but which to acid solutions adds a sweet taste in humans and probably in monkeys too. The "sweet" sites are in man completely inhibited by Gymnemic acid. Gymnemic acid has previously in the literature been described as impairing also the bitter taste. The assumption that Gymnemic acid has a weak impairing effect

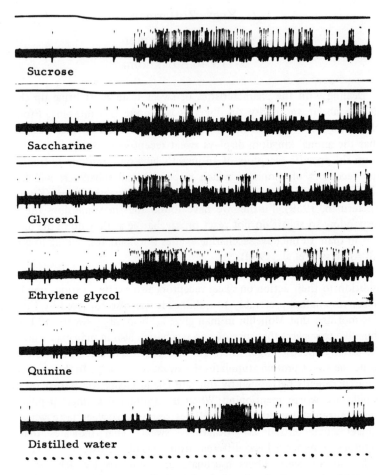

Fig. 1. Records from a small strand of the chorda tympani of Macacus rhesus which contained few active fibers. Note particularly the large spikes, and also those of intermediate size which project both above and below the baseline. All solutions made up in Ringer's solution. Time: 10 per second. (Gordon et al., 1959).

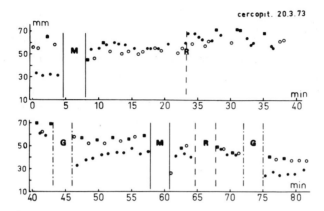

Fig. 2. The peak responses of the summated activity during stimulation with 0.3 M NaCl (■), 0.5 M sucrose (○) and 0.03 M citric acid (●) in a Ceropithecus monkey are plotted. M, indicates the application of miraculin; R, rubbing of the tongue; and G, the application of gymnema. (Hellekant et al., 1974).

upon the bitter receptors in man can explain the fact that after its application on the human tongue, sucrose has no taste whatever. We have recently reported that Gymnemic acid does not wipe out the response to sucrose in the monkey but it wipes out the augmented response of acid after the application of Miraculin (Fig. 2). Beidler said the other day at the New Delhi Congress that such an effect was expected because the binding of the pentose group of the miraculin molecule to the sweet sites must be much weaker than that of sucrose. All these suggestions should be verified by experiments on single or few fiber preparations in man.

As I wrote 15 years ago the existence of a certain number of specific taste fibers in the monkeys — you may call them "sweet-best fibers" etc. if you like — makes it easy to understand that we can discriminate the classical basic groups of taste. The across fibers will give further acuity to our taste. I repeat what Pfaffmann said this morning: "You can have both specificity according to the classical notion but also an operation of across fiber pattern mechanisms. These two views are not incompatible; they are most probably working together."

To sum up I would say that among the mammals sweet receptors are widely distributed but they seem to be absent in the cat and I would believe also in some other carnivorous animals like the lion and the tiger — but we do not know as yet. What still is lacking, however, and which is more important for you young people to tackle is the primary mechanism of excitation. I would like to look forward to our next meeting in France in 1977 when some of you will produce data giving us the clue to understand what is really happening in the taste cell when sucrose is applied to the tongue.

References

Andersson, B., Landgren, S., Olsson, L. and Zotterman, Y. 1950. Acta Physiol. Scand. 21, 105.
Bèkèsy, G.v. 1964. J. appl. Physiol. 19, 1105.
Gordon, G., Kitchell, R., Ström, L. and Zotterman, Y. 1959. Acta Physiol. Scand. 46, 119.
Hellekant, G., Hagstrom, E.C., Kasahara, Y. and Zotterman, Y. 1974. Chemical Senses and Flavor 1, 137.
Kitchell, R., Strom, L. and Zotterman, Y. 1959. Acta Physiol. Scand. 46, 133.

PHYLOGENETIC EMERGENCE OF BITTER TASTE

Chairman: R. Bernard
Moderator: J. Garcia

The Evolution of Bitter and the Acquisition of Toxiphobia

John Garcia and Walter G. Hankins

Departments of Psychology and Psychiatry

University of California at Los Angeles

Abstract

Natural aversions to bitter substances have been acquired by a wide variety of species through natural selection. Strong bitter tastes are rejected and milder ones are suspect when first encountered by humans. Similar reactions are noted in monkeys, infraprimate mammals, and birds. More surprising are the observations that many invertebrate and protozoan species reject flavors called bitter by man. For example, oysters, hydra and stentors also actively avoid quinine. Bitter receptors probably first appeared in the coelenterates during the Cambrian age, since modern sea anemones possess sensory receptors and display a pronounced rejection to quinine. Thus the rejection of bitter may represent a phylogenetically ancient natural response tendency. Bitter poisons are found throughout the world in plants, animals and water holes, hence rejection of bitter has survival value. In addition, many animals and plants have developed bitter flavors, secondarily taking advantage of this widespread natural rejection to fend off predators. Others visually mimic the toxic species. Closely related to the bitter rejection is the conditioned aversion whereby animals learn to reject any flavor followed by illness produced either by toxins or by dietary deficiencies. These conditioned rejections by mammals often resemble the behavioral reactions to bitter. Conditioned aversions are also widespread, and may represent an ancient form of associative learning.

The Nature-Nurture Issue

Western man's obsessive desire to attribute his success to either his natural inherited talents or to his culturally nurtured habits is incessantly discussed by students of behavior. Since behavior can neither occur in an environmental vacuum nor be manifested without inherited structure, the question will never be answered to everyone's satisfaction. This persistent dualistic view often obscures the more basic Darwinian axiom that the present day environmental contingencies which shape the coping responses of the successful individual are basically the same environmental contingencies which shaped the anatomical structure of his successful species, otherwise his survival would have been impossible. The two best known proponents of the behavioral effects of current environmental contingencies (Skinner, 1966) and past evolutionary contingencies (Lorenz, 1965) agree on this essential premise though they disagree on which facet should be considered the primary causal agent in the analysis of the present behavior of living organisms.

The monistic action of nature and nurture is marvelously illustrated in the natural aversions many organisms display towards bitter substances and the rapidity with which they acquire conditioned aversions to any flavor which is followed by toxicosis. The cardinal rule for survival of a hungry person in a strange environment is; "Do not eat anything that tastes unpleasant particularly if it is bitter." This rule is based upon the natural wisdom of the body which man, like other species, acquired thru natural selection. The second rule for survival is; "When a novel food is encountered, eat only a small amount and then wait to see if this food is agreeable to your stomach." This rule reflects the capacity of man to acquire tastes for new sources of nutrition and distastes for toxins. It is also reflected in other animals, who display neophobia or a cautious, finicky approach to novel foods. If the first small tentative meal is followed by beneficial effects then more food is ingested on the next occasion. On the other hand, if toxicosis ensues, the animal rejects the food on the next encounter. This specialized form of learning is also widespread throughout the animal kingdom (Garcia, Hankins and Rusiniak, 1974).

The Natural Rejection of Bitter

Morgan's scientific canon against anthropomorphism limits our generalizations concerning inferences of bitter sensations in organisms other than humans old enough to verbally report the quality of the test substance. However, it is difficult to resist the anthropomorphic temptation when one observes the characteristic grimace displayed by neonate humans and nonhuman primates induced by quinine solutions, or for that matter, the characteristic grooming observed in rats when they encounter bitter food. In fact, species from virtually every phylum reject solutions which are also rejected by man who describes them as bitter.

For example, Schaeffer (1905) reports that *Stentor caeruleus,* a protozoan, will not feed on materials which have been soaked in quinine and then rinsed. When the stream of quinine food reaches the oral disc, the stentors contract, or bend away; Parker (1910) has found that both the ostia and ocula of sponges *Stylotella heliophila* close when small amounts (2×10^{-4} M) of strychnine, a bitter compound, contacts them. Quinine in 3 x 10^{-4} M solution inhibits the feeding of wireworms *Agriotes Spp.* and 3 x 10^{-3} M quinine reduces biting by 75% (Crombie and Darrah, 1947). Earthworms *Lumbricus terristris* are apparently much more sensitive, withdrawing immediately from 10^{-5} M quinine hydrochloride applied to the prostomium (Laverack, 1960).

Another fascinating example is the ancient arthropod *Limulus*, or the horseshoe crab which is equipped with chemical sensors on its mandible and on its legs. The leg receptors apparently do not respond to bitter materials, but the mandible receptors are extremely sensitive (Patten, 1894, Barber, 1956). Arthropods provide many other examples. Both adult and larval insects are sensitive to bitter substances. For example, Von Frisch (1934) reported that some bees would reject 8 x 10^{-4} M quinine in 1 M sucrose solution, and Dethier (1937) found that lepidopterous larvae *Porthoteria dispar* "spit out" bitter substances.

The molluscs and echinoderms appear to be extremely sensitive to bitter materials. Measuring the oral tentacular retraction of oysters *Ostrea virginica,* Hopkins (1935) found clear retraction responses to 4.6 x 10^{-6} M quinine sulfate, and Wells (1963) using a behavioral task found that octopii *Octopus vulgaris* could discriminate 1.5 x 10^{-7} M quinine sulfate in sea water. Olmsted (1917) found that sea cucumbers *Synaptula hydriformis* withdrew from 10^{-4} M quinine. Sessile tunicates *Ascidia atra* retract their siphon to 5 x 10^{-5} M quinine sulfate (Hecht, 1918).

Aversive reactions to bitter substances have been reported for all classes of chordates. Quinine stimulation of oral receptors in fish produce avoidance reactions (Sheldon, 1909; Herrick, 1904; Bardach and Case, 1965; Bardach and Atema, 1971). Little work has been done with the sensitivity of amphibia, but some salamanders *Triturus* do respond to bitter taste qualities (Bardach, 1967). Feral pigeons, *Columba livia,* (Duncan, 1960) and rats *Rattus norvegicus* (Cicala and McMichael, 1964) reject quinine hydrochloride in all detectable concentrations.

Toxins occur naturally in many substances which organisms use for food and water. The poison water hole in alkali deserts is the most familiar example, but toxic plants are even more common. Nearly ten percent of plant species contain toxic alkaloids and toxic glycosides are even more prevalent (Kingsbury, 1964). For example, the cassava root which has been exploited by man as a source of starch from primitive times contains fatal concentrations of cyanogenetic glycosides. The pulp must be repeatedly pounded and washed in running or boiling water to release cyanide gas before consumption. The pulp is not eaten at all if it tastes bitter (Montgomery, 1969). It is clear that the foraging organism which has a natural distaste for bitter and which acquires an aversion for the

41

flavor of toxins has a much better chance of surviving and passing on these characteristics to his offspring.

The First Bitter Rejection

Hodgson (1967) believes that the first chemical receptors must have evolved in the coelenterate animals or their close relatives, probably during the early Cambrian era, over 500 million years ago. This supposition is based upon the existence of sensory cells called palpocils, a modified neuron found in present day coelenterates. The first receptors probably were functionally related to the attractive properties of food and those related to the rejection of toxins probably developed as a refinement in the feeding system of the coelenterate.

Feeding in the sea anemone has been studied since the turn of the century. Parker (1896) found that anemones *Metridium* who are alternately fed pieces of crabmeat and filter paper soaked in crab juice would initially ingest both stimuli, but with repetition, the filter paper was not taken in while pieces of meat were. Jennings (1905) repeated parts of Parker's experiments, then went on to demonstrate that the anemone *Aiptasia anulata* "habituates" to novel mechanical stimuli, a drop of water falling on the medium directly over the animal, when the interstimulus intervals were 5-min or less apart. Thus the sea anemone is capable of simple forms of learning.

This raises an intriguing question. Can the sea anemone learn to reject food that is followed by toxicosis in a single trial? Some observations indicate that such learning may be possible. Hodgson (1970) related that he presented amines in gelatin capsules to sea anemones. They soon habituated to this novel way of feeding and ingested the capsules. He then filled a capsule with quinine and fed it to his subjects. A short time later they rejected the dissolving capsule, everting their gastrovascular cavities. After a single bitter experience the sea anemones refused to ingest another capsule. Harrelson and Harrelson (1973) relate similar observations. Sea anemones fed shrimp colored with a bitter food dye refused to eat shrimp again. One trial learning may yet be demonstrated in this animal whose nervous system is a network of interconnected neurons without the ganglionic masses of tissue found in higher organisms.

Elaboration of Bitter Defenses

The widespread existence of bitter toxic plants and the corresponding development of a general avoidance of bitter tastes in the herbivores and omnivores naturally lends selective advantage to any plant or animal possessing a harmless bitter flavor mimicking a toxin. For example, the skin of toads is covered with glandular cells which secrete bitter materials. The secretion of *Dendrobatus* is so toxic that Indians use it to poison arrow tips. The secretion of *C. Dorsata* is relatively innocuous but bitter enough to afford protection against predation (Stecher, 1968; Noble, 1931).

Secondary elaboration of defensive mimicry comes about when one species comes to resemble the bitter or toxic species in its nongustatory features. Brower and his associates (Brower, 1969; Brower, Pough and Meck, 1970; Cook, Brower and Alcock, 1969; Platt, Coppinger and Brower, 1971) present an elegant example of the subtle relationships that arise in the elaboration of a bitter defense and its mimics which develop among toxic plants, insects and birds. Milkweeds *Asclepiga curassavica* and other plants of the Asclepiadaceae family synthesize toxic cardiac glycosides which apparently do not serve any metabolic function in the plant. However, this toxin causes mammals to vomit; and as a result cattle and other herbivores learn to avoid the bitter plants increasing the probability of the plants' survival. In contrast, the larvae of the Danianae, a large group of insects which includes Monarch butterfly *Danaus plexippus* feed on the milkweed and assimilate its toxin without ill effects. The toxins are retained by the adult after metamorphosis so both the larvae and the adult butterfly possess the toxin and apparently a mild bitter taste. When the bluejay, a member of the insectivorous family Corvidac, eats such a Monarch butterfly it becomes ill and vomits. Brower (1969) gives a graphic description of how the bluejay rapidly acquires an aversion for the taste of the butterfly and then later learns to avoid the distasteful butterfly on sight. Thus the visual pattern of the butterfly provides a secondary defense. Some Monarch larvae feed on nontoxic plants so these butterflies possess neither the toxin nor its unpalatable flavor. However, their visual resemblance to the toxic butterflies provides them with defense against the wary bluejays who have become ill after eating the toxic members of the species. Other species (e.g., *Limenitis archippus)* which resemble the Monarch obtain a degree of protection through mimicry. However, when the bluejay is driven by hunger it will seize any available butterfly in its beak and release only the unpalatable ones. The ultimate consummatory decision rests on taste not vision.

Toxicosis in Carnivorous Predators

Large feral members of the dog and cat families prey upon other mammals which, unlike *Danianae* do not contain a bitter poison distributed throughout their tissues, yet these predators also exhibit behavior remarkably similar to that of the bluejay when the flesh of their prey is artificially poisoned. One meal of hamburger laced with six grams of lithium chloride in capsule form was sufficient to cause coyotes *Canis latrans* to reject hamburger the next time it was presented (Gustavson, Garcia, Hankins and Rusiniak, 1974). Similar results were obtained in a cougar *Felis concolor* fed deer meat containing the same toxic dose (Gustavson and Garcia, 1974). Both species displayed the "disgust reactions" characteristic of their species to the test meal. The coyotes either urinated on the hamburger, buried it or rolled on it as they do with putrid and fecal material. The cougar shook all four paws in rotation and retched as it walked away. Rats display similar disgust reactions to bitter food and to food associated with toxicosis. They dig the unpalatable food out of the dish then vigorously groom themselves and rub their muzzles on the floor.

Gustavson *et al.,* (1974) tested the effect of the conditioned flesh aversion upon the coyote's response to living prey. One or two trials with a given flesh (lamb or rabbit) followed by lithium illness was enough to inhibit the attack upon that prey when it was presented several days later. The coyotes immediately attack the alternate prey indicating that the aversion is specific to the prey which was arbitrarily linked to toxicosis. Brett (1974) conducted similar tests on several large redtail hawks *Buteo Borealis* which, unlike the coyote, do not taste their prey during the attack since they kill with their taloned feet. One trial with a dead bitter mouse followed by lithium illness will inhibit consumption of the prey, but will not reliably prevent the attack on a living black mouse.

While there is some evidence that toxicosis will cause both the avian and mammalian predators to hesitate in their attack on the basis of visual and/or olfactory cues, it is clear that taste is the critical sensory analyzer for consumption. Toxicosis causes the taste of the prey to become unpalatable. When the visual and olfactory cues become associated with the unpalatable food, the predator loses interest in the prey. In fact feral predators will actively avoid the now unpalatable prey just as they naturally reject bitter food.

Aversive conditioning provides man with a behavioral method of controlling predation which protects the domestic prey animal and conserves the feral predatory animal as well. The method is extremely effective because the conditioning takes advantage of the plasticity within the food regulation mechanisms established by natural selection of the predator species. In flavor-toxicosis conditioning, nurture mimics nature.

This research was supported in part by USPHS grant Number RO1 NS 11618.

References

Barber, S. 1956. J. Exp. Zool. 131, 51-73.

Bardach, J. 1967. In The Chemical Senses and Nutrition (M. Kare and O. Maller, eds.) John Hopkins Press, Baltimore, 19-43.

Bardach, J. and Atema, J. 1971. In Handbook of Sensory Physiology, IV (L. Beidler, ed.), Springer-Verlag, New York, 293-336.

Bardach, J. and Case, J. 1965. Copeia. 2, 194-206.

Brett, L.L. 1974. Unpublished data and notes, Department of Psychology, University of California at Los Angeles.

Brower, L.P. 1969. Sci. Amer. 220, 22-29.

Brower, L.P., Pough, H.F. and Meek, H.R. 1970. Proc. Nat. Acad. Sci. 66, 1059-1066.

Cicala, G. and McMichael, J. 1964. Canad. J. Psychol. 18, 28-35.

Cook, L.M., Brower, L.P. and Alcock, J. 1969. Evolution. 33, 339-345.

Crombie, A. and Darrah, J. 1947. J. Exp. Biol. 24, 95-109.

Dethier, V. 1937. Biol. Bull. 72, 7-23.

Duncan, C. 1960. Ann. Appl. Biol. 48, 409-414.

Frisch, K. von. 1934. Z. Vergl. Physiol. 21, 1-156.

Garcia, J., Hankins, W. and Rusiniak, K. 1974. Science. 185, 824-831.

Gustavson, C. and Garcia, J. 1974. Psychol. Today, 8, 60-72.

Gustavson, C., Garcia, J., Hankins, W. and Rusiniak, K. 1974. Science. 184, 581-583.

Harrelson, J. and Harrelson, S. 1974. Department of Psychology, California State University at Los Angeles. Personal Communication.

Hecht, S. 1918. J. Exp. Zool. 25, 261-299.

Herrick, C.J. 1904. Bull. U.S. Fish Comm. 22, 237-272.

Hodgson, E.S. 1970. Biology Department, Tufts University. Personal Communication.

Hodgson, E.S. 1967. In The Chemical Senses and Nutrition (M. Kare and O. Maller, eds.), John Hopkins Press, Baltimore, 7-18.

Hopkins, A. 1935. Bull. U.S. Bur. Fish. 47, 249-261.

Jennings, H. 1905. J. Exp. Zool. 2, 447-472.

Kingsbury, J. 1964. Poisonous Plants of the United States and Canada. Prentice Hall, New Jersey.

Laverack, M. 1960. Comp. Biochem. Physiol. 1, 155-163.

Lorenz, K. 1965. Evolution and Modification of Behavior. University of Chicago Press, Chicago.

Montgomery, R. 1969. In Toxic Constituents of Plant Foodstuffs. (I. Lienter, ed.), Academic Press, New York, 143-158.

Noble, G.K. 1931. The Biology of the Amphibia. McGraw-Hill, New York.

Olmsted, J. 1917. J. Exp. Zool. 24, 333-379.

Parker, G.H. 1910. J. Exp. Zool. 8, 1-42.

Parker, G.H. 1896. Bull. Mus. Comp. Zool. Harvard. 29, 107-119.

Patten, W. 1894. Quart. J. Micro. Sci. 35, 1-96.

Platt, A.P., Coppinger, R.P. and Brower, L.P. 1971. Evolution. 25, 692-701.

Schaeffer, A. 1910. J. Exp. Zool. 8, 75-132.

Sheldon, R. 1909. J. Comp. Neurol. 19, 273-311.

Skinner, B.F. 1966. Science. 153, 1205-1213.

Stecher, P.G. 1968. The Merck Index (Merck and Co.) Rahway, New Jersey.

Wells, M. 1963. J. Exp. Biol. 40, 187-193.

Temporal Patterns of Liquid Intake
and Gustatory Neural Responses
Bruce P. Halpern
Department of Psychology and Section of Neurobiology and
Behavior,
Cornell University, Ithaca, New York 14853 U.S.A.

Terrestrial vertebrates who drink external liquids use two mechanisms: licking and sucking. Licking involves repeated, small, high frequency samples of a liquid; sucking, slower, relatively large-volume intake. Licking is a common intake pattern (6), while suction drinking is observed in animals which include humans. In rodents (1, 10, 11), licking occurs at 5-8 licks/sec, with the higher rates observed at the beginning of licking and when licking from a drinking tube. The usual volume per lick is estimated to be 5 μl (4, 14). However, with an opaque liquid one finds that some of the liquid removed from the drinking tube by a rat in a high-restriction drinking situation (11) is lost before reaching the mouth (Figure 1). The drinking tube (SS) was connected to a reservoir containing 5.9% (W/V) microcrystalline cellulose (Cellex MX, Biorad) dispersed in H_2O (n = 1.3330). Rat was on a 20 hr water deprivation schedule. Arrows (Figure 1) point to drops of liquid removed from the drinking tube by this licking, but lost at the front of the enclosure rather than ingested. Liquid also falls below and inside the enclosure. The duration of liquid contact by the tongue at a drinking tube is usually estimated to be 50 msec (1, 11). Measurements from highly trained rats (conditions as for Figure 1) show an increase in lick parameters with number of licks (Table 1). Such licking will continue for many sec without pause. However, if a novel taste stimulus is introduced, pauses will appear within a few licks (9) (Table 2). If a conditioned aversion procedure is used, thus providing high motivational strength and salience for the taste stimuli, rats will cease licking after a single lick of easily discriminated liquids (5, 12).

Human drinking behavior has a different time scale (8, 9). For a single voluntary sip, ca. 1500 msec elapse from lifting a "glass" until the liquid in the glass contacts a lip. Liquid contact is maintained for almost 1000 msec, varying somewhat as a function of liquid used (Figure 2). Several sec elapse while the glass is returned to the table surface. The first swallow occurs after the end of liquid contact. Simultaneous EMG measurements from the upper lip indicate that during a human sip, EMG increases at the time of glass lifting and is near maximum at liquid contact. Precontact-EMG and early contact-EMG are independent of the particular liquid being drunk. However, the latter portion of contact-EMG and the post - contact EMG show significant ($p < 0.001$) differences as a function

Fig. 1. Photograph of rat (Holtzman male, 498g) licking in high restriction environment.

of the liquid being drunk (Figure 2). For this human sipping, on each of six to eight randomized trials for each of the three liquids, the subject was instructed to take a single sip from a drinking glass (containing 174 ml) "sometime" after a "ready" signal was spoken. Before the start of each sip, the glass rested on a table at which the subject sat. Between sips, the glass was removed, refilled, and the appropriate glass placed on the table. All liquids were colorless. The H_2O (deionized distilled water, n = 1.3330, conductivity = 1.36 x 10^{-6} mhos/cm^2) was the solvent for the sucrose solution (A. R. grade, 309 mM/L). The sucrose solution was the solvent for the *Kool-Aid* (3.69 g/L, supplied by General Foods without Red #40). A lickometer in series with a nickle-chromium electrode in the drinking glass and the subject's ground electrode, detected the onset and duration of liquid contact. Swallowing was detected by an acoustically shielded throat microphone. Averaging of EMG was done from liquid contact to 1450 msec before contact, and to 4500 msec after contact. The bottom line in Figure 2 (dash, two dots) is control-trial EMG: a silent electronic signal was used in place of the spoken "ready" (sip initiation), but at the time at which the "ready" signal would normally be given.

A comparison of human sipping and rat licking shows liquid intakes

Table 1.
Median licking parameters for six rats (all conditions as for Figure 1). Times in msec. 6.3 μl/lick.

Duration	Interval[a]	ILI[b]	Rate[c]	Pauses[d]
		First 10 Licks		
35	100	142	7.8	0
		First 20 Licks		
45	107	149	7.1	0
		First 40 Licks		
49	107	163	6.6	0
		First 70 Licks		
51	110	164	6.2	0

a = stop-start interval
b = interlick interval (start-start)
c = licks per second
d = intervals (stop-start) greater than 150 msec

with 10-fold differences in time course. Also, voluntary multiple human sips are separated by approximately 1500 msec. Finally, the effective flow rate of rat licking is less than 40 μl/sec while flow rates during human sipping range from 3 to 20 ml/sec. (Of course, the tongue area is considerably greater). These liquid ingestion differences suggest that different components of gustatory neural input may be important in ingestion decisions made by human versus rat. Rat gustatory neural responses have a low noise, rapidly rising, early phasic component, with a concentration-dependent slope (7). The response reaches maximum magnitude within a few hundred msec after stimulus presentation (2, 4, 7, 12). The early phasic component may provide sufficient neural input for some taste decisions (7,13). Comparable observations on the time course of human gustatory neural responses are not available. Long time-constant

Table 2

Median licking parameters for seven rats contacting NaCl solution (505 mM) for the 1st time. Other conditions as for Figure 1. Times in msec. 5.6 μl/lick.

Dur[a]	Int[b]	ILI[c]	Rate[d]	Pauses[e]	Licks to 1st Pause	Total Ps[f] Duration
			First 10 Licks			
31	98	131	8.0	2	4	900
			First 20 Licks			
35	101	147	7.2	3	4	1267
			First 40 Licks			
45	112	161	6.6	3	4	1460
			First 70 Licks			
49	112	164	6.3	4	4	2804

a = duration of liquid contact (drinking tube)
b = interval (stop-start)
c = interlick interval (start-start)
d = licks per second
e = intervals (stop-start) greater than 150 msec
f = total duration of all pauses

analog-summator (6) recordings of human gustatory neural responses (3) suggest a rising phase similar to the rat. However, the large differences in temporal pattern between human and rat liquid ingestion imply that much of the post maximum neural components are normally available for human sucking decisions. Generalizations from gustatory neural responses of licking liquid-ingestors to the neural responses of suction drinkers such as humans, may neglect major components.

The preparation of this report was supported by National Science Foundation Grant GB-43557.

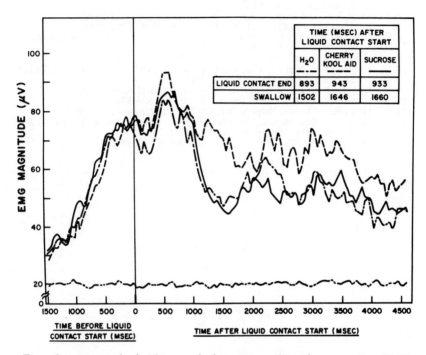

Fig. 2. Averaged (within and between subjects) upper-lip EMG, liquid-contact durations, and times to first swallow-sound, for twelve human subjects.

References

1. Allison, J. 1971. J. Comp. Physiol. Psych., 78, 408-417.
2. Beidler, L.M. 1953. J. Neurophysiol., 16, 595-607.
3. Diamant, H., Oakley, B., Strom, L., Well, C., and Zotterman, Y. 1965. Acta Physiol. Scand., 64, 67-74.
4. Faull, J.R., and Halpern, B.P. 1971. Physiol. Behav., 7, 903-907.
5. Halpern, B.P. and Tapper, D.N. 1971. Science, 171, 1256-1258.
6. Halpern, B.P. 1973. In: W.I. Gay, Ed., Methods of Animal Experimentation Vol. IV, Ch. 4, Academic Press, N.Y.
7. Halpern, B.P., and Marowitz, L.A. 1973. Brain Res., 57, 473-478.
8. Halpern, B.P., Nichols, T.L., and Leibling, L. L. 1973. Annual Report, PRL, p. 35, No. 65, U.S. Army Natick Laboratories.
9. Halpern, B.P., and Nichols, T.L. 1974. Unpublished observations.
10. Hulse, S.H. 1968. J. Comp. Physiol. Psych., 66, 536-539.
11. Marowitz, L.A., and Halpern, B.P. 1973. Physiol. Behav., 11, 259-263.
12. Ogawa, H., Sato, M., and Yamashita, S. 1973. Physiol. Behav., 11, 469-473.
13. Scott, T.R. 1974. Physiol. Behav., 12, 413-417.
14. Stellar, E., and Hill, J.H. 1952. J. Comp. Psych., 45, 96-102.

Instrumental Responding Terminates
Aversive Tastes

Harry R. Kissileff

University of Pennsylvania

Introduction

Previous investigations of preferences and aversions for solutions have utilized either intake (Weiner & Stellar, 1951; Young & Greene, 1953; Richter & Campbell, 1940) in single stimulus or choice situations, or number of choices in a brief exposure (Young & Greene, 1953; Young & Falk, 1955) as measures of the strength of preferences and aversions. There are limitations in these methods of measuring directly what Young (1961) has called hedonic intensity, since intake is influenced by both the stimulus quality of the solution and the nutritive state of the animal (Epstein, 1967); the choice measures are relative, not absolute; and the animal must perform an ingestive act in order to sample the solution. A more direct way of investigating hedonic value which does not depend on ingestive acts is to train the animal to perform an instrumental response to escape from an aversive taste. The magnitude of responding becomes a measure of the aversiveness of the solution. This paper describes and validates a method of training rats to press a lever to terminate an intraoral injection of aversive tasting solutions.

Methods

Five albino rats were successfully implanted with a chronic cheek fistula (Kissileff, 1972). All rats were trained and tested individually in the same chamber. A lever requiring 20 g force to operate it protruded 0.5 cm into the chamber. A 10% or 20% NaCl solution was injected continuously through the fistula at 2 ml/min by a peristaltic pump. When the lever was operated, the injection was terminated for 40 sec and then began again. Operations of the lever during the 40 sec period when the pump was off, were counted but did not postpone onset of the next injection.

In order to shape the rat, a blunt probe (large nail or straightened paper clip) was passed through the opening in the chamber which admitted the lever and was placed just above the lever by the trainer. Shaping proceeded as the rat mastered each step in a sequence of the following actions: attacking the probe when poked by the trainer, orienting toward the lever, attacking the probe and landing on the lever as the probe was withdrawn through the opening. Sessions lasted 30 min to 1 hour per day and

53

consisted of 20 to 30 reinforcements (terminations of injection). Between sessions only, the rats had ad libitum access to lab chow pellets and water.

When the rat turned off all the injections in a session by itself, the lever was recessed 0.2 cm outside the chamber, and the sessions were limited to exactly 30 min per day at the same time daily. Three tests were made to determine whether the rat was pressing the lever to terminate the aversive taste. First, the fixed ratio (FR) required for reinforcement was increased. All further tests were conducted using the FR-4 schedule of reinforcement. Presses made while the pump was off did not contribute to fulfilling the ratio. In the second test, 20% urea, 10% and 30% sucrose, distilled water, and several concentrations of NaCl (5%, 2.5%, 1.1%, 0.5% and 0.3%) were injected instead of 10% NaCl. Each of four rats was tested with several of these solutions at least three sessions in a row. The fifth rat was trained as described, but after the first test, it was tested for only 10 min/day with an injection at 0.5 ml/min using 10-3 M sucrose octaacetate (SOA), 20% urea, 10-3% quinine hydrochloride and 10% NaCl. Finally, the first four rats were tested with both 10% NaCl and distilled water delivered at 0.5 ml/min.

20% SALINE

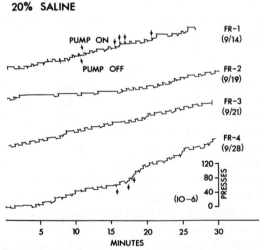

Fig. 1. Pattern of lever pressing and termination of injection for one rat on four fixed ratios of reinforcement. Each trace is one session. Pen in up position indicates pump on. Pen in down position indicates pump off. Pen steps upward with each press of the lever (see scale at lower right).

Results

After a mean of 8.2 (R=7-13) sessions, all rats were pressing and received a mean of 23.8 (R=11-36) reinforcements. The latency to press after the onset of injection was short. (Fig.1). The shorter the latency, the shorter the flat portion of the inverted U which separates successive reinforcements. Spike like configurations (arrows on top line) indicate latencies less than 12 sec. As the fixed ratio increased, pressing clustered in runs of rapid pressing and was not simply spread out randomly. There was also an overshoot (arrows on bottom line) by continued pressing after the injection terminated. Each animal increased the number of bar presses from a mean of 117.4 (R=66-202) at FR-1 to a mean of 146.9 (R=93-210) at FR-4, but the number of reinforcements dropped from a mean of 28.5 (R=22-35.5) at FR-1 to a mean of 19.8 (R=17.5-24.0) at FR-4. It can be concluded that the rats were not simply pressing the lever randomly.

Changing the solution injected showed that concentrated sucrose solutions produced the least, and concentrated NaCl solutions the most, responding to terminate their injection. Water produced an intermediate amount of responding (Fig. 2). In the animal tested for 10 min daily, the number of reinforcements varied with the solutions as follows: NaCl − 9.0, urea − 3.8, SOA − 0.4, quinine − 4.7. A gradual reduction in the saline concentration produced an inverted preference-aversion function with a valley at 0.3% and 0.5%, which coincided with the peak preferences measured using a 2-bottle 24-h test. The technique is therefore sensitive to hedonic intensity of the solution, and the rat has not merely learned a response associated with pain, dehydration or the stimulus on which it was trained.

Finally, the rate of injection contributed to control of responding when water was injected but did not, when NaCl was injected, since there were 58% fewer reinforcements at the slower rate when water was injected, but only 1.6% fewer when 10% NaCl was injected. In summary, rats will perform an instrumental response to terminate intraoral injection of solutions. This behaviour is motivated and controlled by stimulus properties of the solution and to a lesser extent by its rate of injection. The critical difference influencing responding among the solutions is their taste.

Discussion

The results of measuring hedonic intensity using the operant method are in agreement with findings using other methods. They show for the first time that fluids which are unacceptable are also aversive in that they will support escape responding. If further tests confirm these findings, the way is open to exploring directly such issues as the role of alliesthesia in the control of ingestion (Cabanac, 1971), the evaluation of taste vs. internal need in the control of intake (Jacobs & Sharma, 1969), the hedonic value of fluids paired with toxic after effects (Garcia *et al.*, 1974), and the role of taste and motivational level in finickiness following lateral hypothalamic lesions (Teitelbaum & Epstein, 1962) and in states of obesity (Kennedy, 1950). In conclusion, taste is capable of motivating learned instrumental escape responding which can be used as a measure of hedonic intensity of a solution injected into the mouth.[1]

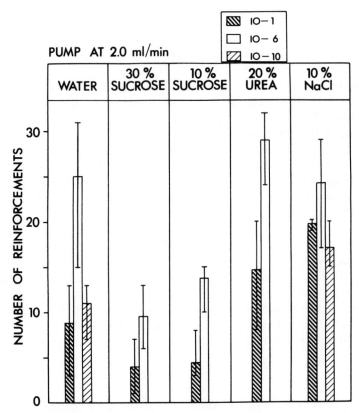

Fig. 2. Number of reinforcements when various solutions are injected. Each bar is a mean (range shown by line above and below top of bar) of at least three sessions responding for the rat indicated by the key at top.

References

Cabanac, M. 1971. Science 173, 1103-1107.

Epstein, A.N. 1967. In "Handbook of Physiology. Sect. 6: Alimentary Canal" (C.F. Code, ed.). Vol. 1, pp. 197-218. Am. Physiol. Soc., Washington.

Garcia, J., Hankins, W.G., and Rusiniak, K.W. 1974. Science 185, 824-831.

Jacobs, H.L. and Sharma, K.N. 1969. Ann. N.Y. Acad. Sci. 157, 1084-1125.

Kennedy, G.C. 1950. Proc. Roy. Soc. Lond. B. 137, 535-549.

Kissileff, H.R. 1972. In "Methods in Psychobiology" (R.D. Myers, ed.). Vol. 2, pp. 125-154. Academic Press, New York.

Richter, C.P. and Campbell, K.H. 1940. J. of Nutrit. 20, 31-46.

Teitelbaum, P. and Epstein, A.N. 1962. Psychol. Rev. 69, 74-90.

Weiner, I.H. and Stellar, E. 1951. J. Comp. and Physiol. Psychol. 44, 394-401.

Young, P.T. 1961. "Motivation and Emotion". John Wiley and Sons, New York.

Young, P.T. and Falk, J.L. 1955. J. Comp. and Physiol. Psychol. 49, 569-575.

Young, P.T. and Greene, J.T. 1953. J. Comp. and Physiol. Psychol. 46, 295-298.

[1]I thank C. Pfaffmann for loan of equipment and B. Farren, C. Woolsey, and M. Paul for help with experiments. Supported by NSF grant #GB 33219.

Response Patterns of
Rat Glossopharyngeal Taste Neurons
Marion Frank
The Rockefeller University, New York, N.Y., U.S.A.

Most single mammalian peripheral taste nerve fibers whose response profiles have been studied were sampled from the chorda tympani nerve of rodents (Erickson, Doetsch, and Marshall, 1965; Ogawa, Sato, and Yamashita, 1968; Frank, 1973). The chorda innervates taste buds of the fungiform papillae on the front of the tongue. Taste buds in the foliate and circumvallate papillae on the back of the tongue are innervated by the glossopharyngeal nerve which is more sensitive to bitter stimuli in rats (Pfaffmann, Fisher, and Frank, 1967).

If hamster or rat chorda fibers are divided into four groups: sucrose-, NaCl-, HCl-, or quinine-best (the best stimulus being that one of the four, at moderate intensity, which elicits the largest response), a very small number of fibers fall into the quinine-best group (1% for hamster, 4% for rat) and these respond nearly as well to HCl. Fibers in each of the other three groups are more numerous. Mean hamster fiber responses greater than one-quarter the size of the mean best response occur in sucrose-best fibers only to sweet (fructose, saccharin); in NaCl-best fibers only to NaNO$_3$ and HCl; but in HCl-best fibers to all tested stimuli (including NaCl, NH$_4$Cl. citric acid, quinine) except the sweet. Thus, among rat and hamster chorda fibers, bitter and acid sensitivities are associated and fibers especially tuned to bitter are not found.

Figure 1 presents response profiles for five sucrose-best (A), NaCl-best (B), HCl-best (C), and quinine best (D) rat glossopharyngeal taste nerve fibers. The glossopharyngeal, freed from other tissue and cut centrally, was split into fine strands giving the differentially amplified signals of single fibers and the circumvallate stimulated via an inserted pipette. A good number of glossopharyngeal-circumvallate fibers are highly tuned to quinine. Of 52 fibers, 10 are sucrose-, 11 NaCl, 12 HCl, and 19 quinine-best. The profiles for the most sensitive are shown. Figure 2 gives relative mean response profiles for the four groups of glossopharyngeal (A) and chorda (B) fibers (data from Ogawa, Sato, and Yamashita, 1968).

Sucrose-best glossopharyngeal fibers are more tuned than chorda fibers in this category: the second-best salt response being larger for chorda fibers. The salt-best fibers all have similar profiles, but many more are found in the chorda: 25 of 49, versus 11 of 52 in the glossopharyngeal.

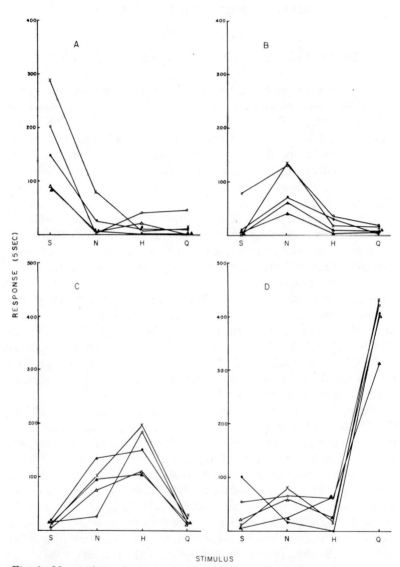

Fig. 1. 20 rat glossopharyngeal-circumvallate nerve fiber response profiles to 4 tastes: 5 for fibers which respond best (largest number of spikes in 5 s) to (A) .3 M sucrose (S), (B) .3 M NaCl (N), (C) .003 M HCl (H), or (D) .001 M quinine hydrochloride.

Acid-best fibers differ: quinine responses are larger in chorda fibers; but, the quinine stimulus used to test them was stronger: .02 M compared to .001 M for the glossopharyngeal.

But the greater differences in fibers of the two nerves are in the quinine-best fibers. They are more common in the glossopharyngeal: 19 of

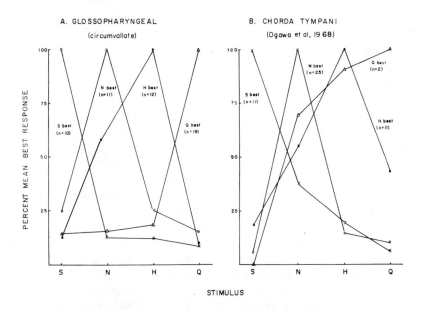

A. GLOSSOPHARYNGEAL
(circumvallate)

B. CHORDA TYMPANI
(Ogawa et al, 1968)

STIMULUS

Fig. 2. (A) Rat glossopharyngeal (GL) and (B) rat chorda tympani (CT) nerve fiber percent mean best response profiles to 4 tastes: .3 M sucrose (S), .3 M NaCl (N), .003 M (HCl (H), and .001 M quinine hydrochloride (Q) for the GL; and .5 M sucrose, .1 M NaCl, .01 M HCl, and .02 M quinine for the CT. Fibers from each nerve are divided into 4 groups: those responding best (largest number of spikes in 5 s) to sucrose, NaCl, HCl, or quinine; mean responses calculated and percent of mean best response plotted.

61

52, versus 2 of 49 in the chorda. They are also much more highly tuned to quinine in the glossopharyngeal: their second-best response to HCl being less than 20% of the quinine response; whereas, in chorda fibers it is more than 90%. The salt response is also much larger in the chorda fibers. In fact, the quinine-best fiber scarcely exists in the rat chorda, but it is the most common and most highly tuned of rat glossopharyngeal-circumvallate fibers.

Table 1 presents correlations between responses of 58 rat glossopharyngeal-circumvallate nerve fibers and Table 2, correlations

TABLE 1. Rank order correlation coefficients between responses of 58 rat glossopharyngeal nerve fibers to 4 tastes[+] and water with circumvallate stimulation.

	NaCl	HCl	Quinine	Water
Sucrose	+0.03	+0.09	+0.10	+0.20
NaCl		+0.45*	+0.24	+0.25
HCl			+0.36*	+0.35*
Quinine				+0.60*

*P < .01, t test.
[+]The molarities for the 4 tastes are: .3 sucrose, .3 NaCl, .003 HCl, and .001 quinine hydrochloride.

between responses of 15 to 23 of these fibers. Sucrose responses are significantly correlated only with fructose responses, and NaCl responses only with HCl responses. HCl responses are also significantly correlated with acetic and citric acid responses, with KCl and NH_4Cl, as well as with quinine and water responses. Quinine responses are also significantly correlated with responses to caffeine and water (that is, water after water). Thus, as in hamster and rat chorda fibers, the HCl-best fiber is less tuned than sucrose- or NaCl-best fibers; but, in the rat glossopharyngeal, it is also less tuned than the quinine-best fiber.

The study of a set of nerve fibers which innervate the back of the tongue, with its stronger bitter sensitivity, has shown that fibers which are

TABLE 2. Rank order correlation coefficients between responses of rat glossopharyngeal nerve fibers to 4 common tastes and 10 other tastes[+] with circumvallate stimulation.

	Sucrose	NaCl	HCl	Quinine	N[+]
Fructose	+0.74*	+0.02	+0.09	-0.13	23
Saccharin	+0.22	+0.25	+0.34	+0.28	21
NH_4Cl	-0.04	+0.21	+0.61*	-0.13	22
KCl	-0.19	+0.43	+0.71*	+0.26	21
Acetic[+]	-0.13	+0.20	+0.78*	+0.07	22
Citric[+]	-0.06	+0.36	+0.73*	+0.04	23
$MgSO_4$	+0.27	+0.14	-0.13	+0.22	15
Urea	-0.26	-0.24	+0.49	+0.59	15
S. O-A.[+]	-0.14	+0.22	+0.45	+0.52	23
Caffeine	-0.24	+0.16	+0.40	+0.60*	22

*$P < .01$, t test.

[+]The molarities for the common tastes are as noted in Table 1. For the other tastes, they are: 1.0 fructose, .1 saccharin, .3 NH_4Cl, .3 KCl, .01 acetic acid (Acetic), .003 citric acid (Citric), .3 $MgSO_4$, 1.0 urea, .0003 sucrose octa-acetate (S. O-A.), and .03 caffeine. N is the number of fibers whose responses were used in each correlation.

highly tuned to bitter do exist and a bitter sensitivity is not necessarily associated with strong sensitivities to acids and salts. In fact, it has added another response pattern: the quinine- or bitter-best fiber pattern to the sucrose-, salt-, and acid-best patterns found typically among rodent chorda tympani nerve fibers.

M. FRANK

Acknowledgement

This work was supported by the National Science Foundation Grant BMS70-00854 AO4 to C. Pfaffmann.

References

Erickson, R.P., G.S. Doetsch, and D.A. Marshall. 1965. J. Gen. Physiol. 49, 247.
Frank, M. 1973. J. Gen. Physiol. 61, 588.
Ogawa, H., M. Sato, and S. Yamashita, 1968. J. Physiol. (Lond.). 199,223
Pfaffmann, C., G.L. Fisher, and M. Frank. 1967. In Olfaction and Taste II. T. Hayashi, editor. Pergamon Press, Inc., Elmsford, N.Y. 361.

Phylogenetic Emergence of Bitter Taste
Chairman's Summary

Rudy A. Bernard

Department of Physiology, Michigan State University ,

East Lansing, Michigan 48824 U.S.A.

The thrust of *Garcia's* paper was that aversion for bitter tasting substances has been acquired by a wide variety of species through natural selection. Because of the widespread link between poisonous substances and bitter taste, rejection of bitter conferred survival value. Closely related to bitter rejection is the taste aversion produced in animals for a food or a flavour followed by illness produced either by toxins or by dietary deficiencies. Dramatic films were shown of a hawk, a coyote, and a cougar avoiding their natural prey after an experimentally induced illness associated with prior ingestion of the prey. This powerful link between taste and the consequences of food ingestion provided the framework for the evolutionary strategy that led to the pairing of rejection with bitterness. Is the rejection of bitter, then, something learned or innate? This issue was raised in several ways during the discussions that followed. *Bardach* offered the observation that Japanese yellow-tail fish, who are visual feeders, rejected quinine after a single touch with their lips. *Laverack* pointed out that the sea anemone's rejection of quinine was not even due to a neural response but rather was a property of the epithelial cells. With reference to the hawk that rejected a bitter flavored black mouse that had previously made it sick *Pfaffmann* asked whether the bitter taste itself was a factor in the rejection and *Weisinger* wondered what taste had to do with it since the hawk and the other predators avoided the prey without tasting it. *Garcia* replied that the bitter flavor was mild, not strong enough to cause rejection, and essentially acted as a cue. Even though the hawk did not have to taste in order to reject the prey, relying instead on external cues, *Garcia* added that when driven by hunger, conditioning to external stimuli broke down and the animal attacked the prey, but did not eat it, showing typical disgust reactions upon sampling the meat. In response to *Stuart's* query about controlling for olfactory cues *Garcia* replied that earlier work with anosmic rats showed that they acquired bait shyness even better than control rats. Clearly other sensory modalities are involved in the conditioned aversion but taste plays the most important and enduring role. This was underlined by *Michel's* experience of a prolonged aversion to orange juice, which had been his favorite drink before it was used to help him take a dose of aureomycin to which he developed a strong reaction. *Garcia* explained that conditioned taste aversion was an automatic changing of palatability, an immediate and primary aversion and not the effect of associative learning. Knowing that the orange juice didn't make him sick did not override the instantly acquired aversion. In earlier work *Garcia* convincingly demonstrated the involuntary character of taste aversion by demonstrating its acquisition by an animal whose experimental illness was imposed under general anesthesia. The final comment on the first paper was made by *Halpern* who took the extreme position that the bitter taste doesn't exist, and has nothing to do with taste aversion. This comment touched on a fundamental issue that was never fully discussed during the symposium. In addition to the perennial question of whether there are 4 basic taste qualities, there is the further question of the applicability to animal experience, from which most of our experimental evidence is derived. This becomes another example of the basic biological problem of species differences, made even more acute when dealing with such an inner experience as taste sensation. Is there a radical discontinuity between the sensory experience of humans and that of animals? In his comments *Halpern* referred to recent work of Sharma and her associates in which squirrel monkeys showed a strong preference for quinine and sucrose octa-acetate over a wide range of concentrations, but not for saccharin even though they showed the usual preference for glucose. The same problem appears when comparing 2 different groups of people as Sharma and her associates reported at the Physiological Congress in New Delhi. They found that a group of Indian laborers, who were poorly fed and from a separate caste, described both sourness and bitterness as increasing in pleasantness with increasing concentration whereas the control group, composed of well nourished, higher caste medical students did not. Both groups made similar judgments concerning the intensity and quality of the stimuli.

To this observer the preceding comments and results stress the importance of distinguishing between taste quality and palatability. Although there may be real differences in sensory quality between species or even individuals within a species, it is more likely that palatability is the variable factor. Quality and palatability seem to have a natural relationship, but in the case of man voluntary behaviour may override it, as when unpleasant medicine is deliberately ingested. In animals palatability is necessarily expressed through ingestive behaviour, but in neither animals nor man can taste quality be directly inferred from ingestive behaviour.

Frank's paper applied the concept of best taste to the classification of rat glossopharyngeal taste fibers. This method is analagous to the best frequency analysis used for auditory neurons and was originally applied to taste by Wang and Bernard. When this type of analysis was applied to rat and hamster chorda tympani fibers, less than 5% showed a quinine-best response and these responded nearly as well to HCl. With rat glossopharyngeal fibers, however, over 35% showed a quinine-best profile with no link to any of the other 3 taste qualities. These data provide a very satisfactory explanation for the earlier observations that the back part of the tongue and the glossopharyngeal nerve are more sensitive to bitter stimuli. The significance of these findings was questioned by *Oakley* when he recalled that aversion to quinine persists in rats with sectioned glossopharyngeal nerves. Since the chorda doesn't seem to have specific quinine fibers, how can the aversion be explained in the context of a labelled line system? *Frank's* rejoinder was that she didn't say anything about labelled lines. To this observer this appears to sidestep the issue, since it is hard to see the usefulness of a best-taste analysis other than as the underpinning of a neo-Müllerian approach. *Zotterman* further alluded to the role of palatability in species and group differences when he commented that carnivorous animals take great pleasure in eating putrefied food that tastes very awkward to us. In Sweden, he added, people eat herring that others find awkward smelling and tasting. A question was also raised concerning the rat's ability to taste PTC, but no definite information was available. In response to a question about the correlation between quinine and sucrose sensitivity in the rat *Frank* replied that it was positive but not significant (.1). *Tucker* then asked why a correlation of .59 in Table 2 wasn't considered significant whereas .60 was. *Frank* replied that it was a question of sample size and that in this case the cut-off point for significance was .60.

Halpern's paper presented a detailed examination of rat drinking patterns and how they differ from those of humans. The rat's discriminatory power was highlighted by its ability to reject an aversive stimulus by licking less than 5 ul for as little as 50 msec. This contrasts dramatically with the human pattern where the time scale for sipping is at the 1000 msec level and flow rates at the ml rather than the ul level. The short time scale of the rat behavioural response is very important theoretically since it clearly establishes that the very earliest component of the neural response is used by the rat for decisions about taste quality. The longer time scale of the human sipping pattern offers the possibility that this species uses the later portion of the neural response for drinking decisions. That such is the case was questioned by *Mistretta* who asked whether human decision time would not be more comparable to that of the rat if the situations were made more comparable. In his reply *Halpern* emphasized that humans don't lick and thereby further advanced his general argument that species differences must be more carefully considered. *Smith* countered that the slower drinking pattern of humans doesn't necessarily imply a longer decision time than in the rat. *Halpern* replied that the shortest reported time for humans was 500-600 msec, although he admitted that truly comparable experiments haven't been performed in both species. His main point, however, was that the human drinking pattern argued for the hypothesis that man has a much longer period of usage of neural information than the rat.

In this observer's opinion Halpern's data inescapably lead to the conclusion that the elaborate cross-fiber patterns generated by Erickson and his followers, in which the first 300 msec of the neural response is systematically omitted from analysis, are not used by the rat for coding taste quality. Based on the analysis of cat single fiber data, however, this observer would like to propose that the later portion of the neural response is required for intensity rather than quality coding and to make the testable prediction that the rat would require more than 50 msec to make decisions concerning stimulus intensity. The rapid decision time for quality identification thus becomes another argument in favor of a neo-Müllerian view.

PHYLOGENETIC EMERGENCE OF SOUR TASTE

Chairman: G. Hellekant
Moderator: L. Beidler

Phylogenetic Emergence of Sour Taste

L.M. Beidler

Florida State University

Tallahassee, Florida 32306

Sour taste is a term derived from human experience and is difficult to apply to other animal species. Therefore, the term 'response to acids' is considered in this paper.

The phylogenetic tree is greatly branched so that man did not emerge from a continuous line that included all lower animals. However, one can examine the response to acids of a variety of animals of differing levels of development to determine whether the ability to respond emerged at any given level.

It is clear from the literature that most, if not all, animals respond to acids. Withdrawal behavior is usually exhibited and is independent of whether or not a nervous system is present.

Amoeba proteus motion is reversibly inhibited at pH 4.7-5.0. Both organic and inorganic acids of proper pH repel Paramecia.[1] Many earthworms avoid soil of pH 3.8-4.5 and neural recordings indicate that the body responds to buffer solutions of the same pH range.[2] Cole et al.[3] measured the percent value closure of the barnacle *Balanus balanoids* as a function of pH of 3 mineral and 7 fatty acids. At equal pH the order of stimulating efficiency increased with the length of the carbon chain, the H^+ concentration of valeric acid being half that of formic acid for the same magnitude of response.

The increase in chain length of fatty acids also increases the effectiveness of response of the sunfish[4]. The following concentrations were effective for a 4 sec reaction time.

Acid	pH
HCl	2.94
Formic	3.50
Acetic	4.19
Propionic	4.50
Butyric	4.81
Valeric	5.13
Caproic	5.50

Similar results were found in the reaction times of the killifish[5] as shown below using closure of the mouth and opercula as a behavioral index.

Acid	pH
HCl	3.66
Acetic	4.30
Propionic	4.46
Butyric	4.64
Valeric	4.72
Caproic	4.69

Electrophysiological studies of the taste buds of the Atlantic salmon, *Salmo salar,* show a similar increase in stimulation as the chain length of aliphatic acids is increased[6]. The Sutterlins recorded from the facial nerve as test solutions were flowed over the snout region on the upper jaw. Behavioral experiments were also undertaken. The fish feed on commercial fish pellets and prefer those treated with caproic acid. Polyurethane foam pellets treated with valeric and caproic acids are also accepted with a greater preference for the latter. These behavioral responses are independent of olfactory cues.

Hidaka studied the effect of salts on the response of the palatal chemoreceptors of the carp, *Cyprinus carpio,* to acid stimulation. The addition of 0.05 M NaCl shifted the response magnitudes-pH curves to a lower pH range. Thus, the behavior of these fish chemoreceptors is very similar to that of human taste receptors as found several decades earlier[8,9] and re-examined by Beidler in 1967.[10]

Chadwick and Dethier[11] studied the behavioral response of blowflies, *Phormia regina,* to tarsal stimulation with 18 fatty acids and one mineral acid. They concluded that the hydrogen ion and the anion play major roles in such stimulation rather than the undissociated molecule. The trend of the efficiencies of stimulation of the acids is similar to that found by others using different species. This can be seen by comparing the values of the mean rejection thresholds.

Acid	Millimolar conc.
Formic	47
Acetic	260
Propionic	130

Acid	Millimolar conc.
n-Butyric	8.2
iso-Valeric	6.6
n-Valeric	6.9
Chloroacetic	6.6
Trichloroacetic	20
Lactic	91
Puruvic	62
Glyceric	112
Malonic	59
Succinic	50
Malic	106
d-Tartaric	39
Fumaric	51
Maleric	37
Hydrochloric	40

The concentration of various acids necessary to evoke taste neural responses in the rat equal in magnitude to that evoked by 5 mM HCl was studied by Beidler.[10] Twenty acids studied showed the following:

Acid	Millimolar conc.
Sulfuric	2.2
Oxalic	3.3
Hydrochloric	5.0
Citric	5.5
Tartaric	5.9
Nitric	5.9
Maleic	6.4
Dichloroacetic	9.0
Succinic	10.0
Malic	10.0
Monochloroacetic	10.4
Glutaric	11.0
Formic	11.6
Adipic	14.0
Glycolic	15.0
Lactic	15.6

Acid	Millimolar conc.
Mandelic	25.0
Acetic	64.0
Propionic	130.0
Butyric	150.0

The above results are similar in many respects to the human taste thresholds as published by Taylor et al.[9] They included not only results from their own experiments but also those published by others. Thus, comparisons must be made with caution.

Acid	Millimolar conc.
Acetic	2.8
Adipic	2.3
Aspartic	2.15
Azelaic	2.0
Benzoic	2.8
Benzene Sulfonic	1.6
Butyric	3.5
Citric	0.97
Chloroacetic	1.7
d-Tartaric	1.1
Formic	1.8
Glutaric	2.25
Glycollic	1.4
Lactic	2.8
L-Malic	1.3
L-Tartaric	0.75
Maleic	0.9
Malonic	1.3
Mandelic	3.8
Mesaconic	1.0
Oxalic	1.0
Phenylacetic	4.75
p-Hydroxy benzoic	2.0
Phthalic	0.65
Propionic	3.3
p-Toluene Sulfonic	1.5
Pyrotartaric	1.5

Acid	Millimolar conc.
Pyruvic	2.0
Sebacic	1.9
Suberic	1.45
Succinic	1.6
Tartronic	0.95
Tricarballylic	1.16
Trimethylacetic	5.75
i-Valeric	3.7

From the data listed above, one may conclude that:

1. All species studied show a response to acid stimulation. Thus, response to acids did not emerge at any specific phylogenetic level.

2. Many species show common properties in their response to acids.

a. Equal responses do not occur at equal pH's.

b. The magnitude of response at equi pH increases with chain length within an homologous series of fatty acids.

c. The magnitude of response depends upon both the hydrogen ion and anion concentration.

d. The undissociated molecule plays little or no role in acid stimulation.

e. An increase in ionic strength by addition of a salt to the acid solution increases the effectiveness of stimulation. The response-pH curve is shifted toward higher pH's.

The similarities observed with many diverse species can be understood since the membranes of all cells contain proteins, phospholipids and varying amounts of polysaccharides. Consideration of H^+ binding to protein explains the relationship of pH or molar concentration with response. The response increases with H^+ membrane adsorption. However, since H^+ binding is more effective if an anion is also bound to prevent membrane charging, the acid response is dependent on the fatty acid chain length and other properties of the acid. In addition, it is known that an increase in ionic strength shifts the H^+ titration curve of proteins to higher pH's which accounts for the unusual effects observed as Na acetate is added to acetic acid or NaCl added to HCl.[10] In summary, there is a great similarity to acid responses of various animal species and the ability of protein to bind H^+. Protein binding of H^+ has best been studied experimentally with wool[12] and has a theoretical basis as well[13,14].

From the above literature, one may conclude that animals of all levels respond to acids and that there is a similarity in the data that can be explained by H^+ binding to the cell membranes. Thus, sourness, if one may refer to the acid response as such, emerged with the appearance of taste itself. Electrophysiological studies of single fiber responses indicate the presence of acid response in all species studied and little specificity is present. To date, the single fiber specificity discovered for sugars has not been seen for acids.

References

1. Jennings, H.S. 1906. Behavior of the Lower Organisms, New York: Columbia Univ. Press.
2. Laverack, M.S. 1961. Comp. Biochem. Physiol. 2, 22-34.
3. Cole, W.H. and Allison, J.B. 1933. J. Gen. Physiol. 16, 895-903.
4. Allison, J.B. 1932. J. Gen. Physiol. 15, 621-628.
5. Allison, J.B. and Cole, W.H. 1934. J. Gen. Physiol. 17, 803-816.
6. Sutterlin, A.M. and Sutterlin, N. 1970. J. Fisheries Res. Bd. of Canada 27, 1927-1942.
7. Hidaka, I. 1972. Jap. J. Physiol. 22, 39-51.
8. Beatty, R. and Cragg, L. 1935. J. Amer. Chem. Soc. 57, 2347-2351.
9. Taylor, N.W., Farthing, F.R. and Berman, R. 1930. Protoplasma 10, 84-97.
10. Beidler, L.M. 1967. Olfaction and Taste II, Oxford: Pergamon Press, pp. 509-534.
11. Chadwick, L. and Dethier, V. 1946. J. Gen. Physiol. 30, 255-262.
12. Alexander, P., Hudson, R.F. and Earland, C. 1963. Wool: Its Chemistry and Physics, New Jersey: Franklin Publishing Co.
13. Tanford, C. 1961. Physical Chemistry of Macromolecules, New York: John Wiley & Sons.
14. Gilbert, G.A. and Rideal, E.K. 1943. Proc. Roy. Soc. 182A, 335-346.

Origin of Receptor Potential In Chemoreception

K. Kurihara, N. Kamo, T. Ueda and Y. Kobatake

Faculty of Pharmaceutical Sciences

Hokkaido University, Sapporo.

There is much evidence (1,2) suggesting that the mechanism of generation of taste receptor potential is hard to explain in terms of the ionic theory proposed by Hodgkin and Huxley. We hence proposed (2,3) that taste stimulation is initiated by a change in electrical potential at the membrane-solution interface. This notion can be applied to other chemoreceptor systems such as olfactory and chemotactic systems.

1. Model Membrane (2,3) A Millipore filter paper impregnated with the total lipids extracted from bovine tongue epithelium was used as a model for the taste receptor membrane, and the membrane potential arising between two solutions across the membrane was measured in the presence of salts, acids and distilled water. It was found that the changes in the membrane potential induced by these stimuli simulated well the taste receptor potentials observed with living taste cells intracellularly or the taste nerve responses. For example, the relations between the magnitude of the potential changes induced by HCl and acetic acid and pH of the solutions (Fig. 1) closely resemble those dervied from the nerve responses in rat (4). The relation between the magnitude of potential change and NaCl concentration followed the Beidler taste equation. The model membrane reproduced the characteristic effect of $FeCl_3$ on taste cells (5); treatment of the membrane with $FeCl_3$ brought about a reversal of polarity of the steady potential. Application of NaCl to the $FeCl_3$-treated membrane induced a variation of a potential change in the opposite direction to that normally displayed.

Response of a model membrane of single phospholipids was quite different from that of the total lipids (6). For example, phosphatidylcholine-membrane did not respond to salts, but did well to distilled water, while phosphatidylethanolamine-membrane responded to salts but not to distilled water. These findings together with some other experiments suggested that each or similar species of phospholipid forms small clusters or domains on the surface of the total lipid-membrane or living receptor membrane (6).

The model membrane is not permeable to ions since the pores of a Millipore filter are filled up with the lipids. The zeta-potential of liposomes made of the same lipids as used for the model membrane was measured in the presence of various taste stimuli. The dependence of the zeta-potential on salt concentration closely resembled that of membrane potential. The characteristic effect of $FeCl_3$ was also observed in the measurement of the zeta-potential. It was concluded that the membrane potential of the model membrane is attributed to the electric potential at the membrane-solution interface.

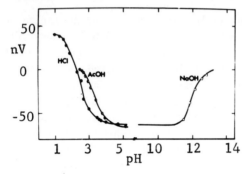

Fig. 1. Changes in membrane potential induced by HCl and acetic acid (model membrane).

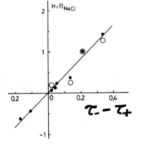

Fig. 2. Relation between magnitude of water response and ($\tau_- - \tau_+$).

interface.

    2. Mechanism of Generation of Water Response (7) The mechanisms of generation of water response was also interpreted in terms of changes in electric potential at the membrane-solution interface. The mechanism was investigated by use of the model membrane and by measuring activities of the frog glossopharyngeal nerve. When the model or receptor membrane is adapted to a certain electrolyte solution, the ions are dissolved in (or adsorbed on) the surface of the membrane, and an application of distilled water to the membrane produces a variation of the emf caused by diffusion of the ions at the membrane surface into the bulk solution. This emf is mainly responsible for generation of the water response. When the electrolyte involved is 1:1 type, the diffusion potential is expressed by the following equation

$$\triangle\Psi = RT/F \; (\tau_- - \tau_+) \; \ln C_2/C_1$$

where C_1 and C_2 are the concentration of the electrolyte component in the bulk solution and that at the membrane surface, respectively τ_- and τ_+ are transference numbers of anion and cation of the electrolyte. The following results observed with the model membrane and the frog's tongue are consistent with the above equation. 1. The magnitude of the water response of the membranes adapted to 100 mM of various 1:1 type salts was linearly related to $(\tau_- - \tau_+)$ for the salts (Fig. 2). 2. The water response increased with increase of the salt concentration in the adapting solution (increase of C_2). 3. Increase of C_1 (addition of electrolyte to a solution which is applied to the membrane) suppressed the water response irrespective of species of electrolytes added. The diminution of the screening effect on the electric double layer at the membrane surface was also responsible for the generation of the water response when the membranes were adapted to some specific electrolytes such as $CaCl_2$, HCl. The cations of these electrolytes were found to bind strongly to the negative group at the liposome surface.

The response of the frog to Ca^{++} was examined under the condition where the water response is suppressed, and it was concluded that the water response of the frog is different from the response to Ca^{++}.

3. *Membrane Potential and Chemoreception in Slime Mold* (8). The plasmodium of *Physarum polycephalum* perceives various kinds of chemicals and moves toward or away from them. Tactile movement and membrane potential were measured quantitatively by using a double chamber method (Kamiya, 1942); a single plasmodium was placed between two compartments through a narrow ditch, and differences in membrane potential and in pressure between the two compartments were measured. By increasing the concentration of various chemicals in one compartment, the membrane potential started to change at a certain threshold concentration (C_{th}) for each chemical (Fig. 3). Chemotactic movement of the plasmodium took place at the same threshold concentration, These results held both for attractants (glucose, galactose, $Th(NO_3)_4$, $Ca(H_2PO_4)_2$, ATP, c-AMP, etc.) and for repellents (various

inorganic salts, e.g., KCl and $CaCl_2$, sucrose, etc.).

The threshold concentration decreased remarkably with increase of the valence of cation, z, and was proportional to z^{-6}, i.e., the Shultze-Hardy rule known in the field of colloid chemistry was found to be applicable. The plasmodium distinguished the species of monovalent

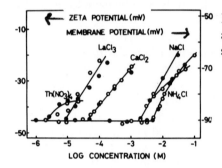

Fig. 3. Membrane potential (●) and zeta-potential (○) in the slime mold.

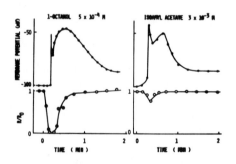

Fig. 4. Changes in membrane potential and resistance induced by odorants (Nitella).

cations in the following order: H, Li, K, Na, Rb, Cs, NH_4. Plots of log C_{th} against the lyotropic number of anion fell on different straight lines for each monovalent cation species. Plots of log C_{th} against the reciprocal of the absolute temperature followed different straight lines for different substances. The slopes of the lines were, however, the same as each other and gave a value of 12 kcal/mol for the apparent enthalpy irrespective of attractants and repellents.

As described above, the thresholds for the tactile movement were identical to those for changes in membrane potential. It was confirmed that the membrane potential also stems from the electric potential at the membrane-solution interface as follows; the Spherical microplasmodia with about 0.5-1.0 mm diamater were obtained by the submersed culture method and electrophoresis of the microplasmodia was carried out in the presence of various chemicals (9). The dependence of the zeta-potential of the microplasmodia on concentration of chemicals (salts, phosphate compounds and sugars) was identical with that of the membrane potential (Fig. 3).

    4. *Olfactory Response in Nitella Excitable Systems* (10). The elicitation of neural responses caused by odorants is not restricted to the olfactory organs, but other excitable systems are also able to respond to odorants (11). In the present study, excitable protoplasmic droplets isolated from Nitella (12) and internodal cells of Nitella were used as a model for olfactory receptor system. Odorants induced depolarization in the internodal cell (Fig. 4) when the concentration exceeded a certain threshold, C_{th}, for each odorant. This depolarization produced impulses which propagated along the internodal cell. This was observed when the cell was arranged appropriately. The membrane potential changes in two steps (Fig. 4). The height of the first peak was independent of the odorant concentration, while the second peak depended on the odorant concentration. The first peak was accompanied by a remarkable decrease in the membrane resistance as in the case of electrical excitation, but no decrease in the resistance was associated with the second peak of the potential. Depolarization of the protoplasmic droplet induced by odorants was also not accompanied with decrease in the membrane resistance. Plots of log C_{th} (threshold for the potential change in the internode or the droplet) vs log T (the threshold for humans) gave a straight line of unit slope. The threshold for octanol was remarkably lowered by decreasing Ca^{++} concentration in the external medium of the droplet. The decrease of Ca^{++} concentration in the medium increases the instability of the surface membrane of the droplet (12). The extremely high sensitivity of the olfactory receptor cell to odorants may be attributed to the instability of the receptor membrane.

    In all chemoreceptor systems dealt with in the present study, changes in membrane potential induced by chemical stimuli were found to stem from changes in electrical potential at the membrane solution interface. This is not unreasonable according to Teorell and Meyer and Sievers. The membrane potential inherently consists of two phase boundary potentials and intra-membrane diffusion potential. When the diffusion potential is small (2,13), membrane potential is predominantly attributable to the phase boundary potentials.

References

1. Beidler, L.M. 1967. Olfaction and Taste II, 506.
2. Kamo, N., Miyake, M., Kurihara, K. and Kobatake, Y. Biochim. Biophys. Acta, in press.
3. ibid., in press.
4. Beidler, L.M. 1971. Handbook of Sensory Physiology, IV-2, 200.
5. Tateda, H. and Beidler, L.M. 1964. J. Gen. Physiol., 47, 476.
6. Miyake, M., Kamo, N., Kurihara, K. and Kobatake, Y., in preparation.
7. ibid., in preparation.
8. Ueda, T., Terayama, K., Kurihara, K. and Kobatake, Y. J. Gen. Physiol., in press.
9. Hato, M., Ueda, T., Kurihara, K. and Kobatake, Y., in preparation.
10. Ueda, T., Muratsugu, M., Kurihara, K. and Kobatake, Y., in preparation.
11. Tucker, D. 1971. Handbook of Sensory Physiology, IV-1, 151.
12. Inoue, I., Ueda, T. and Kobatake, Y. 1973. Biochim. Biophys. Acta, 295, 653. Inoue, I., Ishida, N. and Kobatake, Y. 1973. ibid., 330, 27. Ueda, T., Muratsugu, M. and Kobatake, Y., ibid., in press.
13. Kamo, N. and Kobatake, Y. 1974. J. Colloid Interfacial Sci., 46, 85.

Acid Responses in Frog Taste Cells

Masayasu Sato and Norio Akaike

Department of Physiology Kumamoto University Medical School Kumamoto, Japan

Responses to various chemicals including acids were recorded from taste cells inside fungiform papillae of the frog tongue with an intracellular glass microelectrode. Various salts, acids, quinine hydrochloride, and sucrose produced depolarizing responses in taste cells, but responses to sucrose were poor. Depolarization by the first three kinds of stimuli occurred in all the cells examined, though their magnitude differed according to kind of stimulus and from one cell to another.

The effectiveness of hydrochloric, acetic and lactic acids at an equimolar concentration for taste cell response was studied. Among the three acids examined, lactic acid produced the largest response, but the effects by acetic and hydrochloric acids are not significantly different from each other. Therefore, the order of effectiveness of acids for frog taste receptors is not quite the same as that determined in other vertebrates.

When responses to various stimuli in about 20 cells were compared, the response profiles across cells for acids are quite different from those for NaCl and quinine hydrochloride, while the profiles for acetic and lactic acids were similar to that for hydrochloric acid. These results indicate the absence of cells specifically sensitive to one stimulus kind and suggest that discrimination of stimulus quality could be made by the across – cell response pattern at the taste cell level.

On the Phylogenetic Emergency of Sour Taste
Report of discussion by Chairman G. Hellekant
Dept. of Physiology, Veterinarhogskolan,
Stockholm, Sweden

Beidler reviewed earlier studies on the reaction to acids in a number of species. These studies show that at equal molarity the inorganic acids are more efficient stimuli than organic but at equal pH the organic are better than the inorganic. Beidler also reviewed the data indicating that adding the salt of an acid to the acid in question, which drastically changes the pH by buffering, will not affect its stimulating properties. Increase of chain length in a molecular series of acids causes an increase of the response at equal hydrogen concentration.

All these observations can be explained by the fact that adding hydrogen ions to a protein changes the charge of the protein. At one point the positive charge density will be so high that the next hydrogen ion will have problem to bind. This can be avoided by adding the salt of the acid, eg. its Na-ion to the solution, which then will prevent the build up of charge on the protein.

Beidler concluded that there is no emergence of the taste to acids because even the lowest animals reacts to acids, and that the differences in response from one acid to another can be explained by known physical and chemical prints in a quantiative way.

Discussion

Farbman suggested that the negatively charged carbohydrate which bathe the microvilli in the taste pore may have a similar effect as the lipoprotein. *Beidler* answered that he did not think that the small amount carbohydrate which is possible in the taste pore could have any substantial binding capacity and that it might be debris from the taste cells. *Miller* wanted to know how Beidler visualized the mechanism of the Na-ion compared to that of the hydronium ion. *Beidler* answered that he did not think that the Na-ion stimulates very much. He suggested that more attention should be paid to the transduction mechanism than to the receptor mechanism because most taste cells respond to all stimuli used. *Michell* suggested that the undissociated acid may cross the membrane and may play a role, on which *Beidler* referred to studies which indicate that the amount of undissociated acid plays no role. Finally *Zotterman* reminded the audience that Liljestrand in experiments with buffered acid solutions in man found a better correlation between intensity of sensation and pH of stimulus than between intensity and concentration of stimulus.

Kurihara described experiments using lipid monolayer as a model for a receptor membrane. The lipid was extracted from bovine tongue epithelium or taste papillae and spread on the surface of an acid solution. The interaction between the acid and the lipid was observed by measuring the surface pressure change. Kurihara found that these changes were in parallel to data obtained by Beidler from rat chorda tympani proper nerve. Kurihara pointed out that it is generally accepted that the binding occurs at the microvilli of the taste cells. For several reasons, he finds it difficult to explain the generation of receptor potential in a taste cell by the theory of Hodgkin-Huxley. One is: the microvilli are impermeable to protons, another, the resistance of the taste cell does not change during depolarization by acids. Kurihara described further the results obtained by the use of lipid from bovine tongue on millipore filter paper. The model was on one side exposed to a control solution, and the other to an acid solution. The model responded to acids, water and salt. Similar results were obtained when a platinum plate was inserted between the two sides of the model, which made the model impermeable to ions. Finally Kurihara suggested that the membrane potential is determined by the phase boundary potential which changes during stimulation without any change of membrane resistance.

Discussion

Unidentified voice: How specific is your lipid response?

Kurihara answered that they have used five types of phospholipids. They all responded differently to the stimuli. *Hellekant* wanted to know how this general model for the taste membrane could account for the fact that we experience different taste qualities. *Kurihara* answered that the different phospholipids, do not mix and he thinks that the proportion between them differs from cell to cell.

Sato reported results obtained with microelectrode recordings from the inside of fungiform papillae of frog. The resting potential was about 30 mV. NaCl, quinine and acids depolarized the cells. Sugar was fairly ineffective. Hydrochloric acid was at equimolar concentration not a less effective stimulus than acetic acid. The largest response among the acids was obtained with lactic acid. This is different from the observation in rats in which species HCl is a more efficient stimulus. The response profiles between the acids were more similar than those between NaCl and the acids. A decrease of resistance of the receptor cell membrane during stimulation was observed.

Discussion

Pfaffmann asked Sato about the variation in sugar response in frog on which *Sato* answered that the frog shows seasonal variations. Thus the sugar response is better in the summer, though it is always poor. *Hellekant* wanted Sato's and Kurihara's comments on the fact that Sato found a change of membrane resistance during acid stimulation while Kurihara reported none. *Sato* commented that the change of resistance was related to the type of stimulus. Some gave a larger change than other. *Kurihara* commented that Sato's data showed no change in membrane resistance with regard to acids. *Miller* wanted Sato's comments on the possibility that the current that depolarizes the taste cell may be a capacity current as opposed to a transmembrane current. *Sato* answered that when they inserted an electrode in the cell and measured the resistance between the outside and the inside they observed a change of membrane resistance during taste stimulation. *Miller* replied that his question was in line with observations from Sato's lab. which indicates that tetradotoxin had little effect on the response to sodium chloride. He felt that this implies that the depolarization which Sato reports was not caused by current through the membrane but might be caused by a capacity current. *Sato* answered that they think that their data should be interpreted as a change of permeability or resistance across the membrane.

ONTOGENESIS – TASTE AND SMELL

Chairman: I. Darian-Smith
Moderator: B. Oakley

The Developing Sense of Taste

Robert M. Bradley

Dept. of Oral Biology, School of Dentistry,

University of Michigan, Ann Arbor, Mich. U.S.A.

and

Charlotte M. Mistretta

Dept. of Oral Biology, School of Dentistry & Center for

Human Growth & Development, Univ. of Michigan

Throughout the vertebrate phyla sweet tasting substances are generally preferred while bitter substances are rejected or avoided. Such widespread phenomena suggest that these preferences and aversions are innate, and certainly their survival value for the organisms is apparent. But considerable evidence demonstrates that to some extent food acceptance and rejection can be modified by experience with the environment. Genes and environment never act in isolation, but always interact. To understand the interactions which lead to the establishment of food acceptance and rejection we must study the development of preference-aversion behaviour. And to understand the observed behavioral development, it is necessary to study the development of the sensory systems which function primarily in feeding behavior — taste and smell. The developing olfactory sense has not been studied in any detail and previous investigations of the development of the sense of taste have been exclusively anatomical. Using fetal, newborn. and adult sheep we have been studying the development of taste from structural, behavioral and neurophysiological approaches. The sheep has a lengthy gestation ($\sim$ 147 days) and is born, like the human, with all senses functional.

Morphology

As in the human (2,6), taste buds develop *in utero* in the sheep. Taste buds first appear on the tongue of the fetal sheep at about 7 weeks of gestation (3,4). These presumptive or early buds have a nerve supply. By 14 weeks the buds possess a taste pore and are morphologically similar to adult taste buds. The 7 week developmental period is comparable in length to that of the human (2,6) and contrasts sharply to that in rodents (8,11). At the time of earliest appearance, one presumptive taste bud can be found per fungiform papilla. At about 100 days of gestation in the sheep 2 taste buds are often found in one papilla; in a newborn lamb a single fungiform contains 3–4 taste buds. and in the adult as many as 8 taste buds have been counted in one papilla. In the human, also, there is an increase in the number of taste buds per fungiform over development. It is not known how this increase is accomplished.

Electrophysiology

To determine whether fetal taste buds are functional we recorded neurophysiological responses from the chorda tympani nerve while stimulating the tongues of fetuses aged 96 days to term with a variety of chemicals. For comparative purposes recordings were also made in lambs and adult sheep. Single and multi-unit preparations were studied. An intravenous injection of sodium pentobarbital into the maternal jugular vein was used to anesthetize mother and fetus. The fetus was delivered onto a heated table and the umbilical circulation remained intact. Records of activity in the chorda tympani nerve were made using conventional electrophysiological techniques. For details on surgical and electrophysiological methods, see (3). All chemical stimuli were dissolved in distilled water and kept at room temperature during experiments. Stimuli were applied to the anterior third of the tongue using 20 ml plastic syringes and were rinsed from the tongue with distilled water.

Integrated responses were recorded from fetuses aged 96 days to term. The fetal chorda tympani nerve responded to lingual stimulation with NH_4Cl LiCl, NaCl, KCl, HCl, acetic acid, citric acid, glycerol, Na saccharin, glycine, Na glutamate. and quinine hydrochloride. In all multifiber preparations, stimulation of the tongue with salts or acids elicited a higher frequency neural response than chemicals which taste sweet or bitter to man. Similar responses were recorded in newborn lambs and adults. The order of effective stimulation of a series of 0.5 M monochloride salts (NH_4Cl, LiCl, NaCl, KCl) was the same in fetuses, lambs, and adults. When stimulating the tongues of fetuses, lambs, and adults with a concentration range of NH_4Cl (0.1 to 0.75 M), increasing concentrations elicited neural responses of increasing magnitude.

Single unit responses were recorded from fetuses aged 109 days to term. As in the multi-unit preparations, salts and acids were more effective stimuli than chemicals which taste sweet or bitter (Fig. 1). Typically in fetuses, lambs, and adults glycerol was a more effective stimulus than glucose fructose, or sucrose. As seen in recordings from adults of many species, no two chorda tympani fibers had identical response characteristics.

Records of neural responses in the fetal chorda tympani nerve

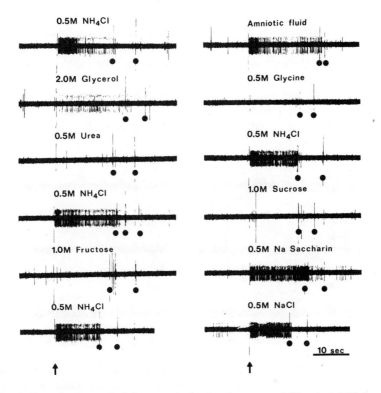

Fig. 1. Impulses recorded from a single chorda tympani fiber in a 128 day fetus. Arrows at the bottom of each column denote time of stimulus application. Filled circles denote water rinses. From Bradley & Mistretta (3).

demonstrated that the peripheral gustatory system is functional over the last third of gestation. The response characteristics of the fetal chorda tympani were similar to those of lambs and adults. This is not surprising, for over the age-range studied the fetal taste buds have a taste pore and appear to be morphologically mature. Yet perhaps the equation of maturity with the appearance of a taste pore is purely arbitrary. Can the taste buds function before a pore is present? Do response characteristics change when the pore develops? It is important to extend the age-range studied and record from the taste system in fetuses aged 50-100 days of gestation, during the period when taste buds are maturing. Since the chorda tympani nerve is easily damaged in such young fetuses, it is not possible to dissect the peripheral nerve. Therefore, we have begun to record from the nucleus of the tractus solitarius obviating the necessity for nerve dissection.

The ewe and fetus are anesthetized as in experiments for peripheral nerve recording. The fetus is delivered onto a heated table and the head secured between modified ear bars. The fetal cerebellum is exposed and

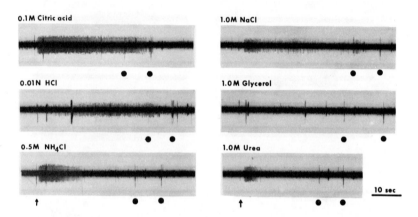

Fig. 2. Impulses recorded from a unit in the medulla of a 140 day fetus. Arrows at the bottom of each column denote time of stimulus application. Filled circles denote water rinses. Note response latency and adaptation to NH_4Cl, $NaCl$, and urea.

aspirated, to reveal the floor of the fourth ventricle. There is no atlas of the developing fetal sheep hindbrain. We have, therefore, established coordinates from histological sections of fetal brains at different ages. A tungsten microelectrode mounted in a micromanipulator is positioned over the obex, and advanced to the appropriate coordinates. When neural responses to chemical stimulation of the tongue are recorded, the brain is lesioned electrolytically at that point and the electrode position confirmed with subsequent histology.

To date, recordings have been made in the medulla of 7 fetuses aged 114 days to term. Some units recorded from the region of the tractus solitarius are similar to those in the chorda tympani in that salts and acids elicit a high frequency neural discharge (Fig. 2). However, the latencies are longer in the brainstem recordings and responses to salts usually adapt before the chemical is rinsed from the tongue (see responses to NH_4Cl and NaCl, Fig. 2). In other fetal medulla units, not yet precisely localized histologically, spontaneous activity is inhibited by citric acid, NH_4Cl, and KCl, but not by NaCl, LiCl or sugars (Fig. 3). We emphasize that these results are preliminary. Using this technique we are attempting to record from the nucleus of the tractus solitarius in fetuses aged 50-100 days of gestation. By correlating neurophysiological responses and light microscopic data on developing structure, it should be possible to identify structural elements of the taste bud which are essential for function.

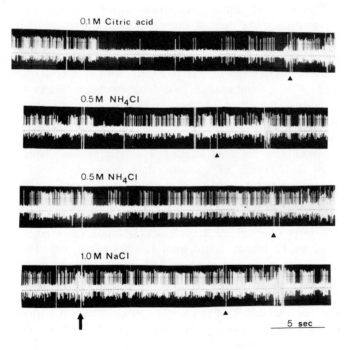

Fig. 3. Impulses recorded from a unit in the medulla of a 116 day fetus. The arrow denotes time of stimulus application. Triangles denote water rinse. Note inhibition of spontaneous activity with citric acid and NH_4Cl.

Swallowing

Knowledge that the taste bud matures morphologically *in utero* and that the peripheral gustatory system is functional for at least part of gestation does not demonstrate that the fetus utilizes its sense of taste. In humans and other mammals with lengthy gestations all sensory systems are functional at birth, having developed *in utero*. But does the fetus use its sense organs to monitor its environment? Since the fetal environment is a liquid, amniotic fluid, the sense of taste may be important in detecting the chemical composition of the amniotic fluid. Fetuses swallow amniotic fluid and this provides a means for performing a preference test *in utero*. If the amniotic fluid is flavored to taste pleasant, a fetus which can perceive taste stimuli should drink more of it and less if it is flavored to taste unpleasant. Human fetuses reportedly swallow more amniotic fluid after an intra-amniotic injection of Na saccharin (7) and swallow less after an intra-amniotic injection of a noxious tasting substance (9).

In preparation for intrauterine fetal preference tests we have studied fetal swallowing in 12 chronic preparations with fetuses aged 101 to 136 days of gestation at the time of surgery. The surgical procedure has been described in detail (5). Using chronic sheep preparations it has been possible to continuously measure fetal swallowing over a period of weeks. A cannulated electromagnetic flow transducer with an internal diameter of 3, 4, or 5 mm, implanted in the fetal esophagus allows continual measurement of fluid flow in either direction. A record of integrated flow provides volume swallowed at any given time.

Using this preparation, the intermittent, seemingly random character of fetal swallowing has been revealed. The fetus swallows large volumes (from 20 to 200 ml) of amniotic fluid during 2–7 brief episodes throughout the day. One of these episodes is illustrated in Fig. 4. The swallowing bouts are from 1–9 minutes duration and occur at unpredictable intervals. In between episodes of large-volume swallowing the fetus often swallows smaller volumes (1-10 ml). The average volumes swallowed by 12 fetuses ranged from 126-502 ml per day. For a given fetus on any one day, as much as 750 ml of amniotic fluid may be swallowed.

The variability of total volume swallowed per day coupled with the seemingly random timing of swallowing episodes makes it difficult to use swallowing as an indication of fetal taste responses. We have injected the extremely bitter chemical Bitrex into the amniotic fluid of two fetal preparations. Although we made daily, repeated injections and the amniotic fluid on withdrawal tasted bitter to us, both fetuses continued to swallow large volumes of amniotic fluid. A lamb of 2–7 days old will stop sucking on a bottle filled with milk when bitter chemical is injected into the milk stream. So the newborn lamb can detect bitter. Perhaps swallowing *in utero* is not under conscious control or perhaps the fetus is not responsive to the bitter amniotic fluid. Experiments are still in progress to decrease swallowing with Bitrex injections.

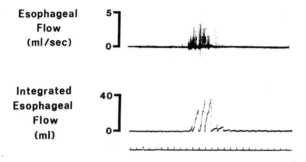

Fig. 4. Record of swallowing in a 128 day fetus on postoperative day 6. Time is marked in minutes on the bottom trace. Positive deflections on the esophageal flow trace result from swallowing. Integrated record provides volume swallowed during the episode. The integrator automatically resets to 0 every 60 secs.

Discussion

To summarize, it is known that taste buds begin to develop and become morphologically mature *in utero* in mammals with relatively long gestations. It is also known that the peripheral taste system is neurophysiologically functional *in utero* and that during swallowing episodes amniotic fluid flows over the taste buds. Amniotic fluid is not constant in composition over gestation. Gradual changes and daily fluctuations in constituents occur (10). The fetus produces large volumes of urine throughout the day (1,12) and, presumably, sometimes has its mouth open, sometimes closed. So that taste receptors are exposed to varying stimuli during development.

It is not yet possible to say whether the fetus can distinguish taste stimuli *in utero* and, therefore, is able to monitor the chemical composition of its environment. This makes it difficult to assess the entensiveness of the fetal taste experience. Certainly, though, to exclude the fetal period and assume that mammalian taste experience begins at birth is short-sighted. As we understand how the environment can affect developing sensory systems it becomes important to ask whether taste stimuli *in utero* play a role in influencing the formation of neural connections in the taste pathway. The intrauterine environment may provide a level and quality of stimulation ideally suited to optimal development. Therefore to understand the origins of taste sensitivities, of preference and aversion, the fetal period should not be ignored.

(Supported by NIH Grant HD07483).

References
1. Alexander, D.P., Nixon, D.A., Widdas, W.F. and Wohlzogen, F.X. 1958. J. Physiol. (Lond.) 140, 14-22.
2. Bradley, R.M. 1972. In Third Symposium on Oral Sensation and Perception. J.F. Bosma, ed., Thomas, Springfield, Ill., 137-162.
3. Bradley, R.M. and Mistretta, C.M. 1973. J. Physiol. (Lond.) 231, 271-282.
4. Bradley, R.M. and Mistretta, C.M. 1973. In Fourth Symposium on Oral Sensation and Perception. J.F. Bosma, ed., DHEW Publication No. (NIH) 73-546, Bethesda, Md., 185-205.
5. Bradley, R.M. and Mistretta, C.M. 1973. Science 179, 1016-1017.
6. Bradley, R.M. and Stern, I.B. 1967. J. Anat. 101, 743-752.
7. DeSnoo, K. 1937. Monatsschr. Geburtsh. Gynackol. 105, 88-97.
8. Farbman, A.I. 1965. Develop. Biol. 11, 110-135.
9. Liley, A.W. 1972. In Pathophysiology of Gestation. Vol. II. Fetal-Placental Disorders. N.S. Assali, ed., Academic Press, N.Y., 157-206.
10. Mellor, D.J. and Slater, J.S. 1971. J. Physiol. (Lond.) 217, 573-604.
11. Mistretta, C.M. 1972. In Third Symposium on Oral Sensation and Perception. J.F. Bosma, ed., Thomas, Springfield, Ill., 163-187.
12. Robillard, J.E., Thayer, K., Kulvinskas, C. and Smith, F.G. 1974. Am. J. Obstet. Gynecol. 118, 548-551.

Regenerative Phenomena and the Problem of
Taste Ontogensis

Bruce Oakley and Marylou Cheal
Zoology Department University of Michigan
Ann Arbor, Michigan 48104

In the first section of this report we present some results on the recovery of taste function following nerve interruption. We conclude by analyzing alternative processes controlling the formation of different kinds of taste fibers and receptors.

Recovery of Taste Function in Gerbils

Each fungiform papilla of the gerbil contains a taste bud and extensive nerve plexus (Fig. 1-left). When fungiform taste buds are denervated, they will degenerate. However, under the trophic influence (1) of the regenerating nerve fibers (Fig. 1-right) new taste buds will reform. We examined this recovery process by combining anatomical observations of regenerating taste nerves and buds with electrophysiological monitoring of the recovery of taste function. Anatomical studies of regenerating rabbit foliate and circumvallate taste buds have been carried out by other workers (2,3).

The chorda tympani taste nerve was cut or crushed at its junction with the lingual nerve in more than 100 Mongolian gerbils (*Meriones unguiculatus*). Proximal to this injury site the chorda tympani nerve passes through the auditory bulla on its way to the brain. The enlarged bulla of the gerbil made the chorda tympani readily accessible for electrophysiological recording with a pair of wire electrodes (100μm dia.) placed against this undisturbed portion of the nerve. Thirteen different chemicals, including salts, acids, sweeteners, and quinine hydrochloride were used to test taste responsiveness. Acute whole nerve recording under pentobarbital anesthesia was carried out at daily intervals starting eight days after nerve interruption. Although it was not generally possible to resolve single units in whole nerve recording, in some instances at early regenerative stages action potentials of 1 or 2 units could be discerned. Immediately after electrophysiological recording the gerbil was perfused with a formalin fixative and the tongue stained using Winkelman's silver (axons) or Heidenhain's iron hematoxylin (taste buds) procedures.

Interruption of the chorda tympani at the site of union with the lingual

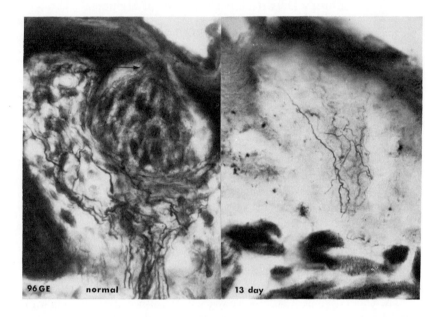

Fig. 1. Axons in normal and re-innervated fungiform papillae of gerbil. The Winkelman silver strain used here is suitable only for fibers. Taste buds are infrequently stained. Left: In this instance cells of the taste bud and axons of a normal gerbil were stained with silver. Distal processes of taste bud cells are indicated by the arrow.

Right: Regenerating fibers without a visible taste bud in a fungiform papilla 13 days after crushing the chorda tympani nerve. Scale is 10 μm; section thickness is 25 μm.

nerve silenced the electrical activity and after several days caused ipsilateral degeneration of taste buds in the fungiform papillae.

Functional recovery proceeded in the following stages (Fig. 2); Stage 1, spontaneous activity, unaltered by stimulation; Stage 2, discharges elicited only by pressure; Stage 3, 1 or 2 fibers responded to taste solutions; Stage 4, 3 to $\approx$6 fibers responded to taste stimulation; Stage 5. numerous fibers responded to taste solutions – responses to one or two chemicals predominated. The most effective chemicals varied for different gerbils.; Stage 6, numerous fibers responded – response profile (relative responsiveness to the test chemicals) was similar to the normal nerve. Cutting, rather than crushing, the nerve delayed the entire recovery process by one or two days.

In our experiments we recorded taste responses as early as eleven days after crushing. For example, on the drawing of the gerbil tongue in Figure 3 there are two sites (open circles) where NH_4Cl activated the same single

Fig. 2.

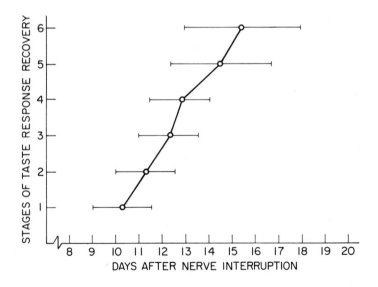

Fig. 2. RATE OF FUNCTIONAL RECOVERY. The open circles represent the mean number of days required to reach a given stage of recovery after crushing the chord tympani nerve. Standard deviations shown. N = 30, at least 3 gerbils/stage. See text for definition of stages.

fiber. Other chemicals had little effect on this fiber. The six most mature taste buds are indicated by solid dots. The photomicrograph in Figure 3 shows the most mature taste bud within the area responsive to NH_4Cl at the tip of the tongue. Spike counts to NH_4Cl stimulation are also shown. This particular experiment illustrates several general points: (a) To arrive at the more anterior receptive sub-field, the nerve fiber must have grown at a minimum rate of 2 mm/day. (b) Taste buds may reappear at the tip of the tongue prior to reappearance in all papillae near the entry of the chorda tympani nerve into the tongue. (c) newly refunctioning taste buds are often smaller than normal. (d) In the few testable instances in our gerbils it was evident that axonal branches tend to innervate similar kinds of taste receptor cells, as has been demonstrated in the normal cat tongue (4).

In mammalian taste buds there is a rapid turnover of cells (8-12 day average life span) (5,6). It has been suggested that as taste cells age, move to the center of the taste bud and die, they change their relative chemical responsiveness (7); young cells respond to one set of chemicals, old to another. However, a primary observation of our electrophysiological experiments with regenerating taste fibers is that many kinds of fibers are observed at early stages of regeneration. Responsiveness to a variety of chemicals reappears rapidly. [A given fiber might be specific (narrow response profile) but the effective chemical cannot be predicted in advance.] We interpret this result to mean that young receptor cells are not restricted to responding to one or a few kinds of chemicals.

Specification of Receptors, Fibers, and Connectivity

We would like to know (a) how different receptor types arise, (b) how different fiber types arise, and (c) how their interconnections are specified. Each of these features must be determined at some stage in development. At this time it may be useful to outline the problem and consider alternative mechanisms.

In one hypothesis of interaction between taste fiber and taste receptor cell percursor, a taste fiber grows into the tongue and innervates several fungiform papillae, and ultimately connects at random with assorted types of taste receptor cells whose taste specificity is pre-determined. The fiber type is thus determined by the types of receptors innervated. Specified receptor percursors are randomly innervated by any of a homogeneous population of taste fibers. However, this hypothesis is inconsistent with data from the cat tongue where taste fibers of the chorda tympani divide and innervate at least 2 fungiform papillae. Impulse activity recorded from the parent axon indicates that the two sets of receptors have similar

Fig. 3.

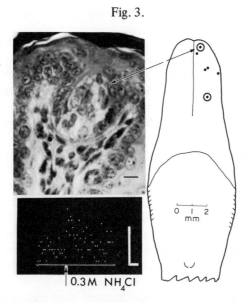

Fig. 3. TASTE RECOVERY AFTER 11 DAYS. Small dots on the diagram of the gerbil tongue indicate the six most full regenerated taste buds. Open circles show the two sites on the tongue where NH_4Cl elicited responses from a single taste fiber. The photomicrograph shows the most mature taste bud at the anterior site. Scale equals 10 μm; section thickness is 10 μm. The electrophysiological record gives the number of impulses in 0.5 sec intervals with whole tongue stimulation; calibration is 10 impulses and 10 seconds.

chemical responsiveness (4). The hypothesis of random innervation of receptor types must be rejected because it cannot explain this similarity of response profiles.

The hypothesis of random innervation of different types of taste receptor precursors is an example of the modulation hypothesis put forward and developed by Weiss (8). The receptor (or peripheral tissue) would modulate the central connections of the fibers so that they are appropriate for the innervation pattern in the tongue. Modulation has been tested and found not to occur in re-innervation of fish muscle (9,10). Recent experiments implying modulation in cutaneous fibers are still open to the possibility of one form or another of selective peripheral re-innervation instead of modulation (11,12). With the rejection of random innervation of different types of cat taste receptors, we must conclude that for the taste system, modulation by itself, cannot account for the establishment of taste connections between the tongue and the second order afferents.

A second hypothesis is that, prior to connection, the fibers are not all alike but are labelled as to type and impose their type upon any of the identical pluripotent receptor cell precursors they chance to innervate. However, Oakley (13) showed that the electrophysiological taste response in cross-regeneration was tissue-specific and not nerve-specific. For example, the rat IXth nerve, which normally gave weak responses to NaCl and strong responses to quinine, gave vigorous responses to NaCl and weak ones to quinine when forced to innervate the front of the tongue. That is, it adopted the normal pattern of sensitivity of the front of the tongue. But, according to the second hypothesis, the nerve fibers should dictate the chemical specificity of the receptor cells they reform regardless of location on the tongue. To say that the quinine response declined because the IXth nerve fibers failed to find suitable cells in the front of the tongue is to say that not enough percursors were labelled as quinine receptors, i.e., that the receptor cell precursors were not all alike, but were already specified.

In principle, the taste differences between taste buds from the anterior and posterior regions of the tongue could result from differences in an absorbent or filtering material in the taste pit of a bud, rather than from differences in the receptor cells themselves. However, it has been shown that the receptor cells within a single taste bud have different taste specificities (14). Consequently, assuming stable nerve fiber properties, a major portion of the difference between taste specificity of the front and back of the tongue must be in the taste receptor cells themselves. Determination of taste quality must occur in the receptor cells prior to re-innervation. Even though the critical nerve crossing experiment was carried out in adult animals, it would not be surprising if similar events occur in the initial specification of taste cells in embryogenesis. Owing to the continuous turnover of taste receptor cells in adult mammals, the taste system cannot rely for adjustment of connections solely upon factors unique to embryogenesis. Adult receptor cell populations and fiber-cell connections are renewed every few weeks (5,6).

Two important alternatives remain. A third hypothesis is that both the taste fiber and epithelial cell precursor of the taste receptor cell are prelabelled, i.e., their ultimate taste specificity is determined prior to connection. Connectivity is specified by a process of matching between fiber and cell. It is consistent with matching that taste preferences and aversions of rats with cross-innervated taste buds do not differ from those of controls (15). Matching between fiber and receptor cell would entail cell recognition (16,17) through, for example, cell-surface interactions.

A fourth alternative postulates a two-stage process in which an unlabelled fiber connects at random with a labelled receptor, which then dictates that all future peripheral connections must be with receptor cells of the same type. Since adjustment of central connections would also be required, the concept of modulation is part of this hypothesis. This remains as a possible alternative.

At the present time, the analysis indicates: (a) that the relative taste responsiveness of the epithelial cell precursors is specified prior to re-innervation and (b) that at some stage matching between pre-specified fiber and receptor is a probable feature of the process of re-innervation.

Supported by U.S. Public Health Service Grant NS-07072.

References

1. Guth, L. 1971. In L.M. Beidler (ed.), Handbook of Sensory Physiology, Volume IV, Chemical Senses, Part 2, Springer-Verlag, New York, 63-74.
2. Fujimoto, S. and Murray, R.G. 1970. Anat. Rec. 168, 393-413.
3. Iwayama, T. 1970. Z. Zellforsch., 110, 487-495.
4. Oakley, B. Chem. Senses Flavor, in press.
5. Beidler, L.M., and Smallman, R.L. 1965. J. Cell Biol., 27, 263-272.
6. Conger, A.D., and Wells, M.A. 1969. Radiation Res., 37, 31-49.
7. Beidler, L.M. 1962. Prog. Biophysics Biophysical Chem., 12, 109-151.
8. Weiss, P. 1936. Biol. Rev., 11, 494-531.
9. Mark, R.F. 1965. Exp. Neurol., 12, 292-302.
10. Sperry, R.W., and Arora, H.L. 1965. J. Embryol. Exp. Morph., 14, 307-317.
11. Baker, R.E., and Jacobson, M. 1970. Develop. Biol., 22, 476-494.
12. Jacobson, M., and Baker, R.E. 1969. J. Comp. Neurol., 137, 121-142.
13. Oakley, B. 1967. J. Physiol. (Lond.), 188, 353-371.
14. Ozeki, M., and Sato, M. 1972. Comp. Biochem. Physiol. 41A, 391-407.
15. Oakley, B. 1969. Physiol. and Behav., 4, 929-933.
16. Garber, B.B., and Moscona, A.A. 1972. Dev. Biol., 27, 217-234.
17. Sperry, R.W. 1965. In R.L. De Haan and H. Ursprung (Eds.), Organogensis, Holt, Rinehart and Winston, New York, 161-186.

Developmental and Electrophysiological Studies of Olfactory Mucosa in Organ Culture

A.I. Farbman and R.C. Gesteland, Depts. of Anatomy and Biological Sciences,

Northwestern University, Chicago, Ill., U.S.A.

Methods

Presumptive olfactory mucosa was dissected from the heads of rat embryos of 11-12 days gestation. The specimens were explanted onto pieces of Millipore filter resting on a stainless steel grid platform in a culture dish. The medium was Waymouth's 752/1 supplemented with 20 mg% ascorbic acid and 15% human cord serum. The explants were grown at 100% humidity in an atmosphere of 5% CO_2 in air at 34-35°C for periods up to 8 days. The medium was changed every other day. Standard techniques were used for light and electron microscopy.

The physiological responsiveness of the explants was tested by measuring receptor currents evoked by stimulation with odorous vapors. These electro-olfactograms (EOG's) were recorded with glass pipettes (tip diameters of about 20 μ) filled with mammalian Ringer-agar. Some preparations were tested *in situ* on the Millipore filter in the plastic culture dish, e.g. see Fig. 1. Others were inverted on the Millipore so that the apical ends of the receptor cells faced the advancing electrode. A chlorided silver wire in the culture medium served as the indifferent electrode. A moisture-saturated stream of air flowing at a constant rate of 130-160 ml/min from a 3 mm diameter teflon tube was directed at the explant. Vapors were introduced into this stream for stimulus periods indicated by the horizontal bar(s) on Figs. 4 and 5.

Results and Discussion

At the time of explantation the olfactory region of the 11 day embryo is distinguishable as a pair of placodes resting in shallow horse-shoe shaped depressions on the antero-lateral aspects of the embryo head. Histologically, the epithelium is pseudostratified columnar with 3-4 layers of nuclei and many mitotic figures near the free surface. Examination of the cells with the electron microscope revealed that the future receptor had a nucleus located in the deeper layers and a narrow apical process which contained a small number of microtubules. Cilia were observed rarely although some cells had a few centrioles in the apical process. This suggested the early stages of ciliogenesis. The future supporting cells had a nucleus in the most superficial nuclear layer and a broad apical portion.

In the 12th embryonic day the nasal pit was deeper, the olfactory region was more extensive and the first evidence of olfactory nerve was seen in the mesenchyme. Cytologically, the apical portions of receptor cells had more microtubules, more centrioles, and more cilia. Although receptor cells were in varying stages of cell differentiation, none appeared morphologically mature.

In culture, the explants became flattened and the epithelium was organized around a cavity. This "nasal cavity" was usually lined with a thick olfactory epithelium along one side and a respiratory type along the other (Fig. 1). Bundles of nerve fibers were seen in the mesenchyme near

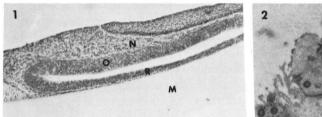

Fig. 1. Explant of 12 day olfactory region grown 5 days in culture. O, olfactory epithelium; R, respiratory epithelium; N, nerve; M, Millipore filter.

Fig. 2. Electron micrograph of a ciliated olfactory dendrite from an explant from a 12 day embryo after 5 days in culture.

the olfactory epithelium and could be traced directly to the epithelium. Most explants contained plates of hyalin cartilage in the mesenchyme.

The differentiation of olfactory epithelium in the explants mimicked in most ways that seen *in vivo*. Within the epithelium, in all cultures, there were several mitotic figures, mostly found near the free surface. Olfactory receptor cells were seen in several stages of differentiation. Many exhibited the morphological signs usually associated with olfactory receptor cells, namely: a ciliated apical process (Fig. 2); a dendrite containing large numbers of microtubules, some mitochondria and vesicles, but virtually no ribosomes (Fig. 3); a perikaryon located in the deeper nuclear layers; a narrow axonal process which joined with several others to form bundles within the epithelium. These axon bundles joined with others beneath the epithelium.

ribosomes (Fig. 3); a perikaryon located in the deeper nuclear layers; a narrow axonal process which joined with several others to form bundles within the epithelium. These axon bundles joined with others beneath the epithelium.

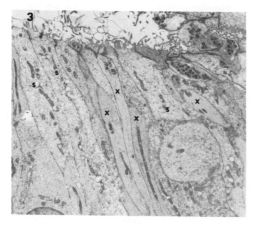

Fig. 3. Low magnification electron micrograph of the apical part of the olfactory epithelium showing receptor cells, x, and supporting cells, s. The epithelium was taken from a 12 day embryo and grown 5 days in culture.

It is clear then that in our culture system olfactory epithelium undergoes differentiation in a manner similar to that seen *in vivo*. These findings raised the question of whether the receptor epithelium can exhibit physiological activity in response to odorous stimuli.

We first explored the surface of the explant with a recording electrode while a stimulus was repeated. In a typical preparation, little response is seen from most of the surface. However we usually found a region with primarily negative stimulus-evoked potentials, and within this region a

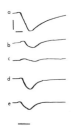

Fig. 4. Tracings recorded from an explant after 7 days in culture. Stimuli were: a, methyl benzoate; b, amyl acetate; c, anisole; d, pyridine; e, n-butanol. Calibration (Figs. 4 & 5) vertical, 200 μV; horizontal, 2 sec. s, duration of stimulus puff.

point of maximal amplitude. The traces in Fig. 4 were recorded from the surface of an inverted explant in response to several stimuli delivered at an intensity of 0.23 saturation. There was no response from a stimulus delivered from an empty vial. Wave form varies with stimulus substances, some of which evoke a prominent initial positive deflection (traces b,c,d,e). All responses had the same general features with minor differences in detail from one preparation to the next.

It is unlikely that the stimulus-evoked voltages are due to a generalized tissue response to chemical substances. If this were the case we would expect relatively uniform responses over the tissue surface and constant polarity for any substance. In Fig. 5 the responses to methyl benzoate at 0.18 saturation recorded at three different locations on a single explant are shown. The explant was oriented as in Fig. 1 and the electrode was advanced from the surface. The initial response was a positive going wave

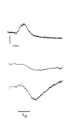

(top trace, Fig. 5). When the electrode was advanced deeper into the tissue the response became smaller (middle trace) and then reversed its polarity to become negative (bottom trace). The only reasonable explanation for this is that the recording electrode in the top trace was closest to the basement membrane side of the epithelium and in the bottom trace was closest to the apical ends of the receptor cells.

Fig. 5
See text.

(Supported by NIH Grant No. NS-06181 and NSF Grant No. GB-30520).

Cell Renewal in the Olfactory Epithelium
of the Mouse

D.G. Moulton

Monell Chemical Senses Center and Department of
Physiology, University of Pennsylvania and
Veterans Administration Hospital, Philadelphia,
Pennsylvania, U.S.A.

Under certain conditions olfactory receptors can regenerate following their experimental destruction (e.g., Graziadei and Metcalf, 1971; see also Moulton, 1974). This, however, does not necessarily imply that programmed cell death and replacement is a normal property of this tissue. In the mouse, evidence for such cell renewal comes from autoradiographic studies involving tritiated thymidine. These show that basal cells divide continuously. 90 per cent of the products migrate peripherally where the majority come to rest in the layer occupied by receptor cell nuclei (Moulton and Fink, 1972). Calculations of the rate of entry of new cells into this compartment indicate that receptors must die and be replaced if tissue homeostasis is to be maintained.

The potential functional implications of such turnover are extensive. It strengthens the ability of the tissue to repair damage and balance attrition. But it may also render it susceptible to influence of altered activity levels in the sympathetic nervous system and of hormonal and nutritional levels, in so far as these factors are known to alter turnover rate in certain other tissues (see, for example, Creamer, 1967). It also raises the question of how the olfactory epithelium maintains the constancy of its ability to discriminate and detect odors.

The extent to which these questions are significant presumably depends on the rate and extent to which cell renewal occurs. To gain further information on this point we have used the plant alkaloid colchicine to block cells undergoing division in metaphase.

Methods

Twenty-four male Swiss-Webster mice were used. They were 4–5 months old, had a mean weight of 46 g, and were caged individually. To control for possible diurnal variations in proliferative activity experiments were spaced across the 24 hour cycle. To achieve this mice were divided into four groups, each group containing two subgroups of three mice. All mice were injected with colchicine (0.2 mg per 100 gm body weight) according to the following schedule: Group A was injected at 6.00 hrs; group B at 12.00 hrs; group C at 19.00 hrs, and group D at 24 hrs. In each group of 6 mice 3 were killed 3 hrs after injection while the remaining 3 were killed 5 hrs after injection.

Pieces of the olfactory epithelium were dissected out, fixed in Bouins-Duboscq fixative, washed until most of the picric acid was removed, dehydrated and embedded in Luft's epon. Sections two microns thick were cut with glass knives on a rotary microtome. The sections were stained with 1% acid fuschin containing a few drops of glacial acetic acid.

Counts of the number of blocked metaphases and total cells per field were made separately for each of the three compartments of the epithelium: supporting cell and basal cell. (The nuclei of the three cell types are concentrated at different depths in the mouse epithelium: those of the supporting cells form a discrete layer at the periphery, those of the basal cells occupy the base, while the intermediate region contains the receptor cell nuclei. For the purposes of analysis we consider each layer as forming a separate compartment.) The estimate of mean turnover time for the total population was derived from differences between the ratio of blocked cells per 1000 cells counted for the 5 and 3 hour periods. This gives an estimate of the rate of entry of cells into mitosis over a two hour period from which turnover time (time required to replace all cells in the population) can be determined. In estimating turnover times at specific time periods during the 24 hour cycle, the same method was used except in the few cases that it yielded negative values. In those instances the ratio of the sum of blocked cells counted at 3 and at 5 hours per 1000 cells counted was taken as a measure of turnover rate.

Results and Discussion

As can be seen in Fig. 1 blocked metaphases were found in all three compartments. The highest frequency, however, occurs in the basal cell layer which is reflected in the shorter turnover time. This confirms the findings of the autoradiographic studies on this species (Moulton et al., 1970; Moulton and Fink 1972). Since it is difficult to set a sharp boundary between basal and receptor cells the counts of blocked cells in the receptor cell compartment may include counts of basal cells. Although blocked mitoses appear throughout the depth of this compartment it is nevertheless possible that the turnover time may be longer than that implied in Fig. 1. But since the magnitude is in any case high we cannot conclude that active turnover — as opposed to growth — characterizes

112

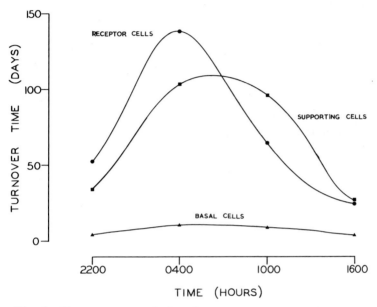

Fig. 1. Turnover rate of the three compartments of the olfactory epithelium estimated for four periods during a 24 hour cycle.

receptor cells in the absence of input from the products of basal cell division. The boundary between receptor and supporting cell compartments is more clearly defined. But if supporting cells are being replaced by active division *in situ* the rate is considerably slower than that seen in the basal cell layer.

The turnover time of the entire population is 28.6 ± 6.6 days. The turnover time of the receptor cells probably lies within a similar range for the following reasons: The supporting cells account for a small and relatively insignificant fraction of the total population of the epithelium and their rate of proliferation is relatively slow. [In the rabbit the ratio of supporting to receptor cells is 1:98 (Clark, 1956)]. The basal cells, although much less numerous than receptor cells, divide relatively rapidly. 90% of the daughter cells, however migrate into the receptor cell compartment (Moulton and Fink, 1972). Finally, the receptor cells themselves are by far the most numerous of the cells in the epithelium. Taking these facts together it appears that turnover time of the receptor cells should approximate that of the entire epithelium.

Diurnal variation in the proliferative activity of olfactory epithelial cells is to be expected. Circadian rhythms occur in mitotic activity and DNA synthesis of most cell renewing tissues. (The few reports of exceptions have been criticized for inadequate design.) In the basal epithelium of the rat tongue, for example the daily fluctuation in mitotic index is about 300% (Gasser et al., 1972). But vulnerability to this rhythm is not necessarily uniform across different cell populations: in the colon, for example, the more differentiated cells are the more susceptible (Chang, 1971). One obvious implication of the fluctuations seen in Fig. 1 is the need to obtain measures of cell proliferation at representative times over a 24 hour period if accurate estimates of turnover time are to be derived.

References

Chang, W.W.L. 1971. Amer. J. Anat. 131, 111-120.
Clark, W.E.LeG. 1956. Yale J. Biol. Med. 29, 83-95.
Creamer, B. 1967. Brit. Med. Bull. 23, 226-230.
Gasser, R.F., Scheving, L.E. and Pauly, J.E. 1972. J. Cell. Physiol., 80, 437-442.
Graziadei, P.P.C. and Metcalf, J.F. 1971. Z. Zellforsch. 116, 305-318.
Moulton, D.G. 1974. Ann. N.Y. Acad. Sci. (in press).
Moulton, D.G. and Fink, R.P. 1972. In Schneider, D. (editor). Olfaction and Taste IV. Stuttgart Wissenschaftliche Verlagsgellschaft MBH., 20-26.

Ontogenesis — Taste and Smell

I. Darian-Smith

Department of Physiology, University of Melbourne,

Parkville, 3052, Australia

In this session of the Symposium we have been concerned with chemoreception in the developing mammal. I comment on each of the papers as they were presented.

Dr. Mistretta and Dr. Bradley provided a description of the functional capacities of primary gustatory afferents and second order neurones in the foetal sheep from the 50th day of gestation until term — a technical tour-de-force. These techniques will now make possible a detailed description of the changing response characteristics of individual taste fibres in the developing foetus. The same data should also allow a precise definition of the capacities of populations of taste fibres for transmitting stimulus information about the chemical environment of the taste buds from the earliest stages of neuronal specification until the final maturation of these fibre populations.

One problem which complicates most studies of foetal tissue function, namely the behavioural status of the foetus following the surgical intervention necessary for the experiment, the influence of anaethesia on foetal function, etc., will need to be resolved to ensure that these studies reflect the sensory processes in the normal foetus. This troublesome factor has, in fact, been partly overcome in the chronic preparation which Dr. Bradley and Dr. Mistretta have developed for their investigation of patterns of fluid flow in the foetal oesophagus. However, I think it unwise to design experiments in terms of 'perceptual responses' in the foetus. There is much more to perceptual behaviour than a functional maturation of the sensory apparatus and central sensory pathways. We are still very confused in our analysis of perceptual behaviour in the human neonate, and still more so in dealing with what the foetus senses.

Drs. Oakley and Cheal have tackled the problem of regeneration of the receptor axon/receptor cell complex in tastebuds of the mammalian fungiform papilla following local interruption of the chorda tympani. In these experiments previous histological studies have been complemented with an electrophysiological examination of functional recovery of this receptor complex. In parallel with the reappearance of taste buds 10 — 12 days following section of the chorda tympani at its point of union with the lingual nerve, Oakley and Cheal were able to record activity in individual taste fibres. They have found that at this early stage of regeneration individual fibres have response characteristics representative of the several different elements which constitute the fibre population in the normal chorda tympani; that is, each of the chemoreceptive fibre types, with its characteristic response profile for the series of chemical test stimuli used, regenerates at approximately the same rate.

These observations are pertinent to the more general problem of the determinants of the functional relationships betwen sensory nerve fibres and their receptor cells, including the inductive action of the sensory fibre on the cells which form the sense organ, the nature of the precursor cells which interact with the sensory fibre terminal, and also the sustaining 'trophic' action of the sensory nerve fibre in maintaining the differentiated receptor cells. The general conclusions arising from the present experiments on taste buds are similar to those reached in experiments by Burgess and his colleagues (1974) on cutaneous afferents innervating clusters of specialized epithelial (Merkel) cells. In each example the experimental findings match well with a model having the following properties: (a) each regenerating sensory fibre retains a functional specificity in the sense that the regenerated receptor cell/sensory axon complex signals the same stimulus information as it did before injury; (b) only certain cells in the tissue invaded by the regenerating fibre respond to the inductive action, presumably chemical, of the fibre. Thus, taste buds develop only from epithelial cells of the appropriate papillae (Zalewski, 1970) — and Merkel's cells in hairy skin differentiate mainly at sites within the epithelial layers at which these same specialized receptor cells have previously developed (i.e. previous to degeneration of the terminal); (c) some mechanism exists which ensures that cell regenerating fibre makes physical contact with the appropriate precursor receptor cell in the reinnervated tissue — perhaps a particular matrix of Schwann cells, which is 'recognised' by the invading fibre.

We still do not know whether the mechanisms of receptor regeneration have features in common with those factors which maintain a stable afferent fibre/receptor cell in say, the taste organ, where receptor cells are replaced every 12 — 15 days. Nor are the mechanisms of receptor cell induction in the embryo understood.

Dr. Farbman has examined the development of olfactory cells in tissue culture, the explant being taken from rat embryos at about the twelfth day of gestation. In tissue culture cell differentiation continued to a stage at which identifiable olfactory receptor cells with dendritic and axonal extensions could be identified after 5 — 8 days in culture. Tissue serum is essential to maintain this tissue growth and differentiation, but the necessary component of the serum has not been identified. Possibly it is nerve growth factor. The same serum constituent is also necessary for successfully culturing taste buds from embryonic tongue tissue. Preliminary electrophysiological studies indicate that the electrolfactogram evoked by a number of odorous substances of the cultured olfactory mucosa is similar to that of normal adult olfactory mucosa, but, of course, it is appreciated that these electrical responses provide a very crude index of organized functional activity.

Dr. Moulton has examined the turnover of the cellular elements of the olfactory mucosa of the rat. His estimate is that there is a complete replacement of these cells approximately every 28 days. Does this include the dendritic component of the primary olfactory neurone? Dr. Moulton argued that regular replacement of the component cells of the olfactory mucosa renders it particularly susceptible to humoral, nutritional and various local changes in the nasal passages.

References

Burgess, P.R., English, K.B., Horch, K.W. and Stensaas, L.J. 1974. J. Physiol. 236, 57-82.
Zalewski, A.A. 1970. Exper. Neurol., 26, 621-629.

CHEMORECEPTION IN AQUATIC ANIMALS

Chairman: M. Sato
Moderator: J. Bardach

Chemoreception of Aquatic Animals

J.E. Bardach

University of Hawaii

Hawaii Institute of Marine Biology, P.O. Box 1346,

Kaneohe, Hawaii 96744

Chemoreceptors have evolved to adapt to the information transmittal properties of chemical signals. In fact "Mechanisms for detecting chemical changes in the environment occur in all groups of animals and they may be considered in an evolutionary sense as the most primitive distance receptors" (Kohn, 1961). We will examine external chemoreceptors, the signals they have become specialized to receive, their mode of functioning and central nervous system and behavior correlates of chemoreception in aquatic animals because of information that can accrue to comparative smell and taste physiology and especially because terrestrial animals have evolved from aquatic ancestors.

Such a synthesizing effort may seem premature, with many lucunae in pertinent information — from structure to function, from biochemistry to behavior — on all chemoreceptor-associated phenomena in the aquatic milieu. By attempting it, nevertheless, we hope both to point to certain areas of research that appear more fruitful than others and to some particularly glaring voids in our understanding of the chemical senses and their function. Various possible entries into the subject and various modes of its organization are conceivable; we propose the relatively simplistic one of going from environment-receptor interaction processes, including receptor specificities, through the theme of receptor-integrator relations such as coding, stimulus filtering, etc., to integration processes proper in the central nervous system, leading wherever possible to considerations of behavior, governed or influenced by chemical signals. Information available and traditional orientation of ISOT symposia will make us concentrate our discussion to relatively few groups, mainly the fishes, the higher molluscs and crustacea.

One good way to zone in on the first subthemes, those of receptor-environment interaction and receptor specificity would be to look at, a) what substances in water have been reported to serve as signals, b) what particular properties of water — as opposed to air — such as its solvent properties, prevalence of currents and the like govern contact between stimulus molecule and receptor, and c) certain anatomical

Table 1. Chemical signals of aquatic animals.

Taxonomic Category	Nature of signal (mostly pure compounds)	Function of signal	Source
Escherichia coli (bacteria)	serine (also cystine, alanine & glycine); aspartate (also glutamate); D-galactose (also D-glucose & D-fructose)	chemotaxis	Adler,1969
Protozoa	K, Li, Na, NH_4	avoidance	Prosser,1973
Dictyostelium discoideum (slime mold)	"Acrasin"; cyclic 3'5' – adenosine mono-phosphate	aggregation of amoeboid cells	Konijn et al. 1967
Hydra	glutathione	feeding response	Lenhoff, 1961
Cnidaria in general (Coelenterates, corals, etc.)	proline, tyrosine, glutamine	feeding response	Prosser,1973
Polythoa (soft coral)	proline, glutathione (in synergy)	feeding	Reimer,1971
Bdelloura candida (flatworm)	methylamine hydrochloride, trimethyl and triethyl amine oxide	attraction to host Limulus	Boylan,1974 pers. comm.
Dugesia dorotocephala (Planaria)	extract from brine shrimp, mol. wt. over 1000; 5-10 residues	feeding	Ash et al. 1973
Echinaster (starfish)	glucose, leucine, phenylalanine (high conc.)	initiate filter pumping (feeding)	Ferguson, 1969
MOLLUSKS			
Numerous gastropods (snails, slugs, limpets, etc.)	univalent cations, Mg ++ quinine sapid substances	feeding (close up) avoidance food search (some distance)	Kohn,1961
Buccinium (snail)	skatol sucrose, glycine	feeding (close up)	Kohn, 1961
Nassarius (snail)	skatol sucrose, glycine	feeding (close up)	Kohn, 1961
Nassarius obsoletus	glycine, lactate, betaine	feeding resp.	Carr, 1967
Aplysia (sea hare)	glutamic acid	feeding ?	Bailey & Laverack, 1963
	glutamic acid, aspartic acid alanine, cystine, leucine proline	feeding withdrawal mating	Jahan-Parvar,1972 & this symp.
Helisoma duryi (freshwater pulmonate snail)	polypeptides from crushed tissue, mol. wt. ca. 10,000	alarm substance	Snyder,1967
Pecten (scallop)	steroid saponin from starfish	predator avoidance	Mackie et al 1968
Crassostrea (oyster)	acid, quinine, sucrose, salt	?	Dwivedy, 1973
Octopus	acid, quinine, sugar	?	Wells, 1963

Taxonomic Category	Nature of signal (mostly pure compounds)	Function of signal	Source
Octopus	mixture of L-proline, L-alanine, L-arginine & glycine more stimulatory than mixture of L-glutamic acid & L-glutamine	feeding?	Bardach & Henderson, unpubl.
CRUSTACEA			
Carcinides maenas (crab)	L-glutamic acid, D-aspartic acid, D-leucine, L-proline, glutaric acid & keto glutaric acid	feeding?	Case & Gwilliam, 1961
Cancer (another crab)	aminobutyric acid, taurine, glutamic acid, serine	feeding?	Case, 1964
Pachygrapsus (shore crab)	aminobutyric acid, taurine, glutamic acid, serine	feeding?	Case, 1964
Pachygrapsus (shore crab)	4, 4-dimethyl butyrolactone	mating pheromone derived from molting hormone	Kittredge et al 1964
Panulirus (spiny lobster)	dopachrome (indole-5,6-quinone)	predator avoidance	Kittredge et al 1964
Homarus americanus (Maine Lobster)	DL-alanine, glutamic acid, L-proline, succinic acid, malic acid, L-tyrosine	feeding?	McLeese, 1970
" "	L-cystine, L-proline, L-asparagine, hydroxy L-proline, taurine, L-glutamic acid, D-asparagine, glycine, L-tyrosine; other amino acids less active.		Shepheard, 1974
" *antennule*	several units with different sensitivity spectra e.g., arginine unit also responds to methionine, lysine, leucine	feeding?	Ache, 1972
Homarus gammarus (European Lobster)	L-aspartic acid, L-threonine, L-serine, l-glutamic acid, L-valine, L-methionine, L-isoleucine, L-leucine, L-tyrosine, L-phenylalanine, L-lysine, L-histidine., taurine, L-proline, glycine, L-alanine, L-arginine, glycine betaine HCI, TMAO[a] HCI, TMA[b] HCI, homarine, hypoxanthine, inosine, AMP[c], L (+) lactic acid (as mixture in different proportions; see source for details) [a]Trimethylamine oxide hydrochloride [b]Trimethyamine hydrochloride [c]Adenosine-5'-monophosphate	feeding?	Mackie, 1973
Barnacles (species ?)	L-glutamic acid, L-proline, betaine, taurine protein ?	feeding? settling of larvae	Crisp, 1967

123

Taxonomic Category	Nature of signal (mostly pure compounds)	Function of signal	Source
Limulus, King crab (not a crustacean)	glycine	feeding ?	Barber,1956
FISHES			
Petromyzon marinus (lamprey)	isoleucine methylester	prey location	Kleerekoper & Mogensen, 1959 Kleerekoper 1971 (pers. comm)
Negaprion (shark)			
Sphyrna (shark)	glutamic acid, glycine, cystine, serine	feeding?	Gilbert et al. 1964
Gynglymostoma (shark)	trimethylamine oxide, glycine	food location?	Mathewson & Hodgson 1971
Salmo salar (Atlantic salmon) smell	15 amino acids with L-alanine, L-histidine & L-serine as most effective	feeding?	Sutterlin & Sutterlin 1971
taste (N. facialis)	K, Na, Mg, Ca, chlorides, dissolved Co_2, H_2SO_4, caproic, fumaric acids	feeding?	Sutterling & Sutterlin, 1970
taste (N. glossopharyngeus)	Ca, Mg, Na, K. chlorides, proline, glucose, short aliphatic acids, caproic to formic (least)	?	Sutterlin & Sutterlin, 1970
Oncorhynchus (Pacific salmon)	L-serine	predator avoidance	Idler et al. 1956
Salmo gairdneri (Rainbow trout)	homoserine, L-methionine, L-glutamine, L-alanine, L-asparagine, L-cystine, L-leucine, −L-threonine,		Hara et al. 1973
Salvelinus fontinalis (Brook trout)	L-serine, L-valine, L-arginine, glycine, L-histidine, L-lysine, L-cystine-cystine,isoleueine, GABA, soserine, L-tyrosine, D-alanine,		
Coregonus clupeaformis (Whitefish)	D-serine, L-phenylalanine, B-alanine (the 3 species in this family responded differently with species-specific sensitivity arrays; olfactory bulb)		
Ictalurus catus (White catfish) smell	L-glutamine, L-methionine, L-alanine L-asparagine, D-methionine, L-cystine as most effective	?	Suzuki & Tucker, 1971
	peptides, amino acids in characteristic mixture, likely to include: lysine, glutamic acid, alanine & cystine	alarm substance in skin-slime	Tucker & Suzuki, 1972
Ictalurus natalis taste *Ictalurus nebulosus* (yellow & brown bullheads), also *Urophycis & Microgadus* taste (cod family)	monovalent chlorides, mono & disaccharides, quinine, inorganic acids, several amino acids with cystine and L-alanine as most effective; cholesterol, inosities	feeding ?	Bardach et al. 1967

Taxonomic Category	Nature of signal (mostly pure compounds)	Function of signal	Source
Cyprinus carpio (carp) taste	Seven fiber types with different response spectra to: NaCI, acetic acid, quinine, sucrose; saliva (glycero phospholipid & sphingolipid)	feeding?	Konishi & Zotterman, 1963
Ictalurus punctatus (channel catfish) taste	extreme sensitivity to L-amino acids, & analogs, esp. alanine, arginine, serine	food search?	Caprio, this symp.
Fugu pardalis (puffer)	L-amino acids, ribonucleotides	feeding ?	Hidaka, this symp.
Caribbean reef fish smell	Hypotauryl-2-carboxyglycine (extracted from mollusc tissues)	elicits food search	Thomas,1970
Lebistes reticulatus (guppy) smell	Estrogen, either 17 beta estradiol or estriol	attracts male to female	Wilson, 1970
Alosa pseudoharengus smell	heat stable, mostly nonvolatile polar compunds smaller than mol. wt. 1000	home stream odor	Boylan et al. 1973

facilitations to enhance stimulus-receptor contact.

Before tabulating information on chemical signal substances in water (Table 1), arranged by phyla, the chemicals involved and the life processes the latter subserve, e.g., feeding, reproduction, homing, etc., stress on a few facts is in order:

1) A distinction between smell and taste is more tenuous in aquatic animals than in those on the land. Still such a distinction has been postulated for molluscs (Kohn, 1961) and it exists in the aquatic vertebrates (fishes and amphibia) where one still finds greater overlap in smell and taste capabilities than among reptiles, birds and mammals. 2) Sensitivity spectra seem specific to taxonomic groups in water as in air, as are, of course, pheromones. 3) Great chemosensory acuities exists in certain aquatic organisms.

Table 1 summarizes, albeit not exhaustively, the chemical signals to which aquatic animals respond and the behavior they elicit; it illustrates several things: 1) Amino acids and their analogs prevail certainly as feeding stimuli and L-amino acids are more effective than their D-isomers. This is not surprising as food means animal protein for most of the animals tested which are, by choice or chance, predominantly carnivorous. Furthermore, amino acids occur in water, free and in suspended particular matter. Degens (1970) indicates that the following are most prevalent in the first 3,000 meters of the ocean; serine, glycine, alanine, ornithine, leucine, proline, valine, in this descending order. In surface samples we find representative prevalence relations of amino acids to be as follows: glutamic acid, glycine, serine, leucine, alanine, aspartic acid, cystine (Vernberg & Vernberg, 1972). Amino acids are released quite freely by marine and freshwater invertebrates as shown in Table 2 (from Johannes & Webb, 1970) with glycine, serine and ornithine being most prevalent. Evidently — even though their sample of eight different species is small for our purpose — those that diffuse most readily from animal to exterior and those that are most prevalent in sea water are alpha amino acids. They also re-occur fairly prominently in Table 1, certainly in the feeding stimulant and food search category. In fact in Table 1 31 of 47 entries in the "nature of signal" category encompassing 39 different species or groups of species, contain amino acids. In these 31 entries the following — in descending order — are the 6 most prevalent: alanine, glutamic acid, proline, glycine and serine, trailed by cystine. Hara (this symposium) postulates an optimum of five carbon atoms, among other properties, one of which is the alpha position of the amino group, for amino acids to stimulate the olfactory bulb of fishes (he generalizes from experiments with the family Salmonidae in which one finds both fresh water and marine species). Some such correspondence is evident in Table 1 and certain of its sources but more detailed investigations will be necessary for these correlations to underpin physiological and ecological generalizations in an unequivocal fashion.

2) It is characteristic of most signals in Table 1 that they are relatively small molecules, that many of them are strongly polar and that they are relatively soluble in water. Few, if any natural aquatic chemical signals so far reported are volatile. Some of them reveal peculiar adaptations such as dopachrome to which certain crustaceans seem highly sensitive. The substance is a breakdown product of melanin which, in turn, is the main constituent of the ink of octopus, a prime predator of the crustaceans in question (Kittredge et. al., 1974).

3) Another set of facts which demands attention through further research relates to the pheromone category. Though far less information exists on reproduction-related chemical signals in water than on land or in the air, it appears that cyclic or steroid molecules play an important role here (estrogens in fish, steroids and cyclic compounds as sex pheromones in aquatic fungi [Wilson, 1970] and cyclic adenosine monophosphate as an aggregant in slime molds). That the latter is also a taste substance for a fish (Hidaka, this symposium) and bears mentioning because it points to the acquisition in evolution of signal properties by substances that existed many hundreds of millions of years ago. It may thus be advisable for

Table 2. Composition of dissolved free amino acids released by some aquatic invertebrates (mole per cent) [1].

	Artemia salina, adult (yeast) [2]	Artemia salina, nauplii (yeast) [2]	Clymenella torquata [3]	Acanthopleura granulata [4]	Nerita versicolor [5]	Acropora muricata [6]	Daphnia magna [7]	Chaoborus punctipennis [8]
cysteic acid			5.5	3.3	1.3	3.1		2.2
taurine			9.6	7.7	32.8	3.9	3.5	1.2
meth. sulf.			2.0	2.7	0.9	2.5	0.8	
aspartic acid	7.9	4.5	6.0	4.1	1.3	3.9	5.7	14.7
threonine	8.3	4.1	4.3	2.2	1.7		5.2	6.8
serine	15.1	14.2	20.5	1.6	2.1	6.7	25.6	1.1
glutamic acid	1.0	0.8	2.5	34.9	4.0	11.6	2.7	
proline	3.0	1.5	4.7	6.2	3.9	2.2	3.6	
glycine	10.3	50.2	9.2	15.5	38.0	54.5	15.9	10.8
alanine	8.3	5.9	9.9	5.3	4.6	2.4	8.4	8.2
valine	4.8	2.0	2.5	1.5	1.8	0.8	2.9	4.3
cystine	0.8			2.6				tr.
methionine				0.3				
isoleucine	2.8	1.2	1.6	1.0	1.3	0.3	2.4	3.4
leucine	4.2	1.6	1.9	2.0	2.3	0.4	2.8	4.9
tyrosine	5.0	0.8		0.5	0.7	0.4	1.9	2.1
phenylalanine	2.9	0.4	1.3	0.9	0.8	0.02	1.3	2.6
ornithine	10.7	7.8	11.8	1.1		3.4	10.2	3.1
lysine	7.7	2.2	3.8	2.7	1.4	1.5	2.7	7.2
histidine	4.5	2.1	2.3	1.0		1.8	3.0	5.5
arginine	2.8	0.6	0.7	2.9	1.1	0.4	1.4	6.1
unidentified ninhydrin positive	trace	trace						15.9

1   The food of animals from cultures is shown in parentheses under their names. Other species were incubated at *in situ* temperatures immediately after their removal from their natural environment, except for *Nerita versicolor*, which was held without food for three hours prior to incubation.

2   hypersaline crustacean

3   estuarine annelid

4   marine chiton

5   marine gastropod

6   marine scleractinian coral

7   freshwater crustacean

8   freshwater dipteran larva

From: Johannes and Webb, 1970

physiologists interested in the nature of chemical stimulation to "cast their net" far more widely than they have been wont to do.

The relatively broad spectrum of sensitivities to amino acids certain aquatic animals display and their apparent species specific arrays of relative acuities (Fujiya & Bardach, 1966; Tucker & Suzuki, 1972; Hara et. al., 1973), notwithstanding occurrence of one or the other "special substance" (Kleerekoper & Mogensen, 1963; Konishi et. al., 1966) further suggest that biologically meaningful scents are often mixtures of a "fingerprint" type (Mackie, 1973; Hara et. al., 1973; Bardach & Villars, 1974). Whether or not specific sensitivities have been encoded genetically in fishes and invertebrates as they have been in snakes (Burghardt, 1967) will have to be elucidated. Mackie (1973) has clearly shown for the lobster that a synthetic mixture of amino acids and similar compounds, mimicking in composition an extract of squid muscle was the "most" effective stimulant, to which different types of chemo-sensory cells responded. Such scent mixtures seem important for behavior aside from food search and feeding (e.g., homing).

A "division of labor" between smell and taste prevails at least in certain catfishes and some representatives of the marine family Mullidae, the goatfishes (Holland, 1974), with smell subserving reproductive and social, and taste mainly feeding behavior. This division requiring recognition of complex scents both by nose and taste-sensors may explain the concurrent evolution of amino acid sensitivity in both organ systems (Caprio, this symposium).

4) Some complex chemical signals involve substances of larger molecular weight than those discussed so far. Certain of these, for instance home stream or home site markers (the shad, Table 1) and the fright substance from snails (Table 1) are probably fingerprints again, while others such as certain gamones of algae (Wilson, loc cit) are single large molecules. It would seem likely that it is not the entire molecule but certain building stones of it which stimulate the receptor membrane sites and that these — quite possibly amino acid radicals prominently among them — again combine into fingerprint patterns.

Heavier substances act either on contact where solubility and diffusion properties are not so important or they are carried actively by currents which are of paramount importance in the effective distribution of signals and the contact between stimulus and receptor. Wilson (1970) discusses size of waterborne signals, their fadeout times, diffusivity and tradeoffs between molecular weight and optimal modes of emission of signals into the water as opposed to their being sent into the air and, last but not least, the importance of currents. It need hardly be labored nor reinforced by quotations that brooks, streams and rivers spread chemical signals and perhaps not either that water in lakes and the sea is hardly ever still. What needs to be said, though, is that currents bridge distances, and can evoke rheotaxis in response to chemical stimulation in certain crustaceans, e.g., the crab *Carcinus maenas* (Pardi & Papi, 1961) but, the eddies that form at their edges, with the shores, or with other waters of different densities through their respective temperature and/or salinity properties, can also create confusion (Bardach, et. al., 1969), certainly when quick gradient detection may be of the essence.

Planktonic animals which aggregate still swim in patterns that suggest they engage in gradient search (Kittredge, et. al., 1974), within the small, "active space" (Wilson, 1970) of substances to which they react, an active space that is related to their size and swarming habit. Larger animals often have anatomical enhancements for the detection of a chemical gradient and the location of a stimulus-source. These specializations are legion and belong certainly, in part, into the realm of comparative anatomy. Some highly efficient ones deserve mention here nevertheless, such as the nose trumpet of an eel (Holl, et. al., 1970), the taste bud distribution of catfishes (Bardach & Atema, 1971) and the sensor studded sheet into which an octopus can extend its protean shape (Figs. 1a, b,c). These and other anatomical features facilitate "hits" by signals. What these latter do, what nerve-carried messages they elicit and what the fate and function of these messages are, will be treated — again selectively — by the papers which follow this introduction.

Fig. 1a. Distribution of tastebuds on the skin of the catfish, *Ictalurus natalis*. Centimeter grid. Each black dot represents 100 tastebuds. Higher concentrations on lips and barbels are indicated by small dots and solid black respectively. Top: lateral view; Bottom: dorsal and ventral view. (From Bardach & Atema, 1971).

Fig. 1b. *Rhinonurrena ambonensis:* A. anterior portion lateral view, B. dorsal view (in direction of broken-line arrow), C. ventral view (in direction of solid-line arrow). a. nasal funnel, b. rostrum, c. mandibular barbels, d. nasal sack, e. anterior nares, f. posterior nares, g. gill chamber, h. gill opening. (From Holl, et. al., 1970).

Fig. 1c. Ventral view of *Octopus maya* illustrating water volume or surface that can be sampled for chemical stimuli by the animal with radially spread arms. (Photo. Haw. Inst. Mar. Biol.).

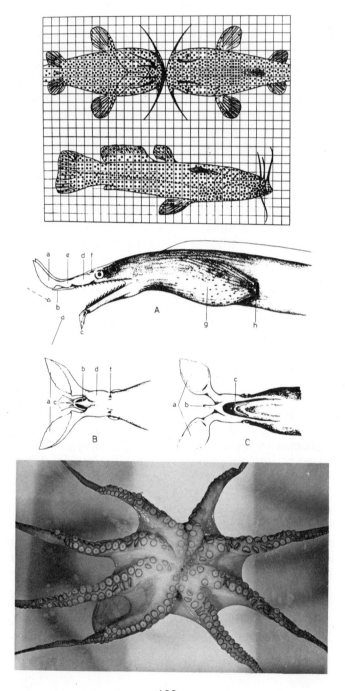

References

Ache, B.W. 1972. Comp. Biochem. Physiol. 42A, 807-811.
Adler, J. 1969. Science 166, 1588-1597.
Ash, J.F., McClure, W.O. and Hirsch, J. 1973. Anim. Behav. 21, 796-800.
Bailey, D.F. and Laverack, M.S. 1963. Nature, Lond. 200, 1122-1123.
Barber, S.B. 1956. J. Exp. Zool. 131, 51-73.
Bardach, J.E. and Atema, J. 1971. Taste in fishes. In Handb. of Sensory Physiology. L. Beidler, Ed. Jul. Springer Verl. Berlin, Heidelberg and N.Y. vol. IV (2), 293-336.
Bardach, J.E. and Villars, T. 1974. The chemical senses of fishes, 49-104. In Chemoreception by Marine Organisms, P. Grant, Ed. Acad. Press London and New York.
Bardach, J.E., Johnson, G.H. and Todd, J.H. 1969. Annals of the New York Academy of Sciences, 163, 227-235.
Burghardt, G.M. 1967. Science 157 (3789), 718-721.
Case, J. 1964. Randall. Biol. Bull. 127, 428-446.
Case, J. and Gwilliam, G.F. 1961. Biol. Bull. 121, 449-455.
Carr, W.E.S. 1967. Biol. Bull. 133 (1), 90-127.
Crisp, D.J. 1967. Biol. Bull. 133 (1), 128-140.
Degens, E.T. 1970. In Symp. on Organic Matter in Natural Waters. U. Alaska 1968. Ed. by D.W. Hood, College, Alaska, Inst. of Mar. Sc., U. of Alaska, 77-1 06.
Dwivedy, R.C. 1973. Transactions of the ASAE, 16 (2), 367-370, 373.
Ferguson, J.C. 1969. Biol. Bull. 136 (3), 374-384.
Fujiya, M. and Bardach, J.E. 1966. Bull. Jap. Soc. Sci. Fish, 32 (1), 45-46.
Gilbert, P.W., Hodgson, E.S. and Mathewson, R. 1964. Science, N.Y. 145, 949-951.
Hara, T.J., Carolina Law, Y.M. and Hobden, B.R. 1973. Comp. Biochem. Physiol. 45A, 969-977.
Holl, A., Schulte, E. and Meinel, W. 1970. Helgolander wiss. Meeresunters. 21, 103-123.
Holland, K. 1974. Behavioral and electrophysiological investigations of the taste function of the barbels of a Hawaiian goatfish, *Parupeneus porphyreus*. M.S. Thesis, Univ. Hawaii.
Idler, D.R., Fagerlund, U.H.M. and Mayoh, H. 1956. J. Gen. Physiol. 39, 889-892.
Jahan-Parwar, B. 1972. Amer. Zool. 12, 525-537.
Johannes, R.E. and Webb, K.L. 1970. In Symp. on Organic Matter in Natural Waters U. Alaska 1968. Ed. by D.W. Hood. College, Alaska, Inst. of Mar. Sc., U. of Alaska, 257-274.
Kittredge, J.S., Takahashi, F.T., Lindsey, J. and Lasker, R. 1974. Fishery Bull. 72 (1), 1-11.
Kleerekoper, H. and Mogensen, J.A. 1959. Z. Vergl. Physiol. 49, 492-500.
Kleerekoper, H. and Mogensen, J.A. 1963. Physiol. Zool. 36, 347-360.
Kohn, A.J. 1961. Am. Zoologist 1, 291-308.
Konijn, T.M., van de Meene, J.G.C., Bonner, J.T. and Barkley, D.S. 1967. Proc. Natl. Acad. Sci. U.S. 58, 1152-1154.
Konishi, J. and Zotterman, Y. 1963. Taste functions in fish, 215-233. In Olfaction and Taste I, Zotterman, Y. (Ed). Proc. 1st Int. Symp., Wenner Gren Center, Stockholm 1962, Macmillan, N.Y.
Konishi, J., Uchida, M. and Mori, Y. 1966. Jap. J. Physiol. 16, 194-202.
Lenhoff, H.M. 1961. In The biology of hydra. Lenhoff, H.M. and Loomis, W.F. (Eds). Miami: Miami Univ. Press.
Mackie, A.M. 1973. Mar. Biol. 21, 103-108.
Mackie, A.M., Lasker, R. and Grant, P.T. 1968. Comp. Biochem. Physiol. 26, 415-428.
Mathewson, R.F. and Hodgson, E.S. 1971. In "Report of the Director, Lerner Marine Laboratory, American Museum of Natural History, Bimini, Bahamas to Office of Naval Research. Oceanic Biology Branch." 81-90. Contract No. ONR552 (7).
McLeese, D.W. 1970. J. Fish. Res. Bd. Canada. 27, 1371-1378.
Pardi, L. and Papi, F. 1961. Kinetic and tactic responses, 365-399. In The Physiology of Crustacea. Vol. II. Sense Organs, integration and behavior. Waterman, T.H. (Ed), Academic Press London and N.Y.

Prosser, C.L. 1973. Comparative animal physiology, W. Saunders Co. Phila., London and Toronto. 688.

Reimer, A.A. 1971. Comp. Biochem. Physiol. 40A, 19-38.

Shepheard, P. 1974. Mar. Behav. Physiol. 2, 261-273.

Snyder, N. 1967. Cornell Univ. Agr. Expt. Sta. Mem. 403, 1-222.

Sutterlin, A.M. and Sutterlin, N. 1970. J. Fish. Res. Bd. Can. 27,(11), 1927-1942.

Sutterlin, A.M. and Sutterlin, N. 1971. J. Fish. Res. Bd. Can. 28 (4), 565-572.

Suzuki, N. and Tucker, D. 1971. Comp. Biochem. Physiol. 40A (2), 399-404.

Thomas, S.E. 1970. Natural products from marine invertebrates. Ph.D. Thesis. Univ. of the West Indies, Kingston, Jamaica, W.I.

Tucker, D. and Suzuki, N. 1972. In Olfaction and Taste IV, Proceed of the 4th Int. Symp. D. Schneider, Ed., 121-134.

Vernberg, W.B. and Vernberg, F.J. 1972. Environmental Physiology of Marine Animals. Springer-Verlag, New York, Heidelberg, Berlin.

Wells, M.J. 1963. J. Exp. Biol. 40, 187-193.

Wilson, E.O. 1970. In Chemical Ecology, (Sondheimer, E. and J.B. Simeone, Eds). Academic Press, N.Y. and London, 133-155.

Chemoreception in Gastropods

B. Jahan-Parwar

Worcester Foundation for Experimental Biology

Shrewsbury, MA 01545

Because of their large and identifiable neurons the gastropod molluscs have been used in recent years in a rapidly increasing number of studies on various aspects of the cellular neurobiology. Only few investigators, however, have realized the advantages of the gastropod nervous systems for the study of chemoreception[1,2,8,9,11,16]. This paper provides a summary report on several series of published and unpublished studies from my laboratory on chemoreception in the marine gastropod *Aplysia*, which demonstrates that this gastropod offers excellent conditions for the cellular analysis of chemoreception. It will include: 1) behavioral evidence for the ability of *Aplysia* to find food in Y mazes, and data on the molecular nature of stimuli that elicit different behavioral responses[11,12]., and 2) electrophysiological evidence for the presence of the chemoreceptors on three sets of sense organs, the anterior and posterior tentacles and the osphradium, and their central connections[5,9,10,11].

Materials and Methods

Behavioral

The subjects (Ss) were *Aplysia californica* weighing between 150-200 gm. Eighteen were used in maze experiments. They were kept in individual Y-mazes (60 cm long, 3.5 liter capacity) through which seawater was continuously flowing at the rate of 200 ml/min in each arm and out the foot. The chambers were placed in an environmental room with constant temperature (14°C) and light-dark cycle (12L − 12D). During the test-off period, Ss were kept behind removable doors near the ouput end (foot) of the Y chambers. At the beginning of the daily test trials, seaweed (the natural food for *Aplysia*), or various concentrations (100%, 10% and 1%) of a standard seaweed extract (SSWE) was presented into one randomly selected arm of each maze; the doors were removed and responses were recorded for 30 minutes with a time-lapse movie camera. A correct response was rewarded with one gm. dried seaweed soaked in seawater. Ss were not fed otherwise. The seaweed used in these experiments was dried *Rhodymenia palmata* which is commercially available as "Dulse". The SSWE was prepared by letting 5 gm. Dulse stand in 1000 ml seawater for 30 minutes and then passing the resulting solution through the Whatman

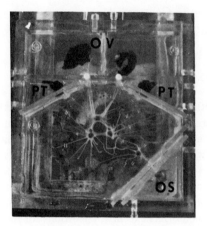

Fig. 1. The preparation chamber containing the major chemosensory organs and the CNS of an Aplysia: OV, oral veil; PT, posterior tentacle; OS, osphradium. Some of the nerves are placed over the bipolar electrodes imbedded in the floor of the chamber and can be stimulated electrically. Each side of the chamber represents 10 cm.

#1 filter. The test solutions were added at a constant rate of one drop/sec (~4 ml/min) to the stream of seawater through the test arm. Each daily test trial was preceded by a 30 minute long control trial with plain seawater.

Electrophysiological

The major chemosensory organs of *Aplysia*, the oral veil (consisting of the anterior tentacles, the lips and the anterior head region in between), the posterior tentacles and the osphradium, connected to the CNS were isolated and placed in individual compartments of a seawater filled chamber (Fig. 1). The compartments were sealed to prevent leakage of solutions from one into the other and maintained at 14°C. With this arrangement, it was possible to maintain the preparation in functional state for a long period of time (days) and search with microprobes for the central neurons that respond to chemical stimulation of each sense organ.

Central responses were recorded using the conventional intracellular microelectrode techniques. In some experiments, the central ganglia were removed and afferent chemosensory induced responses were recorded via hook-electrodes from nerves and branches innervating various regions of the sense organs. These and the bioassay techniques are described in more detail elsewhere[11].

Results

Behavioral

The ability of *Aplysia* to locate food by distance chemoreception was

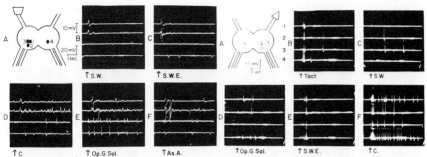

Fig. 2. Responses of four cells in the caudal dark cell cluster region of the cerebral ganglion following chemical stimulation of the left hemi-oral veil. SW, seawater; SWE, standard seaweed extract; C, Coumarin; op.g. Sol., opaline gland solutions; As-A, aspartic acid.

Fig. 3. Responses of four cerebral cells to tactile and chemical stimulation of the right posterior tentacle. Note here that the natural stimuli, the standard seaweed extract and the opaline gland solution are less effective than in the oral veil (Fig. 2).

investigated in several series of Y-maze experiments. In one experiment (60 trials, 6 Ss) the full strength SSWE was the test substance. Responses were 96% correct. Similar results were obtained with the 10% SSWE (30 trials, 96% correct). 1% SSWE produced 65% correct response (36 trials). In these experiments the first sign of sensing food, stretching of the tentacles, was usually observed within the first 2 minutes of the trial. In the latter experiment, for example, the amount of food attractants that can reach the Ss within the first 2 minutes of the trial is estimated to be, at least, one billion times lower than the amount given off by the seaweed into the standard solution. The results of 126 control trials provided no indication for the arm (side) preference. These results suggest that *Aplysia* are capable of distant chemoreception.

Effect of Amino Acids

Since amino acids have been implicated as natural chemosensory stimuli for both vertebrates and invertebrates, we bioassayed 20 natural L-amino acids (Nutritional Biochemicals Co.). These were tested first in

10^{-2} M solutions. The threshold was then determined for the effective substances. Seven of the examined amino acids (asparagin, aspartic acid, glutamic acid, methionine, phenylalanine, serine, and tyrosine) were effective in 10^{-2} M concentrations in producing the typical feeding behavior, the mouth opening response. The glutamic acid with the effective concentration ranging between $10^{-6} - 10^{-7}$ M proved to be the most effective stimulus. The effective concentration of the aspartic acid ranged between $10^{-5} - 10^{-6}$ M. The food sensing behaviors, stretching of the tentacles, head waving, were produced by much lower concentrations $(10^{-8} - 10^{-9}$ M) of both these amino acids. The threshold concentration of the other feeding eliciting amino acids was 10^{-3} M and higher. The amino acid analysis of the SSWE revealed that all amino acids that elicited feeding response were present in the extract.

Four of the examined amino acids had different behavioral effects on *Aplysia*. Alanine, cystine and leucine caused withdrawal response, while proline appeared to activate the mating behavior. The threshold concentrations of all these substances were higher than 10^{-3} M. The other natural amino acids did not have any distinct behavioral effect.

Electrophysiological

A systematic search was undertaken with the intracellular probes for the central neurons that respond to chemical stimulation of the oral veil (over 200 recordings in 28 cerebral ganglia), the posterior tentacles (139 recordings, 24 cerebral ganglia) and the osphradium (319 recordings 25 abdominal ganglia). Natural stimuli such as food ingredients were more effective than the other stimuli in producing a central response.

Responsive cells to stimulation of the oral veil (97 neurons) and the posterior tentacles (36 cells) were scattered all over the dorsal and ventral surfaces of the cerebral ganglion. The majority of these (64 OV, 28 PT), however, were found in two symmetrical pairs of darkly pigmented cell clusters in the caudal quandrants of the cerebral ganglion. The neurons in the ipsilateral clusters produced strongest and shortest latency responses (Fig. 2 and 3). This suggests that the chemosensory pathways from each hemi-oral veil and posterior tentacle primarily project to the ipsilateral caudal dark cell cluster region of the cerebral ganglion.

Local stimulation of the bluish brown pigmented anterior tentacular cup (groove) area with food ingredients was more effective than other oral veil regions in activating the cerebral neurons. This suggests that the food receptors of the oral veil are predominantly located in the anterior tentacular cup area. This finding was supported by 2 lines of evidence: 1) The responses of the most sensitive cerebral neurons to the oral veil

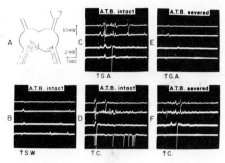

Fig. 4. Responses of four caudal dark cell cluster neurons to chemical stimulation of the left hemi-oral veil before (B,C,D) and after (E,F) cutting the nerve branch that innervates the anterior tentacular cup area.

stimulation with food substances, the caudal dark cell cluster neurons, were abolished after cutting the anterior tentacular branch that innervated the cup area (Fig. 4), while cutting the nerves and branches that innervate the other oral veil regions appeared to have little or no effect; 2) A systematic extracellular unit recordings from the various oral veil nerves and branches revealed the presence of a large number of food sensitive units in the anterior tentacular cup nerve, while only few were found in the other nerves.

The results of the threshold studies of food units in chemosensory pathways from the anterior and posterior tentacles suggest that the most sensitive food receptors are located in the anterior tentacular cup area. The anterior tentacular chemoreceptors were also responsive to other qualities, but their threshold was 100-1000 times lower for the food stimulants. This suggests "partial specificity" of these receptors to food stimuli.

Sixty including 18 previously identified abdominal cells, were found to respond to local stimulation of the osphradium with the SSWE and 2 amino acid constituents of seaweed, the glutamic acid and aspartic acid. Solutions containing the sexual glands of *Aplysia* were not effective in activating these cells. Four cells responded to coumarin and 2 to hypotonic seawater. The threshold of the osphradial food receptors appeared to be higher than those of the anterior tentacles.

The majority of responsive cells (56) were situated in the right abdominal hemi-ganglion with a large concentration near the entrance of the branchial nerve that connects to the osphradium. Many of these cells send axons into the branchial nerve, as it is indicated by their antidromic invasion following electrical stimulation of this nerve. These neurons continue to respond to chemical stimulation of the osphradium after synaptic transmission is blocked in the abdominal and osphradial ganglia with high Mg^{++} (Fig. 5). This suggests that these cells are either centrally

137

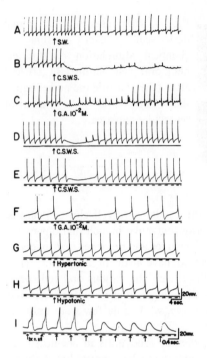

Fig. 5. The responses of an abdominal ganglion cell to chemical stimulation of the osphradium. This cell had an axon in the branchial nerve, as indicated by the antidromic spikes in I and continued to respond to the concentrated seaweed solution (B,D) and glutamic acid (C,F) after high Mg^{++} blockage of the synaptic transmission in the abdominal (solid line) and osphradial (broken line) ganglia. The humps preceding the somatic spikes in I are antidromically evoked A-spikes.

located primary chemosensory neurons or are ephaptically connected to the osphradial chemoreceptors. Even if the latter possibility was the case, the responses of these cells would represent in a one to one manner those of the osphradial chemoreceptors.

Discussion

The results of the Y-maze experiments provide strong support for the previous assumption that *Aplysia* is capable of locating seaweed (food) from a distance[7,11,13,14]. The low effective concentration of the glutamic and aspartic acid for eliciting the food sensing and feeding response raises the question of whether these substances serve as chemosensory cues for *Aplysia* in food location. The fact that both these amino acids are present in free state and high concentration in all the seaweed species that *Aplysia* feed on[4,6,15] supports this possibility.

The electrophysiological experiments provide the first direct evidence for the presence of chemoreceptors in the anterior and posterior tentacles and the osphradium of *Aplysia*. The results of threshold studies indicating that the anterior tentacular chemoreceptors are the most sensitive food receptors of *Aplysia*, confirm the previous behavioral findings[7,11,13,14]

The partial specificity of these receptors to the biologically significant stimuli (food ingredients) has its parallel in the pheromone receptors of insects[3].

The convergence of the chemosensory pathways from the major head chemosensory organs, the anterior and posterior tentacles, into the caudal dark cell clusters of the cerebral ganglion suggests that these regions are important centers for integrating chemosensory information. The identification of the central projection areas of the major chemosensory pathways and the possibility of identifying the individual neurons and their interconnections in these areas provide excellent opportunities for analyzing the wiring diagram of the neural circuits involved in chemoreception which is the key to the understanding of the mechanism of chemoreception.

Summary

1) *Aplysia californica* is capable of locating seaweed in Y-maze situations. Two amino acid constituents of seaweed, the glutamic and aspartic acids, elicit in low concentration ($10^{-8} - 10^{-9}$ M) the typical food sensing and feeding behavior. Several other amino acids elicit escape and sexual behaviors.

2) Chemoreceptors are present on the anterior and posterior tentacles and the osphradium. Of these, the anterior tentacular chemoreceptors are most sensitive and partially specific to food stimuli. The chemosensory pathways from the anterior and posterior tentacles primarily project into the caudal dark cell cluster region of the cerebral ganglion and those from the osphradium into the right half of the abdominal ganglion. Some of the abdominal cells appear to be centrally located primary chemosensory neurons.

Acknowledgements

This work was supported by the Research Career Development Award K4-HD-5178 and the NIH grants NS08868 and NS11452.

References
1. Bailey, D.F. and Laverack, M.S. 1966. J. Exp. Biol. 44, 131-148.
2. Bailey, D.F. and Benjamin, D.R. 1968. Symp. Zool. Soc. London, 23, 263-268. Acad. Press, London.
3. Boeckh, J. 1967. Z. Vergleich. Physiol. 55, 278-406.
4. Carefoot, T.H. 1967. J. Mar. Biol. Ass. UK., 47, 565-589.
5. Downey, P. and Jahan-Parwar, B. 1972. Physiologist, 5, 122.
6. Fowden, L. 1962. Amino acids and proteins. P. 189-209. In: Physiology and biochemistry of algae. (R.A. Lewin, editor). Academic Press.
7. Frings, H. and Frings, C. 1965. Biol. Bull. Lab., Woods Hole 128, 211-217.
8. Gelprin, A. 1974. Proc. Nat. Acad. Sci. USA, 71, 966-970.
9. Jahan-Parwar, B., Smith, M. and von Baumgarten, R. 1969. Am. J. Physiol. 216, 1246-1257.
10. Jahan-Parwar, B, and Downy, P. 1973. Fed. Proc. 32, 336.
11. Jahan-Parwar, B. 1972. Am. Zool. 12, 525-537.
12. Jahan-Parwar, B., Wells, L.J., Fredman, S.M. 1974. Proc. 26th Intern. Cong. Physiol. Sci., 230.
13. Jordan, H. 1917. Biol. Zbl. 37, 2–9.
14. Preston, R.J. and Lee, R.M. 1973. J. Comp. Psychol. 2 (3), 368-381.
15. Schlichting, H.E. Jnr. and Purdom, M.E. 1969. Proc. Int. Seaweed Symp. 6, 589-594.
16. Suzuki, N. 1967. J. Fac. Sci. Hokkaido Univ. Series VI, Zool., 16, 174-185.

Properties of Chemoreceptors in Marine Crustacea

M.S. Laverack

Gatty Marine Laboratory, St. Andrews,

Fife, Scotland

This paper is really a plea for more attention to be paid to crustacea in order that fundamental principles might be sought. The names of workers involved in, and actively publishing work on crustacea can almost be counted on the fingers of one hand, and yet innumerable topics can be shown to be worthy of further investigation. I will attempt a short review that shows both the potential of the subject and some of its problems.

Structures

It cannot be doubted that in decapod crustacea at least, there are chemoreceptors scattered over the whole body surface. Numerous sensilla may be implicated, but experimental work has not demonstrated that one single *type* of sensillum can be associated with chemoreception. In *Homarus* various aggregations of sensilla may be involved, especially those of the outer ramus of the antennules, the subchela of the pereiopods and the oesophageal wall.

The outer ramus of the antennules carries fine, slender, unbranched hairs known collectively as aesthetascs. Aesthetascs are innervated by large numbers of neurones (Laverack & Ardill 1965; Ghiradella, Case & Cronshaw 1968). The innervation in *Homarus* characteristically ends a short distance above the base which is distinguished by a bulbous appearance (Fig. 1); sensory dendrites do not appear beyond this level. The aesthetasc ends in a fine cone (Fig. 2).

Hairs of different kinds occur on maxillipeds and pereiopods; a catalogue of these has been given for a crayfish by Thomas (1970). It seems clear that many of these are concerned with chemoreception.

The subchelate areas of the pereiopods carry plates that are subdivided into very fine hairs and the whole structure receives heavy innervation.

Environment-receptor interaction

If we may concede that there may be low threshold (olfactory) and high threshold (gustatory) receptors then we might expect there to be two types of ending; the first in connection with the environment via channels in the wall of the hair sensillum; the second having a patent pore that

Figs. 1 & 2. Aesthetasc hairs of *Homarus*. Fig. 1 shows the base, and Fig. 2 the termination of the hairs. Note the intact conical ending (Laverack unpublished). Scales 100μ & 20μ respectively.

opens directly to the environment and with dendritic endings brought close to the surface (cf. the situation in insects).

Among crustacea that have so far been examined we cannot definitely state that both types occur and it seems probable that only one does. Ghiradella et. al., (1968) suggested that the permeability of the aesthetasc hair wall is sufficient to allow access of the dendrites to molecules. Fig. 2 (*Homarus*) shows that there are no terminal pores for aesthetascs. Shelton (1974) claims that in *Crangon* pores do occur. SEM studies shows that intact aesthetascs are long and devoid of a terminal pore but many hairs in sand-dwelling shrimps are abraded and shortened, hence seeming to have pores, especially in late intermoult animals. A pore is claimed for setae on the penis of *Belanus* (Munn, Klepal & Barnes 1974) but abrasion may take place here also.

Summary of structure

Chemoreceptors are 1) often but not always in the form of hairs; 2) not possessed of pores except when broken; 3) sometimes flimsy and permeable (possibly olfactory); 4) otherwise rigid and associated with feeding (possibly gustatory).

Fibre diameters

Sections of the antennulary nerve show that over 90% of sensory axons are ca 0.3μ in diameter, packed in bundles of naked axons without individual glial sheaths (Fig. 3). The numbers are enormous, running into hundreds of thousands for each antennule and each representing a single sense cell (cf. Steinbrecht [1969] on *Bombyx* antenna). Similar fibre sizes and aggregations are found in other presumptive chemoreceptor nerves to the labrum and the oesophagus.

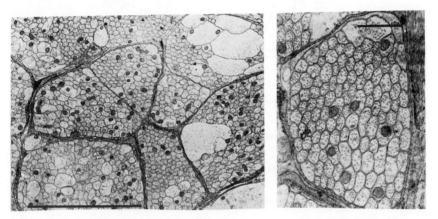

Fig. 3. A) a small area of a TS from the outer ramus nerve of *Homarus;* B) a single bundle of axons (Laverack unpublished). Scales 5μ & 0.5μ respectively.

Osmotic effects

It has been pointed out on behavioural grounds (Lance 1962) that for some crustacea osmotic changes are probable stimulants. Tazaki and Tanino (1973) have demonstrated that lobster antennules respond to osmotic stimuli (varying concentrations of NaCl) (Fig. 4A), and Laverack has recently shown that intact preparations of the legs of undissected *Galathea* motor units can be used as monitors of osmotic responses (Fig. 4B) following dilution of sea water.

Receptor Sensitivity

Table 1 is constructed from information gleaned from several papers. It shows the quantitative analysis of amines from squid (natural food and bait for *Homarus*) (Mackie 1973) compared with the number of receptor units responding in antennules of *H. americanus* (Ache 1972, Shepheard 1974), the threshold of such units, and the response intensity relative to glutamic acid or glycine (for *Cancer,* from Case 1964). From this we may conclude a) it is very difficult to amass much information from such techniques as small bundle recording since the number of identifiable units

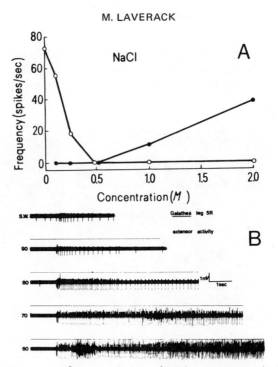

Fig. 4. A) response of two units to changing concentrations of NaCl (*Panulirus*) (from Tazaki and Tanino with permission); B) activity in propus extensor motor fibre following stimulation in various concentrations of sea water of whole 5th leg in intact *Galathea* (Laverack unpublished).

is very small, b) that certain substances are more likely to stimulate than others and that the L-isomer is more stimulatory than the D form, c) that threshold sensitivities of receptors are quite variable, d) that there may be receptors with different distributions.

Thus in *Homarus* there may be far more taurine (threshold 10^{-6}M) than L-glutamic acid (10^{-6}M) than glycine (10^{-4}M) receptors; and the responses relative to population may be very variable from species to species. Curiously also the greatest sensitivity is shown to OH-proline (not recorded for squid extract) but not to L-proline (found in large quantities).

There is some evidence that crustacean chemoreceptors may be either specialist or generalist units. Amongst the former, Ache (1972) reports 1

Table 1: for explanation see text

Squid Extract in order of magnitude	Mackie (1973) mg/g	HOMARUS No. of units responding				HOMARUS Response Relative to Glutamic acid (Shepheard)	CANCER Response Relative to Glycine Case (1964)	Electrophysiological Threshold (Shepheard)
		Ache (1972) %		Shepheard (1974) %				
L − proline	161	1/11	9	4/7	57		100	10^{-3}
TMAO HCL	125							
Glycine betaine	100						106	
Glycine	98	2/11	18	2/5	40	22.9	100	10^{-4}
Taurine	37	7/11	63	7/7	100	13.8	141	10^{-6}
L − alanine	30	2/11	18	3/4	75			10^{-3}
L − arginine	25	3/11	27	3/4	75	10	88	10^{-3}
TMA	10					6	40	
L − lactic acid	10							
Homarine	8							
L − leucine	6.3	6/11	54	1/3	33	6	56	
L − glutamic	5.8	5/11	45	5/9	55	100	137	10^{-6}
Hypoxanthine	5.2							
L − threonine	4.9					0	105	
L − valine	4.1	0/3	0			1.3	100	
OH − proline	?					53	100	10^{-7}
D − glutamic						41	51	10^{-6}
Tyrosine								10^{-4}

unit that responds only to taurine, and 3 to taurine and L-glutamic acid. Shepheard (1974) shows 1 unit responding to L-glutamic acid and 1 to taurine and β-alanine. Other generalist units are stimulated by several compounds.

Some attention has recently been paid to the production of a synthetic mixture for use as bait. None so far concocted is as effective as natural material (cf. mussel and squid extracts). McLeese (1970) showed that several amino acids may invoke walking in *Homarus,* L-proline seeming to be especially potent. From the analysis above it seems that a mixture containing taurine, OH-proline, glycine, leucine, alanine and arginine might be appropriate. Sensitivity towards these compounds, however, does not necessarily indicate attraction and other behavioural results may be expected. Pheromone attraction of females for males has been demonstrated but little is known of the biochemistry of the pheromone (Atema & Gagosian 1973).

References
Ache, B.W. 1972. Comp. Biochem. Physiol. 42A, 807-811.
Atema, J. and Gagosian, R.B. 1974. Mar. Behav. Physiol. 2, 15-20.
Case, J.F. 1964. Biol. Bull. 21, 449-456.
Ghiradella, H.T., Case, J.F. and Cronshaw, J. 1968. Amer. Zool. 8, 603-621.
Lance, J. 1962. J. Mar. Biol. Ass. U.K. 42, 131-154.
Laverack, M.S. and Ardill, D.J. 1965. Quart. J. micr. Sci. 106, 45-60.
Mackie, A.M. 1973. Mar. Biol. 21, 103-108.
McLeese, D.W. 1970. J. Fish. Res. Bd. Can. 27, 1371-1378.
Munn, E.A., Klepal, W. and Barnes, H. 1974. J. exp. Mar. Biol. ecol. 14, 89-98.
Shelton, R.G.J. 1974. J. Mar. Biol. Ass. U.K. 54, 301-308.
Shepheard, P. 1974. Mar. Behav. Physiol. 2, 261-274.
Steinbrecht, R.A. 1969. J. Cell Sci. 4, 39-53.
Tazaki, K. and Tanino, T. 1973. Experientia 29, 1090-1091.
Thomas, W.J. 1970. J. Zool. Lond. 160, 91-142.

Taste Responses to Ribonucleotides and Amino Acids in Fish

I. Hidaka and S. Kiyohara

Mie Prefectural University and Nagoya University, Japan

It has been shown by previous workers (1,2,3) that the chemoreceptors of the fish are able to sense sweet, salty, sour and bitter substances as the tongue chemoreceptors of higher vertebrates do (4). On the other hand, it has also been known that they are sensitive to many other classes of chemicals; e.g., there have been reported many afferent fibers in the sea catfish which responded to extracts of nereis, milk, etc., but responded to neither of the four stimuli, NaCl, sucrose, HCl and quinine (5), though it has not fully been revealed what kinds of constituents of the mixed stimuli are effective. Comparing different fish species, there seems to exist considerable differences in their sensitivity to chemicals. For example, certain fishes respond well to sucrose (1,2), while others do not seem to be sensitive to it (3,5).

Sea water and freshwater may have different effects on the characteristics of the chemoreceptors of the fishes living in them. Our knowledge of what is involved is, however, meager, especially since there have been few reported studies on the marine fish (3,5).

In the present study, the stimulating effects of several chemicals on the lip chemoreceptors of the puffer, *Fugu pardalis,* were examined by recording the electrical responses from the facial nerve supplying the upper lip. The fish were caught by net in Gokasho Bay not far from the laboratory. They were kept in an aquarium filled with artificial sea water for a few days before experiment. The procedure for obtaining the facial nerve preparation was principally the same as that described by Konishi et. al. (5). After removing the eye ball, the ramus maxillaris running under the eye to the upper lip was isolated from the surrounding tissues and placed on a platinum electrode. The multifiber activity of the nerve bundle was recorded with the aid of an electronic integrating device (6). The impulse discharge of single fibers was recorded from small strands with one or a few functionally intact fibers.

147

Stimulating Effects of Chemicals

Salts. NaCl stimulated the receptors weakly. Fig. 1 shows an example of the integrated responses from the whole nerve to application of various stimuli to the upper lip. As seen in the figure, 1 M NaCl failed to elicit a response distinguishable from that to sea water or distilled water. Similar results were obtained with single fibers (Fig. 3), suggesting that the lip chemoreceptors of this fish are less sensitive to NaCl than the barbel or palatal receptors of freshwater fishes (1,7) or even the lip receptors of the sea catfish (5). KCl also stimulated the receptors, but, like NaCl, weakly. No appreciable responses were obtained with sea water or distilled water from any of the chemo-sensitive single fiber preparations tested.

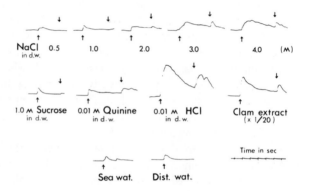

Fig. 1. Responses to various stimuli. (↑): stimulation. (↓): washing with sea water.

Acids. The puffer responded markedly to HCl (Fig. 1) and organic acids. As in the catfish (8), organic acids such as acetic acid were more effective at the same pHs than HCl over the pH range about 2 to 4 in distilled water. Experiments were conducted using three kinds of HCl solutions – prepared in 1) distilled water, 2) sea water, and 3) 0.5M NaCl. Results with five whole nerve preparation are summarized in Fig. 2. In the figure, magnitudes of the responses to the three kinds of acid stimuli at one second after application of stimulus are plotted related to the magnitude for 0.01M HCl in distilled water in each preparation. As the figure shows, the two kinds of acid stimuli containing NaCl were more effective than the one prepared in distilled water over the pH range about 1.7 to 3.0. Single fibers were also found which did not respond to 1.0 M NaCl but responded to mixtures of NaCl and HCl with higher impulse frequencies than to single acid solutions, suggesting that NaCl enhances the response of the acid receptor as in the carp (9).

Sugars. No appreciable response was elicited by sucrose in sea water or in distilled water (Fig. 1). D-glucose and D-ribose were also not effective. It has been reported also in the sea catfish that in none of the twenty-eight

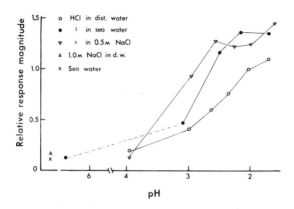

Fig. 2. Effects of salts on the response to HCl.

single fibers tested did sucrose give an appreciable response, while, in the tomcod (3), it gave a positive response in but one fiber out of twelve tested. In contrast to these marine species, some freshwater species (1,2,3) seem to respond well to sucrose.

Quinine. Quinine hydrochloride stimulated the receptors (Fig. 1). But, it was occasionally observed that this substance at high concentrations damaged the receptor activity.

Nucleotides. The puffer responded markedly to squeezed juice of the short-necked clam (*Tapes japonica*), as is seen from the example of the integrated response to a twentyfold-diluted sea water solution in Fig. 1. It was attempted to examine the stimulating effects of several substances contained in the clam tissues. Thus, adenosine-5'-monophosphate (AMP), adenosine-5'-diphosphate (ADP), hypoxanthine inosine-5'-monophosphate (IMP) and uridine-5'-monophosphate (UMP) were tested at concentrations equivalent to those found in the water extract of the clam (10). In these substances, AMP, ADP, IMP and UMP proved to be effective. ADP and UMP were the most effective.

Amino acids. Glycine and L-amino acids were also tested. Alanine, glycine and proline were effective, while, arginine, cysteine, histidine, leucine, methionine, phenylalanine and valine were not effective at 0.01

M. It is noteworthy that cysteine is a very effective stimulus for other species (3), suggesting some species differences.

Others. The puffer responded well to betaine which is also found in the clam tissues, as in other species (1,5). 1% glycogen and 0.1% trimethylamine oxide were both ineffective.

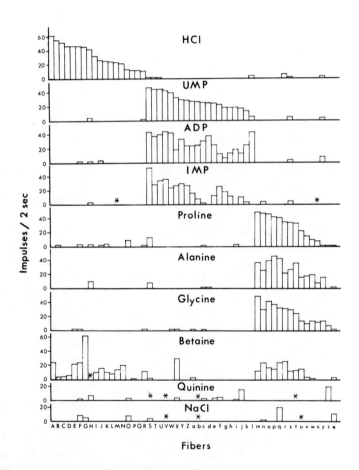

Fig. 3. Response profile of 53 fibers. (*): not tested.

150

Single Fiber Analyses

Impulse discharges in response to 0.005 M HCl, 0.001 M UMP, 0.001 M ADP, 0.001 M IMP, 0.01 M proline, 0.01 M alanine, 0.01 M glycine and 0.01 M betaine, in sea water, and 0.01 M quinine hydrochloride and 1.5 M NaCl, in distilled water, were recorded from fifty-three preparations. In Fig. 3, impulse numbers of these units during the first two seconds after stimulation are illustrated. As seen in the figure, the units tested could be divided roughly into three groups; 1) those responding well to HCl and betaine, 2) those responding well to UMP, ADP and IMP, and 3) those responding well to proline, alanine, glycine and betaine. Some units in each group further responded to other stimuli, but only weakly with a few exceptions, suggesting that the puffer has well differentiated taste units which respond rather specifically to specified qualities of taste substances. The fact that each of the units in the second group tended to respond to all of the three nucleotides uniformly suggests common receptors for the three substances. Similarly, the three amino acids and betaine may share common receptors. It was of interest that most of the units in the first group responded to betaine (pH, 8.0). NaCl stimulated some units, but most of them, weakly as expected from the multifiber activity (Fig. 1).

References

1. Konishi, J. and Zotterman, Y. 1961. Acta Physiol. Scand. 52, 150.
2. Tateda, H. 1964. Comp. Biochem. Physiol. 11, 367.
3. Bardach, J.E., Fujiya, M. and Holl, A. In "Olfaction and Taste II" (T. Hayashi, ed.), Pergamon Press, Oxford, 1967, p. 647.
4. Sato, M. In "Handbook of Sensory Physiology" 4(2) (L.M. Beidler, ed.), Springer-Verlag, Berlin, 1971, p.116.
5. Konishi, J., Uchida, M. and Mori, Y. 1966. Jap. J. Physiol. 16, 194.
6. Beidler, L.M. 1953. J. Neurophysiol. 16, 595.
7. Tateda, H. 1961. Nature 192, 343.
8. Tateda, H. 1966. Mem. Fac. Sci., Kyushu Univ., Ser. E. (Biol.) 4, 95.
9. Hidaka, I. 1972. Jap. J. Physiol. 22, 39.
10. Konosu, S., Fujimoto, J., Takashima, Y.T., Matsushita, T. and Hashimoto, Y. 1965. Bull Jap. Soc. Sci. Fish. 31, 680.

Chemical Attraction of Some Marine Fish

A. Sutterlin

Environment Canada, Biological Station,

St. Andrews, N.B.

The energy flow, due to feeding interactions by fish within different trophic levels in the marine environment, is mediated by a number of sensory systems. Vision is probably especially important in food finding and selection, but chemoreception by itself or in conjunction with vision seems dominant in some marine species. Despite the apparent importance of chemoreception, relatively little work has been done in the ocean, using artificially prepared chemical baits or pure chemicals in an attempt to attract marine fishes. This study examines the effectiveness of natural bait, bait extracts, and a variety of pure compounds as attractants to 4 species of marine fish directly in the field.

Materials and Methods

Flounders (*Pseudopleuronectes americanus*) were observed from the laboratory by means of an underwater T.V. camera situated in the intertidal zone adjacent to the lab. Sea-water solutions of the compounds to be tested were pumped (50 cc/min) from the lab through a plastic delivery tube and released through a spout beneath the television camera. Testing periods for a single compound consisted of 1 hr of pumping during the same phase of flood and ebb tide.

Killifish (*Fundulus heteroclitus*) and Atlantic silversides (*Menidia menidia*) were observed from behind a blind erected at the edge of a brackish brook connecting two tidal ponds. Selected chemicals for testing were pumped (1 cc/min) from behind the blind and released through a hole in a plate which rested on the bottom. The number of fish attracted to the odor was determined by photographs taken every minute.

Comparative trapping methods were used for hagfish (*Myxine glutinosa*). A total of 15 hagfish traps were set approximately 100 m apart on mud bottom (depth 45-65 m). Compounds or bait to be tested were placed in perforated bait jars within the trap. Traps remained down in most tests for a period of 2½ hrs. Freeze-dried codfish extract (FDCE) was prepared by grinding 1200 g of partly frozen codfish fillets into a litre of distilled water. This slurry was stirred for 15 min and the filtrate lypholized to yield aproximately 12.5 g of powder. FDCE was serially extracted with hexanes, chloroform, and ethanol. After stripping and weighing, the residue was redissolved and absorbed on filter paper which was placed in the bait bottles and tested. The ethanol extract was separated by gel filtration (Sephadex G10) with distilled water as eluant. The ethanol extract (50 mg/paper) was applied along the bottom of chromatograph paper (Whatman 3 mm) mounted on a moving kymograph drum. After separation by ascending chromatography (butanol-acetic acid) the edges were developed (ninhydrin). The remaining undeveloped paper was cut into horizontal strips and tested. The active zones were eluted with ethanol and analyzed with an amino acid analyzer.

Results

Of the 17 L-amino acids tested on flounders at 10^{-3} M, glycine was the most effective, attracting over 50 flounders per hr compared to 2 flounders using sea water. To a lesser extent, alanine, methionine, asparagine, cysteine, glutamic acid and leucine were also effective. On a weight basis, FDCE was not as effective as several of the amino acids. Several amines and amino-alcohols were tested but were found non-attractive. Of a total of 510 flounders attracted to 10^{-3} M glycine during 5 different tests, the majority entered the camera field from downcurrent, suggesting that the fish had intercepted the chemical plume somewhere downstream during their movement into and out of the intertidal zone. Most flounders remained in the field of view about 15 seconds and many intermittently circled the spout. Some fish exhibited digging indulations and others snapped at pieces of white clam shells.

Of 21 L-amino acids tested, GABA, alanine, histidine, glycine, β alanine, threonine, leucine, glutamic, isoleucine and valine were attractants to killifish. The first 4 initiated biting responses directed at the release orifice. Schools of silversides could be attracted by only 4 of the 21 amino acids tested (alanine, β alanine, threonine and methionine).

Muscle tissue from codfish, herring and trout were all effective baits for hagfish; shelled clams and shrimp were less effective. Twenty-four amino acids (0.01 GMW of the L form) were wrapped in filter paper and placed in the bait bottles in separate traps and no hagfish were caught. Six combinations taken 4 at a time and 4 combinations taken 6 at a time were tried; no hagfish were caught. All 24 L forms were tried together, each at 0.002 GMW; no hagfish were caught. Mixtures of amino acids in proportions found for free amino acids in various fish muscle gave 0 catch. A variety of amines and peptides associated with fish tissue all failed to catch a single hagfish. During the above testing, at least 1 trap was baited with 1 g of FDCE and catches were consistently over 100 hagfish. Boiling

FDCE for 15 min did not alter its activity. No activity was found in the hexane or chloroform extract, while both the ethanol extract and residue were active, specific activity being greater in the ethanol extract. Gel filtration of the ethanol extract showed activity associated only with low molecular weight compounds. The only active zone in the separation by paper chromatography was between Rf 0.35-0.50, but the activity was much lower than the total activity from the undivided paper. Amino acid analysis of the ethanol extract identified 22 ninhydrin positive compounds and within the active fraction separated in paper, 11 compounds were identified. Reconstituting the mixtures above failed to catch a single hagfish when tested in the traps.

Discussion

Aquarium observations indicate that flounders and other pleuronectiformes, unlike some flatfish, find food by both chemical and visual means (Pipping, 1927; de Groot, 1969). Parker (1911) suggests that killifish usually seek food by visual and chemical means, and that the olfactory organ is an important distance receptor. No information could be found concerning the feeding mechanisms of Atlantic silversides except that their diet is composed of copepods, mysids, shrimp, small squid and marine worms (Leim & Scott, 1966). Hagfish were chosen for a number of reasons. The functional significance of the rudimentary eyes of Myxine are controversial, however the olfactory capabilities based on laboratory observations have been commented on (Greene, 1925; Strahan, 1963). It is generally concluded that they are not predators of large fish, but will attack dead, injured or immobilized fish.

It is probably safe to assume that live, intact organisms (plants and animals) make up the majority of the diet of the 4 species studied here. This is probably less certain for the hagfish, however. The significance behind chemical attraction to specific chemicals (amino acids) or to a complex of chemicals (FDCE) might be explained by considering these fish as opportunistic scavengers. A second possibility is that these odors act not as feeding stimuli but simply as novel stimuli releasing non-specific exploratory behavior. A third possible explanation is that these odors are similar to those released by live organisms which enable fish to locate individual prey or large concentrations of prey (Brawn, 1969; Steven, 1959; and Tester, 1963). Why certain of the species should be attracted by a particular spectrum of amino acids is not especially apparent unless it has something to do with partitioning food resources among certain species.

Amino acids have been implicated in the feeding responses of a number of invertebrates (Lindstredt, 1971), and fish (Tester et. al., 1954;

Hodgson et. al., 1967; Mathewson & Hodgson, 1972; Steven, 1959; and Bardach et. al., 1967).

The failure to find any specific chemical or groups of chemicals that would attract hagfish, or to account for the attractive principle(s) in the FDCE is puzzling. A more conclusive study on eels by Hashimoto (1968) and Konosu et. al. (1968) showed that the attractiveness of clam extracts to eels could be partially accounted for by specific proportions of 7 amino acids. It is interesting to note that of a variety of compounds examined electrophysiologically on the olfactory system of a number of freshwater fish, amino acids stand out as being highly stimulatory (Sutterlin & Sutterlin, 1971; Suzuki & Tucker, 1971; and Hara et. al., 1973). This sensitivity may have evolved as a food-finding mechanism. A second possibility is that this spectrum of compounds may be of navigational significance for non-chemical feeders. It has been shown that the relative proportions of different amino acids distributed vertically and horizontally in oceanic waters differ (Pocklington, 1972). If different ocean currents have different odors, such information could possibly be used in initiating, terminating, or assisting fish in their migrations.

References

Bardach, J.E., Todd, J.H. and Crickmer, R. 1967. Science 155, 1276-1278.
Brawn, V.M. 1969. J. Fish. Res. Board Can. 26, 583-596.
deGroot, S.J. 1969. J. Cons. Int. Explor. Mer 32, 385-396.
Greene, C.W. 1925. Science 61, 68-70.
Hara, T.J., Law, Y.M. and Hobden, B.R. 1973. Comp. Biochem. Physiol. 45A, 969-977.
Hashimoto, Y., Konosu, S., Fusetani, N. and Nose, T. 1968. Bull. Jap. Soc. Sci. 34, 78-83.
Hodgson, E.S., Mathewson, R.F. and Gilbert, P.W. 1967. In Sharks, Skates and Rays. Ed. P.W. Gilbert, R.W. Mathewson, and D.P. Rall. John Hopkins, Baltimore, Md.
Konosu, S., Fusetani, N., Nose, T. and Hashimoto, Y. 1968. Bull. Jap. Soc. Sci. 34, 84-87.
Leim, A.H. and Scott, W.B. 1966. Fishes of the Atlantic Coast of Canada. FRB Bull. No. 155.
Lindstredt, K.J. 1971. Comp. Biochem. Physiol. 39, 553-581.
Mathewson, R.F. and Hodgson, E.S. 1972. Comp. Biochem. Physiol. 42A, 79-84.
Parker, G.H. 1911. J. Exper. Zool. 10, 1-5.
Pipping, M. 1927. Acta Soc. Sci. Fenn. 2, 1-28.
Pocklington, R. 1972. Analytical Biochem. 25, 403-421.
Strahan, R. 1963. Acta Zool. 44, 73-102.
Sutterlin, A.M. and Sutterlin, N. 1971. J. Fish. Res. Board Can. 28, 565-571.
Suzuki, N. and Tucker, D. 1971. Comp. Biochem. Physiol. 40, 399-404.
Tester, A.L., vanWeel, P.B. and Naughton, J.J. 1954. U.S. Fish & Wildlife, Special Scientific Rept. No. 130, 62p.
Tester, A.L. 1963. Pacific Sci. 17, 145-170.

Extreme Sensitivity and Specificity of Catfish Gustatory Receptors to Amino Acids and Derivatives

John Caprio

Department of Biological Sciences, Florida
State University, Tallahassee, Florida 32306 U.S.A.

Introduction

Amino acids have recently been shown to be effective olfactory stimuli in fishes, with electrophysiological thresholds estimated at 10^{-7} to 10^{-8}M[1-3]. Catfish (Siluriformes) depend heavily on their chemical senses for both feeding and social behaviour [4,5] and possess an abundance of external taste buds which are also highly sensitive to the α-amino acids [6]. In the present experiments, the sensitivity, range, and specificity of the gustatory responses to amino acids and derivatives in the channel catfish (*Ictalurus punctatus*) were studied.

Electrical Response Recording

The fish was immobilized and positioned in a plexiglass container with aerated well water perfusing the gills. By removing the eye, a branch of cranial nerve VII (facial) to the maxillary barbel was exposed and placed on a Pt-Ir wire in mineral oil to prevent drying. A continuous flow of well water, about pH 7, was directed to a glass tube positioned around the maxillary barbel. Stimuli dissolved in well water ($1.0 cm^3$ volume) were diluted at least to 60% of their original concentration as determined by photodensitometry of dye solutions. Electrical "summated" activity was amplified by a conventional, capacity-coupled preamplifier, passed through an a.c. to d.c. converter (Sanborn Model 350-1400 amplifier operated on the a.c. linear mode) and displayed on a pen recorder.

Sensitivity and Range of Taste Response

The summated response of the barbel chemoreceptors to a prolonged amino acid stimulus revealed a quickly adapting, phasic response. The threshold concentration for the more effective amino acids was estimated to range between 10^{-9} and 10^{-12}M, the lowest electrophysiological thresholds reported for taste in any vertebrate. The relationship between the neural response and stimulus concentrations can best be described by a power law function, with exponents between 0.08 and 0.14 for the more

157

stimulatory amino acids. Fig. 1 illustrates a response-concentration series for L-alanine, the most potent of the amino acids tested. The electrophysiological threshold was determined by the intersection of the least squares fitted line, corrected for dilution, with the control response value. The control response, i.e., that obtained with the stimulus adjusted for zero concentration, is due primarily to chemical contaminants. Responses fitting the power function were regularly obtained over seven to eight log units. Amino acid concentrations above 0.01M were avoided because of possible deleterious effects on the receptors. A rapid decline of gustatory receptor sensitivity correlating to the duration of fish captivity was also noted [7].

Gustatory Specificity

Gustatory receptors of the maxillary barbel are extremely sensitive to the L-α-amino acids. The L-isomer of an amino acid is always more effective than its enantiomer, although certain D-amino acids are more stimulatory than other L-forms. The taste response is strongly influenced by the position of the amino group (Fig. 2). Amino acids having an amino group in the alpha position (i.e., those derived from proteins) are the most effective.

An α-amino acid may be thought of as consisting of an asymmetrical carbon center surrounded by four functional groups: (1) carboxyl, (2) α-amino, (3) α-hydrogen, (4) side chain, R. The effects on the summated taste response due to molecular changes associated with these groups were studied. All comparisons were made at 0.1mM, and the magnitude of the response (peak height) to each chemical is represented as a percentage of that produced by the standard, L-alanine. Statistical comparisons were performed using the t-test.

(1) Carboxyl: Esterification resulted in no significant loss of activity (Fig. 3), indicating an ionically charged carboxyl group is not a requirement for receptor response. However, amide substitution including peptide bond formation (Fig. 4) or replacement of the carboxyl with alcoholic (Gly>2-NH$_3$-EtOH, p <.001) or sulfonic (β-Ala>Taurine, p <.01) groups resulted in a diminshed taste response.

(2) α-amino: Acylation, alkylation and other structural alterations of the amino group resulted in decreased neural activity (L-Ala>N-Ac-L-ALA, p <.001; Gly>N-CH$_3$-Gly, p <.01; Gly>N-Pthalyl-Gly, p <.001).

(3) α-hydrogen: Replacement of the α-hydrogen by a methyl group produced a decreased taste response (Ala>2-CH$_3$-Ala, p <.001).

(4) Side chain, R: In general, the lower molecular weight amino acids

158

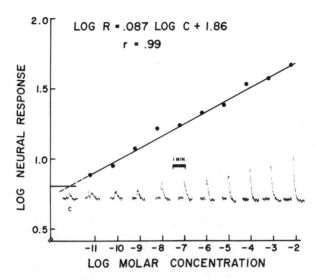

Fig. 1. Relation between summated taste response and L-alanine concentration. Dashed line shows intersection of linear regression line with control response yielding threshold value. Inter-stimulus intervals varied from 2 to 8 minutes.

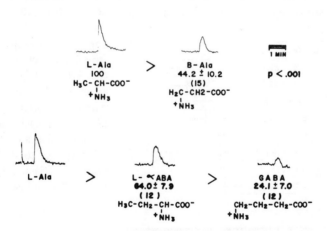

Fig. 2. Effects of amino group position on the summated taste response. Mean±S.D. are listed below chemical abbreviation. Number in parenthesis refers to number of fish tested, (N). This scheme is adhered to in the following figures.

159

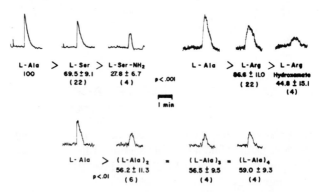

| L - Ala 100 | L-Ala-OMe (HCl) 98.8 ± 5.2 (17) | L-Ala- OET (HCl) 97.6 ± 5.1 (7) | L-Ala - OBenzyl 99.8 ± 6.2 (6) |

Fig. 3. Esterification of the carboxyl group resulted in no significant loss of neural activity.

L-Ala > L-Ser > L-Ser-NH₂ $p<.001$
100   69.5±9.1   27.8 ± 6.7
       (22)   (4)

L - Ala > L-Arg > L-Arg
        86.6 ± 11.0 Hydroxamate
        (22)   44.8 ± 15.1
                     (4)

1 min

L- Ala > (L-Ala)₂ = (L-Ala)₃ = (L-Ala)₄
$p<.01$ 56.2 ± 11.3   56.5 ± 9.5   59.0 ± 9.3
         (6)   (4)   (4)

Fig. 4. Reduction of summated taste response resulting from amide substitution on the carboxyl group and peptide bond formation.

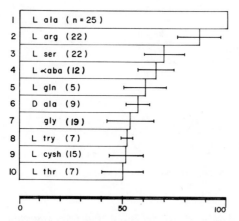

1	L ala (n = 25)
2	L arg (22)
3	L ser (22)
4	L ∝aba (12)
5	L gln (5)
6	D ala (9)
7	gly (19)
8	L try (7)
9	L cysh (15)
10	L thr (7)

0   50   100

Fig. 5. Comparison of stimulatory effectiveness of the 10 most potent amino acids tested on the maxillary barbel at 0.1mM. Relative position of each chemical is represented as a percentage of the response to the standard L-alanine. (N)=number of fish tested. Bars indicate S.D.

were the most effective. Of a total of twenty-eight α-amino acids tested, nine had skeletons of four carbons or less, and seven of these were included in the ten most effective (Fig. 5). Addition of a hydroxyl group also reduced effectiveness (L-Ala> L-Ser, p <.001; L-Phe>L-Tyr, p <.01; L-Pro>L-Hypro, p <.01).

Comments

Olfactory receptors of the white catfish (*I. catus*) are highly sensitive to certain α-amino acids[2]. Preliminary studies of olfactory receptors of the channel catfish also reveal a high sensitivity, but with somewhat different structure-activity relations than reported here for taste [8]. Brown (*I. nebulosus*) and yellow (*I. natalis*) bullheads are able to follow a cysteine hydrochloride gradient solely through the gustatory sense [9], although cysteine stimulates both olfactory and gustatory receptors in these species [10]. These findings present interesting questions concerning the evolutionary development of two separate chemosensory systems, each responding to extremely dilute concentrations of amino acids.

Acknowledgement

I wish to thank Drs. Don Tucker and Lloyd M. Beidler for their encouragement and advice, and to acknowledge support from NIH grants NS-8814 and NS-5258.

References

1. Sutterlin, A.M., and N. Sutterlin, 1971. J. Fish Res. Bd Can. 28, 565-572.
2. Suzuki, N., and D. Tucker, 1971. Comp. Biochem. Physiol. 40A, 399-404.
3. Hara, T.J., 1973. Comp. Biochem. Physiol. 44A, 407-416.
4. Atema, J., 1971. Brain, Behav. Evol. 4, 273-294.
5. Todd, J.H., J. Atema and J.E. Bardach, 1967. Science 158, 672-673.
6. Caprio, J., Comp. Biochem. Physiol. (In press).
7. Tucker, D., 1973. Fish. Res. Bd Can. 30, 1243-1245.
8. Caprio, J., unplublished.
9. Bardach, J.E., J.H. Todd and R. Crickmer, 1967. Science 155, 1276-1278.
10. Bardach, J.E., M. Fujiya and A. Holl, 1967. In Olf. and Taste II, Pergamon Press, New York, T. Hayashi, ed. 647-665.

Functions of the Olfactory System in the Goldfish

(Carassius Auratus)

H.P. Zippel and W. Breipohl

Physiologisches Institut ll
Humboldtallee 7
D-3400 Gottingen

Institut fur Anatomie
Geb. MA 5/52
D-4630 Bochum

Although it is well established that olfaction is of meaningful biological importance for vertebrates, the available comparative data on the functions of the olfactory system, in terms of morphological investigations in conjunction with electrophysiological and behavioural findings, are somewhat limited. Since the space available for this presentation is rather restricted, the following remarks provide only a general outline of the recent findings, a more detailed description of which will be published elsewhere.

Data obtained with animals trained to differentiate between different odour stimuli indicate that learning in this sensory modality is characterized by the difficulty to discriminate synthetic odours *qualitatively*. Using a shock-free training procedure, Zippel (1970) has shown that in single animals, as well as in groups, trained to one odour (Fig. 1), a generalized association initially exists between smell and reward (Tubifex). Following training to a single stimulus, animals have considerable difficulty in discriminating between two odours presented during differentiation training: in the first instance, a level of 80-100% correct responses is achieved within 70-100 trials, whilst a further 30-40 training sessions are required to reach this criterion during differentiation training (Fig. 1). Furthermore, Bieck and Zippel (1971) have demonstrated that in animals trained to one odour the *quantity* of the stimulus offered is of paramount importance, whereas in differentiation training the learned information is primarily *qualitative* in nature irrespective of the quantity of the competing stimulus.

These findings are corroborated by electrophysiological recordings (using glass platinum electrodes, $\emptyset$ 10-20μ) from mitral cells in the olfactory bulb of intact animals. 80% of the cells (n=41) respond uniformly to synthetic olfactory stimuli (amylacetate, camphor, coumarin, eugenol, morpholine, β-phenylethylalcohol, phenol) applied as one drop to the olfactory mucosa, with only 20% showing multiform reactions. Up to 50% of the uniformly responding cells are inhibited to a greater or lesser extent by synthetic odours (Fig. 2A, d-f), 15% are excited, 15% fail to

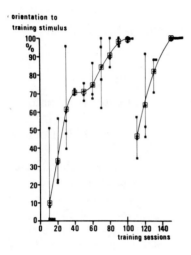

Fig. 1. Training of one fish to a single stimulus, coumarin (sessions 1-100), and subsequent differentiation training (sessions 101-150) with coumarin (positive stimulus) and amylacetate (competing stimulus). Closed squares = individual values; closed circles in open squares = mean values of 10 trials.

respond, and 20% show combined patterns. An increase in concentration is followed by an augmentation of the above effects. In accordance with the findings of Döving and Hyvärinen (1969), computer analysis provides no evidence for frequency coding of single synthetic stimuli in the olfactory bulb; hence in olfactory coding, the crucial factors would appear to be direction, strength and duration of change in the spontaneous activity.

In comparison with synthetic stimuli, natural stimuli (e.g., Tubifex extract) are more readily discriminated, cause a greater excitation (55% with Tubifex) and often have *antagonistic* effects (Fig. 2A, b). These effects are particularly striking in uniformly responding units and as such provide reliable evidence for the coding of biologically important stimuli in the intact olfactory system. Antagonistic effects are recorded less frequently with rotten Tubifex extracts (Fig. 2A, c).

The effects of mechanical stimuli on mitral cells have been described by a number of investigators but not examined systematically. Recent experiments, stimulating the olfactory mucosa with quantitatively defined water currents (laminary: 1,5,10,20,40 ml/min), revealed that even the smallest current – which is undoubtedly within the biological range – has a pronounced effect on the neuronal impulse rate. *None* of the 33 cells hitherto registered failed to show a reaction to laminary stimulation. In this case also, inhibition is the predominant (60%) effect (Fig. 2B, b and c); only 25% of the neurons are excited and 15% show combined patterns

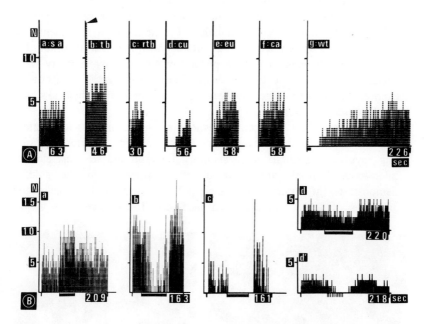

Fig. 2. Post-stimulus-histograms of cell activities following application of odour and mechanical stimuli. A: all recordings from the same cell; stimulus given at beginning of each histogram. a: spontaneous activity; b-f respectively, all applied as one drop: Tubifex extract; rotten Tubifex extract; 10^{-5} M coumarin; 10^{-5} M eugenol; 10^{-5} M camphor; g: short, turbulent stream of water.

B, a-d: responses of 4 different cells to laminary water currents (5 ml/min). Duration of stimulation indicated by black bar below abscissa; d: laminary and d': repetitive stimulation of the same cell; N = number of action potentials per second.

(Fig. 2B, a). In a number of cases, the effects of stimulation are observable long after the stimulation itself has ceased (Fig. 2B, c). On- as well as off- and on/off-effects have been observed. In many cases, turbulent application of water currents causes a strong inhibition (Fig. 2A, g), whereas in other cells repetitive stimulation is most effective (Fig. 2B, d and d').

On the basis of these results, it cannot be excluded that in a large number of other investigations on aquatic animals, which involved the addition of odour substances to a more or less turbulent water flow through the olfactory mucosa, the observed effects were, in fact, recorded *only* from those cells which reacted positively to, or were minimally influenced by, mechanical stimulation. Furthermore, it is evident that natural as well as synthetic odours interact with mechanical stimuli. For instance, a cell which has been previously inhibited by laminary water currents can be excited by *one* drop of Tubifex extract. This is but one example and other interactions between olfactory and mechanical stimuli will be described elsewhere.

One of the most important questions arising from the above data is whether or not receptors exist which are specialized for different stimuli (e.g., synthetic, natural, or mechanical). Electron and scanning electron microscopical investigations of olfactory receptors in the goldfish demonstrate the comparatively high polymorphism of these cells, two examples of which are presented in figure 3. Moreover, the various

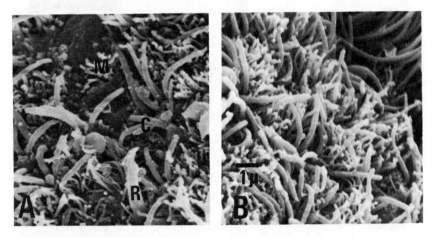

Fig. 3. Scanning electron micrographs of olfactory receptors.
A: different receptor types, C = ciliated, M = microvilli bearing, R = rod shaped.
B: another area of the olfactory epithelium showing a high density of receptors.

receptor cells are not equally distributed over the olfactory organ but differ from one another in quantity and density. However, "the structural differences do not allow a definite classification of sensory cells into functionally different elements" (Breipohl et al., 1973).

Consequently, the question of the relationship between specialized receptor cells and stimuli affecting the olfactory system in vertebrates still remains open. The above findings nevertheless indicate that the olfactory system — at the level of the second neuron and thus at the receptor site itself — appears to have multimodal capacities. It is established that the intact olfactory bulb is subjected to the technical and analytical interferences of centrifugal influences from its contralateral counterpart and higher CNS centres, but the information finally processed by the mitral cells of this organ is effected *incorporating* these influences. In biological terms, the problem thus arises as to what extent the information recorded from an isolated olfactory bulb, having only the input from the receptor cells, is representative of the natural situation.

It is to be hoped that further investigations including behavioural as well as electrophysiological and morphological data will allow a more detailed insight into the functions of the olfactory system.

Footnotes

This work was supported in part by a grant from the Deutsche Forschungsgemeinschaft (SFB 33).

The manuscript has been carefully revised by A. Burt; his kind help is gratefully acknowledged.

References

Bieck, B., Zippel, H.P. 1971. J. Biol. Psychol. 13, 13-16.
Breipohl, W., Bijvank, G.J. Zippel, H.P. 1973. Z. Zellforsch. 138, 439-454.
Döving, K.B., Hyvärinen, J. 1969. Acta Physiol. Scand. 75, 111-123.
Zippel, H.P. 1970. Z. vergl. Physiol. 69, 54-78.

Chemoreception in Aquatic Animals

Moderator's Summary

John E. Bardach

Hawaii Institute of Marine Biology

P.O. Box 1346; Kaneohe, Hawaii 96744

In retrospect of the Melbourne session where aquatic animals were singled out for the first time for special treatment at an ISOT symposium the summing up will have two parts: (1) A terse reiteration of evolution-oriented general common features of chemoreception in the aquatic milieu, and (2) the listing of specific items of interest or new information that emerged from in— and out of—session discussions.

1. Common Features of Chemoreception in Aquatic Animals

a. The overriding importance of amino acids, amines and related compounds clearly emerged, predominantly for food search and feeding, a fact really not surprising if it is remembered that these compounds existed in the waters of the globe certainly before image forming but not necessarily general light sensitivity evolved in animals. L alpha amino acids of 3 to 6 carbon atoms appeared as the most effective signals.

b. Examples of the mollusks, crustacea and fishes discussed in some detail and representatives of other taxonomic groups mentioned cursorily showed distribution of chemoreceptors either well spread over the body surface or, at least placed on various body parts or appendages, a feature presumably due to density/solvent characteristic of water and permitting efficient sampling of the water all around the individual. In mammals and birds, and to some extent also in insects, external stimulus/receptor interaction sites are more restricted. With this difference in chemoreceptor distribution between aquatic and terrestrial animals central integration of chemical stimuli by the former deserves special attention and represents a set of research problems barely begun to be investigated with some rigor. It is recognized that profound differences exist in the CNS of divergent evolutionary lines but noting some results presented at ISOT-V for mammals, electrophysiological scrutiny of the fish brain, proximal to the bulb, would be of much comparative interest.

c. Distinction between smell and taste in aquatic organisms is tenuous and while it can be made on anatomical grounds in vertebrates it is open to debate whether or not the distinction is a truly meaningful one. Similarly debatable is the meaning of a distinction of basic taste modalities even in fishes, let alone the invertebrates. Fiber specialization exists in many aquatic animals not necessarily to one but to several related compounds as do fibers with narrow and wider spectra of sensitivities. Meaningful stimuli are most often multimodal composites of specific compounds.

d. So far major emphasis in research with aquatic animals has centered on food-related chemosensory performance but this symposium, among other reminders, in the literature, has drawn attention to the role of chemoreception in other behavior domains, e.g., social or reproductive cohesion or segregation of groups or individuals, enhancers or releasers of reproductive pairing in the strict sense, navigation, predator avoidance and the like.

2. Specific Items of Interest

JAHAN-PARWAR: The anterior tentacles in *Aplysia* are the most sensitive of the animals' three sets of chemosensory organs — the other two being the posterior tentacles and the osphradium — responding to stimulus release in their vicinity of 10^{-9} M glutamic acid. The three sets of sense organs project to the cerebral ganglion. The osphradium may be a sampler of water quality. Some abdominal ganglion cells appear to be centrally located primary sensory neurons.

LAVERACK: Anatomy of crustacean chemosensory receptors, with a spectrum of different terminal structures, deserves further study. Some fine chemosensory fibers seem to end lining the cuticle. Uncalcified and untanned spots in the antennule, for instance, may represent stimulus-receptor contact points. Sensitivity during moulting was discussed without conclusions being reached beyond presuming greatly reduced crustacean chemical sensitivity at that time.

HIDAKA: The puffer fish has low, while freshwater fishes have far higher, NaCl sensitivity.

ZOTTERMAN, BARDACH: Phospholipid sensitivity was noted, so far, in carp and catfish taste fibers. It may be more widespread.

CAPRIO, HARA: A discrepancy in results of Hara and Caprio related to esterification of the carboxyl end of the amino acid; Caprio finds no effect while Hara finds reduced olfactory stimulation. Caprio worked with channel catfish and Hara with salmonid fishes; Hara recorded from olfactory bulb and Caprio from taste nerves. Further work on this question seems indicated.

TAKAGI: Sensitivity correspondence in aquatic vertebrate olfactory and taste sensory systems exist to salt, acid and bitter stimuli though stimulus intensities applied to olfactory epithelium were high. No olfactory response occurred to sugar. Water-induced waves in the bullfrog were stopped by NaCl solution. The carp olfactory system only responded to water from congeners; no general olfactory sensitivity was noted to water as such.

ZIPPEL: Taste learning in goldfish is more "efficient" than smell learning; it is faster and simultaneously accomplishes quality and quantity discrimination. Mechanical (i.e., differential waterflow) stimulation of olfactory mucosa has important bearing on olfactory discrimination, suggesting that investigations of mechano-sensitivity of fish olfactory mucosa should be refined. Taste, in certain fishes appears dominant chemosensory input, with large brain regions presumably being taste dominated. Atema's (1971) demonstrated division of labor between 7th and 9th nerve nuclei dominating food search and swallowing, respectively seems important here as does Bruckmoser's (1973) investigation of taste fiber projections in the fish brain. In view of Giachetti and MacLeod's finding (this symposium) of a joint smell-taste projection to the mammalian neocortex, attempts should be made to elucidate through electrophysiology taste fiber connections in the fish, anterior to medullary taste nuclei.

OAKLEY: Investigations are in progress (with Doeving) demonstrating that Artic char (*Salvelinus alpinus*) segregate into breeding populations through olfactory cues from the skin mucus of the animals; the nature of the substances involved are under investigation.

BARDACH: Fish, in addition to smell and cranial taste nerve-innervated taste buds have chemical sensitivity in lateral line responding to monovalent cations (Katsuki et. al., 1971) and chemo-sensitive spinal nerve endings, some with taste bud-resembling terminals (Whitear, 1971). The behavioral role of these two chemosensory systems is not known.

References

Atema, J. 1971. Behav. Evol. 4, 273-294.
Bruckmoser, P. 1973. Verh. d. Deutschen Zool. ges. 66 Jahresvers. 219-229.
Katsuki, Y., Hashimoto, T. and Kendall, T. 1971. Jap. J. Physiol. 21, 99-118.
Whitear, M. 1971. J. Zool. London. 163, 237-264.

INVITED LECTURE

The Evolution of Olfaction and Taste

Ya. A. Vinnikov

The Laboratory of Evolutionary Morphology,

Sechenov Institute of Evolutionary Physiology and Biochemistry,

Academy of Sciences of the U.S.S.R., Leningrad

Resumé

The problem of olfaction and taste has two sides. First, a more general aspect deals with the evolution of sense organs at cellular and molecular levels of organization. The second is limited only by particular questions of evolution of chemoreceptors. So, both sides must be considered with a view to elucidate principal laws of olfaction and taste.

Due to information theory as well as to the improvement of the instrumental power of biological research, we can state that adequate "encoding", that is transformation of any specific kind of environmental energy, is performed at the cellular and molecular level of organization of the sense organs.

An evolutionary approach to the study of cellular and molecular organization of the sense organs reveals, as was thought before, only two modifications of sensory cells: primary and secondary. The first type of cells, which in the course of evolution arise initially in Turbellaria and Annelids, are supplied with a peripheral process topped by sensors consisting of motile cilia, containing 9 x 2 + 2 or 9 x 2 + 0 fibrils, and/or microvilli. An additional study is needed for identification of receptor cells in Coelenterata.

The excitation process that arises in the peripheral process is transmitted through the central process to the CNS. Invertebrate primary sensory cells represent principal and universal types of specialized receptor elements which are responsible for all the kinds of reception. In vertebrates they are found only as two relic modifications: namely, photoreceptors of the retina and olfactory cells of the smell organ. Both develop as parts of the nervous system (Fig. 1).

Apart from the primary sensory cells, a new receptor modification arises, a secondary sensory cell in both vertebrates and, seemingly, in some invertebrates (Ctenophora). These cells lack both peripheral and central processes. However, they detect a stimulus by the same apparatus as the primary cells: namely, by motile cilia containing 9 x 2 + 2 fibrils, and microvilli or their derivatives (stereocilia). The excitation is transmitted

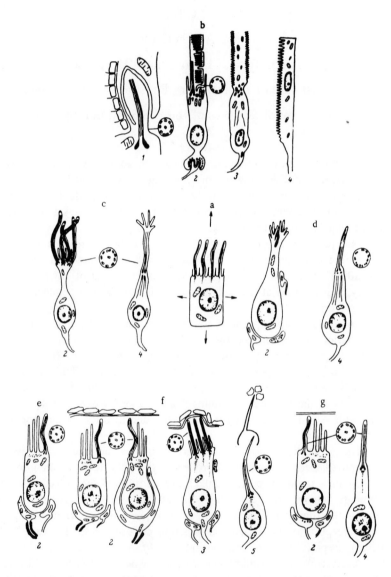

Fig. 1. Scheme of evolution of the receptor cells of the most highly developed representatives of phyletic lines of animals. Cells: a) flame; b) photoreceptors c) olfactory; d) taste; e) lateral line receptor; f) of organ of gravitation; g) of auditory organ. 1. Eyes of Euglena; 2- vertebrates; 3- cephalopoda & mollusks; 4- insects; 5- higher crustaceans.

through the synapses which are formed by terminals of afferent nerve fibers. The receptor cells of the lateral line organs, the inner ear and, which is of special interest here, — the taste buds (Fig. 1), must be related to this group.

Thus, the receptor cell supplied with motile cilia, kinocilia, stereocilia, and/or microvilli — forms the base for the evolution of sense organs (Fig. 1). A mechanism responsible for motion of kinocilia and possibly also microvilli, is related to an interaction between ATP and ATPase contained in the nine pairs of peripheral cilium fibrils or in the fibrils inherent to microvilli. Stereocilia are moved only by an external mechanical force. Since cilia, kinocilia, microvilli and stereocilia are the essential attributes of a receptor cell, and considering their specific role and structure, we have concluded that they must be classified as peculiar motile antennae perceiving energy from the environment. Antennae of the receptor cells are the most ancient sensors (Vinnikov, 1971). Hence, in variance with the neuron which receives and transmits information only by its synapses, the receptor cell is stimulated directly by energy from the environment acting upon its antennae.

The question arises, how do these antennae, whose structures and function have so much in common, detect only strictly specific kinds of energy from the environment? In other words, at what level of their organization and by what mechanism are realized the transforming and detecting functions of the receptors? At present, many data suggest that molecular processes in the plasma membranes of the antennae underlie these functions. Specific protein molecules which are capable of "encoding" environmental energy to information have been found to be localized in the plasma membranes of the receptor cells. These receptive molecules, therefore, function as starting or triggering links in a chain of excitatory processes (Fig. 2).

What facts do we have in favor of this statement? First of all we must appreciate the famous discovery of Wald (1939-1968) who was the first to demonstrate that photoreception is related to the molecular organization of the visual pigment, rhodopsin, which is incorporated in disc membranes of antennae (rods and cones) of photoreceptors. According to Wald (1968) a single photon absorbed induces conformational changes followed by thermal decay of the rhodopsin molecule. These processes underlie the triggering mechanism of excitation of a visual cell. As like other proteins, rhodopsin may differ in its dimensions and molecular weight in various animal types (40,000 in vertebrates, 3,000-4,000 in insects). Dastoli and Price (1966); Dastoli, Lopiekes and Doing (1968) and others have demonstrated special "sweet-sensitive" and bitter-sensitive" protein

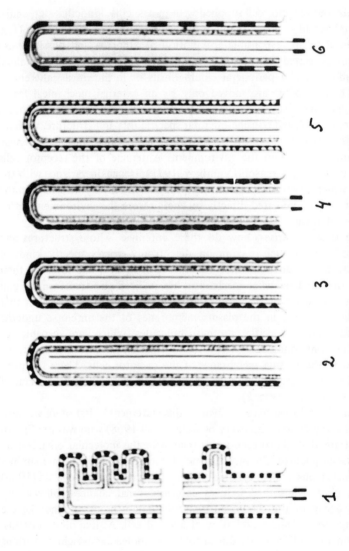

Fig. 2. Scheme of assembly of receptive protein molecules in the plasma membrane of the antennae of receptor cells; 1- photoreceptor — rhodopsin; 2- taste cell-sweet-sensitive protein; 3- taste cell-bitter-sensitive protein; 4- olfactory cell-olfactive protein; 6- receptor cell of the organ of gravitation-hypothetical receptive protein.

molecules (MW = 150,000) in taste bud cells which interact with sweet and bitter food substances. These proteins seem to be localized in the outer layer of the plasma membrane of taste bud cell antennae (Beidler, 1971). They interact with sweet and bitter taste molecules by means of only London hydrogen bonds which is enough to result in a conformational change of the receptive molecule (Ash, 1968-1969; Ash and Skogen, 1970; Riddiford, 1970; Rozental and Norris, 1973).

Finally, the molecular mechanism of hearing is confirmed by our data (Vinnikov and Titova, 1961-64; Vinnikov, 1965, 1970, 1971) according to which acetylcholine esterase and attendant choline receptive proteins within antennae (stereocilia) of auditory cells of the organ of Corti are capable of interacting with acetylcholine of the endolymph. This mechanism gives a possible explanation as to how the auditory cell starts the triggering process of excitation.

The presence of specific receptive proteins in plasma membranes of antennae of other mechanoreceptors, and in particular the gravity receptors, can be taken at present only as a hypothesis.

A quantum of stimulus acting upon a receptive molecule in an antenna results in conformational changes of the molecule which are followed by changes in its bonds with surrounding lipids or other protein molecules of the membrane which in turn, induces changes in mainly the Na^+ and K^+ ion permeability of the plasma membrane that results in depolarization of the cell and generation of the receptor potential. The energetic stimuli (light, sound, etc.) are known to be capable of non-specific action upon a living cell, and the response of the latter is in general similar to that of a receptor cell. In the course of evolution, this non-specific response to a stimulus transforms into the specific response of a specialized receptor cell. If this were not so, the reception should not have reflected the real world. In my opinion, the capability of specific responses, that is adequate encoding, is reached in phylo- and embryo-genesis by biosynthesis of particular molecules of receptive proteins within the receptor cells. The proper genes controlling this synthesis are nothing more than the result of numerous mutations which were at least fixed by the natural selection after many errors were rejected in previous generations. Thus, there are some general principles of evolution of sense organs including those of gustation and olfaction.

So what do we know about the structural and functional organization of the gustatory and olfactory cells? The gustatory cells of annelids, molluscs and arthropods are primary-sensitive cells equipped with flagella. The insect gustatory cells are, as a rule, incorporated in sensillae and their antennae are surrounded with a porous cuticle through which the taste

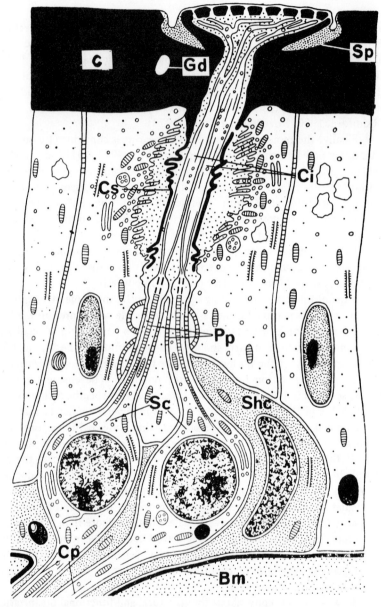

Fig. 3. Diagram of the antennal taste organ (Sensilla basiconica in the beetle (Acilius sulcatus L.). (After Ivanov,1969) Fm — fibrous membrane; c — cuticula; Jm — joint membrane; H — hair; Cs — cuticular sheath; Pp — peripheral process; Ci — cilium; Sc — sensory cell; Cp — central process; Shc — Sheath cell; Bm — Basement membrane; Gd — gland's duct; Sp — secretion product.

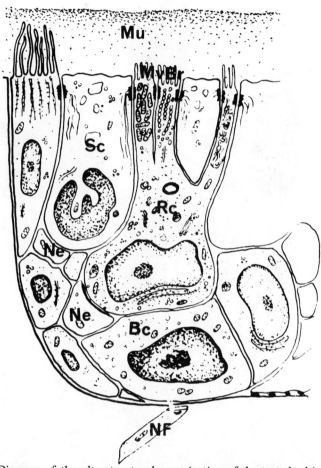

Fig. 4. Diagram of the ultrastructural organisation of the taste bud in frog (After Pevzner, 1970). Mu — mucous; MvBr — microvillar brush; Rc — receptor cell; Sc — Supporting cell; Bc — Basal cell; Ne — Nerve ending; Nf — nerve fiber.

molecules penetrate when an animal contacts food (Fig. 3). It was pointed out that vertebrates' gustatory cells are secondary sensitive ones and their motile antennae represent the peculiar microvillar brushes (Fig. 4) immersed in the mucous (in low vertebrates) or in shifts (in high vertebrates). Otherwise, the olfactory cells of both vertebrates and

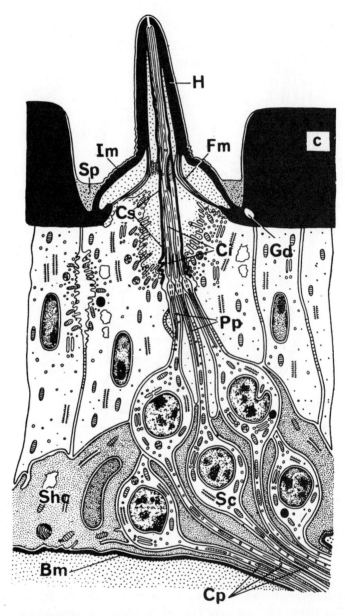

Fig. 5. Diagram of the antennal olfactory organ (Sensilla placoidea) of the beetle (Acilius sulcatus L.), (After Ivanov, 1971). Ci – cilium; SC – cuticular sheath; Pp – peripheral process; Sc – sensory cell; Cp – central process; Bm – basement membrane; c – cuticle; Sp – secretion product; gd – gland's duct; Shc – Sheath cell.

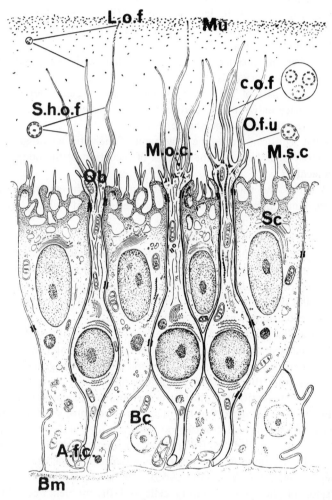

Fig. 6. Diagram of the ultrastructural organisation of the olfactory epithelium. (After Bronshtein, 1972). Mu — mucous; Ob — olfactory bulbs; Sh.o.f. — short olfactory flagella; L.o.f. — Long olfatory flagella; c.o.f. — complex olfactory flagella; O.f.u. — Olfactory flagella with unusual fibrillar set; M.o.c. — microvilla of olfactory cell; A.f.c. — Axon of olfactory cells; Sc — supporting cells; M.s.c. — microvilla of supporting cell; Bc — Basal cell; Bm — Basal membrane.

invertebrates are primary sensitive ones equipped with flagellae (Figs. 5,6). The single flagellum surrounded with a porous cuticle (in insects) or some flagellae and microvilli (in vertebrates) directly face the environment. In some cases, the vertebrate olfactory cells bear only the microvilli on their surface as, for example, In Jacobson's organ (Bannister, 1968; Graziadei and Tucker, 1968, et al.) or a few olfactory cells of some fishes (Bronshtein and Pyatkina, 1973).

A number of questions arise when dealing with the evolution of gustatory and olfactory cells:

1) In what part of the gustatory and olfactory cells does a contact really take place between the odorous or taste molecules and the receptive proteins of the cell plasma membrane?

2) What significance does the movement of the antennae of gustatory[+] and olfactory cells have for the process of reception of the odorous and taste molecules?

3) What is the role of the mucous in which the antennae of the gustatory and olfactory cells are immersed and which contains mucopolysaccharides, proteins and some ions?

Biochemical investigations mentioned seem to convince us more and more that the taste and olfactive receptive protein molecules which are able to change their conformation during an interaction with odorous and taste molecules are really incorporated into the plasma membranes of the gustatory and olfactory cells. In this respect, taste and smell seem to obey the general evolutionary mechanisms of reception.

As far as the gustatory cells are concerned, we can suggest that sweet- and bitter-sensitive molecules are incorporated into the outer protein layer of the plasma membranes of the motile microvilli of the gustatory cells. This follows from the classic investigations of Beidler (1965, 1971) who was the first to introduce the term membrane "receptor sites". The suggestion can be advanced that the membrane "receptor sites" containing the receptive proteins are localized on that part of a microvillus which is embedded into so-called "shifts" rich in mucopolysaccharides, proteins, and Na^+. In my opinion the "shifts" represent ion-exchange structures (Pevzner, 1964, 1969; Vinnikov, 1965). The same may be true for the pore region of the insect gustatory cells, but the problem needs further investigations. Electrophysiological data have shown that vertebrate gustatory cells are multipurpose (Pfaffmann, 1941; Beidler, 1961, 1971

[+]According to recent data, the tubular structures of microvilli contain a contractile substance which is responsible for the movement (Boyd, Porsons, 1969 and others).

and others) if compared with invertebrate gustatory cells which have, for example in insects, a definite specialization (Dethier, 1963).

My laboratory demonstrated the movement of the antennae of vertebrate olfactory cells both in cold-blooded and warm-blooded vertebrates. This movement absolutely differs in comparison with the movement in the ciliary epithelium although it is supported by the same ATP-ATPase mechanism (Bronshtein, 1964, 1973).

The movement of the flagellae of invertebrate olfactory cells is not proved. One can believe that olfactive receptive proteins by analogy with taste proteins must be incorporated into the plasma membranes of the flagella and microvilli of the olfactory cells. Although it follows from the anatomy of the insect organ, in vertebrates such a localization would demand a number of evidences. Bronshtein and Minor (1973) have shown that gradual destruction of the olfactory cell flagella by detergents has led to parallel infringement of its ability to generate an EOG. A temporary cessation of movement of flagellae in hypotonic or isotonic sucrose solution was not followed by disappearance of the EOG during specific stimulation. In these conditions the duration of a negative wave of EOG increased along with the delay in restoration of sensitivity of the olfactory cells after repeated stimulation. The authors have reasonably assumed that the movement of the flagella contributes to restoration of the sensitivity of the receptor membrane by clearing it from "used" molecules of odorants. It is known that when vertebrates left water for land in their phylogenesis, their olfactory organs as well as olfactory cells did not undergo a noticeable structural reconstruction. The data obtained in our laboratory testifying to an alteration of the length of the olfactory cell antennae in Urodela during seasonal alternation of their environment from water to land (Kostanjan, 1971) gives evidence of a very slight reconstruction of the olfactory cells which don't affect the principles of their ultrastructural organization. What mechanisms are responsible for stability of function of the olfactory cells under alteration of ecological conditions? In my opinion, the secretion of the olfactory epithelium in aquatic animals (Breipohl et al., 1973) as well as the secretion of Bowmann's glands in terrestrial vertebrates is a protective mechanism responsible for constant conditions of function of the olfactory cells. In the vicinity of insect olfactory sensillae, special glands were described (Ivanov, 1969, 1971). The secretion which covers the olfactory epithelium from the outside and especially the antennae in both aquatic and terrestrial animals seems to preserve the way of interaction between odorants and receptive olfactive proteins. Because of the secretion in both aquatic and terrestrial animals, the olfactory cell antennae don't make contact directly with water or air possessing the molecules of the odorous substances. The odorous molecules must first immerse in the secretions which were called "olfactory mucous" by Bronshtein and Leontjev (1972). The mucous dissolving the odorous substances is, as a matter of fact, that micromedium in which the odorous molecules and the receptive membrane of the olfactory antennae interact. Thus, the olfactory mucous is the transmission media between the odorous molecules and the olfactory cell. Recently attempts were made to study the chemical composition of the olfactory mucous (Bronshtein and Leontjev, 1972). It appeared that the olfactory mucous of frog contains about 100 mEq Na$^+$ and about 70 mEq K$^+$ and that of guinea pig contains 75 mEq Na$^+$ and about 80 mEq K$^+$ (per kg of wet sample weight). Earlier, by means of electrocytochemical methods, the sodium ions were demonstrated in the secretion as well as in the cells of the Bowmann glands and ATPase in the plasma membranes of the olfactory flagellae (Bronshtein, Pyatkina, 1969). From these data the olfactory mucous can be thought to act as a reservoir for the necessary ion concentration on the surface of the electrogenic membrane of the olfactory cells (Bronshtein, Leontjev, 1972; Bronshtein, Leontjev and Pyatkina, 1973).

According to recent data (Bronshtein, 1973 a, b) the olfactory mucous is of great importance for movement of the flagellae of the olfactory cell. The media containing K$^+$ and Na$^+$ which are known to be necessary for a movement of the flagella and cilia of other cells (Singer and Goodmann, 1966) is optimal for the movement of the olfactory flagella. All these data are in agreement with the recent findings that some parameters of movement of cilia (flagellae) depend on the extent (level) of polarization of their membranes because a direct correlation has been found between the magnitude of membrane potential and the frequency of beating of cilia (Kinosita, 1954; Kokina, 1960; Horridge, 1965). Thus, the character of motion of the olfactory antennae is thought to be controlled by mechanisms responsible for a membrane potential (Bronshtein, 1973 b). The motion of the antennae may, in turn, change the ion concentration in different layers and regions of the olfactory mucous.

Thus, knowledge of general laws of the receptor processes help us to understand the evolution of structural and functional organization of the gustatory and olfactory receptors. Along with their specialization, one can find a number of adaptional mechanisms providing stability of function of these sense organs under changing ecological conditions.

References

Ash, K.O. 1969. Science, 165, 901.
Ash, K.O. and Skogen, J.D. 1970. J. Neuroch. 17, 1143.
Bannister, L.H. 1968. Nature 215: 5125, 275.
Beidler, L.M. 1961. Progr. in bioph. a. biph. chem., 107.
Beidler, L.M. 1965. Cold. Spr. Harb. Symp. on Quant. biol. XXX, 191.
Beidler, L.M. 1970. Taste receptor stimulation with salts and acids. Handb. of Sensory Physiology. IV. Chemical Senses, 200. Springer-Verlag.
Boyd, C.A.R. and Porson, D.S. 1969. J. Cell. Biol. 41, 64b.
Breipohl, W., Bijvank, G.J. and Zippel, H.P. 1973. Z. Zellforsch. 140, 567.
Bronshtein, A.A. 1964. Dokl. Akad. Nauk SSSR, 156, 3, 715.
Bronshtein, A.A. 1973 a. Tsitologia, 15, 7, 841.
Bronshtein, A.A. 1973 b. Tsitologia, 15, 8, 995.
Bronshtein, A.A. and Leontjev, V.G. 1972. J. Evol. Biol. chem. and Physiol., 8,6,580.
Bronshtein, A.A. and Minor, A.V. 1970. Dokl. Akad. Nauk SSSR, 213, 4, 987.
Bronshtein, A.A. and Pyatkina, G.A. 1969. J. Evol. Biochem. and Physiol. 5,3,274.
Bronshtein, A.A. and Pyatkina, G.A. 1973. Materialy IV Vsesoyuznoi Konferentsii po Bionike, 2,18.
Bronshtein, A.A., Pyatkina, G.A. and Leontjev, V.G. 1972. International Biophysics Congress Abstracts. 4, XVI, 28.
Dastoli, F.R. and Price, S. 1966. Science, 154, 3751, 905.
Dastoli, F.R., Lopiekes and Doing, A. 1968. Nature, 218, 884.
Dethier, V.G. 1963. The physiology of insect senses. London and New York.
Graziadei, P.P. and Tucker, D. 1968. Feder. Proc. 27, 583.
Horridge, G.A. 1965. Nature, 205, 497.
Ivanov, V.P. 1969. The Proceedings of the Entomol. Soc. Nauka. L. 53, 301.
Ivanov, V.P. 1971. The chemoreception of insects, Vil'nyus, 75.
Kinosita, H. 1954. J. Fac. Sci. Univ. Tokyo. IV, 7,1.
Kokina, N. 1960. Biofizika, 5,2,134.
Kostanyan, E.G. 1971. J. Evol. Biochem. and Physiol. 7, 1, 96.
Pevzner, R.A. 1964. Kand. thesis, L.
Pevzner, R.A. 1970. Tsitologia, 12, 8, 971.
Pfaffmann, C. 1941. J. Cell comp. Phys. 17, 243.
Riddiford, L.M. 1970. J. Insect. Physiol. 16, 653.
Rozental, J.M. and Norris, D.M. 1973. Nature, 244, 370.
Singer, J. and Goodmann, S. 1968. Exptl. Cell. Res. 43, 2, 367.
Vinnikov, J.A. 1965. Cold Spring Harbor Symp. Quant. Biol., 30, 293.
Vinnikov, J.A. 1970. J. Evol. Biochem. and Physiol. 6, 593.
Vinnikov, J.A. 1971. The cytological and molecular bases of sensory reception, Nauka, Leningrad. (English translation, Springer-Verlag, in press).
Vinnikov, J.A. and Titova, L.K. 1964. The organ of Corti. Its histophysiology and histochemistry. Consult. Bureau, New York.
Wald, G. 1939. Collecting Net, 14, 7.
Wald, G. Nature, 219, 800.

MARKET PLACE SYMPOSIUM

Time Course of the Rat Chorda Tympani Response to Linearly Rising Current

David V. Smith

University of Wyoming, Laramie, Wyoming, U.S.A.

The response of the peripheral taste nerves of vertebrates consists of an initial phasic discharge that shows a rapid exponential decline to a tonic level, which then continues to exhibit a slow exponential decline over many minutes (3,6). This initial transient portion of the gustatory neural response has been implicated in the coding of taste information (1,5), although the mechanisms underlying the production of the transient response and its subsequent decline are not well understood. While it has been suggested that gustatory adaptation in the frog is due to synaptic mechanisms or to properties of glossopharyngeal neurons (4), the time course of the rat chorda tympani response has been attributed to the rate of interaction of stimulus molecules with taste receptor sites (2). The present study examines the sensitivity of the gustatory response to the rate of stimulus application.

Neural responses were recorded from both whole chorda tympani nerve and single fiber preparations. The tongue was stimulated with trapezoidally shaped anodal current pulses delivered through a bathing medium of Ringer solution. Both the intensity of the current and its rate of rise were systematically varied. Anodal current was used so that the rate of stimulus rise could be held constant throughout its rising phase. For comparison to the responses to anodal current, responses were also obtained to NaCl stimulation at different rates of solution flow. Responses of the whole nerve were integrated using a fast time constant (47 msec) so that the time course of the response would not be distorted.

The transient portion of the chorda tympani discharge is extremely sensitive to the rate of anodal current rise (Fig. 1). The transient reached an amplitude that was a power funtion (slope = 0.47) of the rate of stimulus onset and was maintained at that level throughout the rising phase of the stimulus. When the rising current reached its plateau, the neural response fell to a tonic level that was completely unaffected by the rate of current rise. Anodal currents of different intensities presented at the same rate of rise produced transient responses of equal amplitude, with the duration of the transient reflecting the rise time of each stimulus (Fig. 2). The 14 chorda tympani fibers that were sampled each showed a phasic-tonic response to anodal current with the transient response

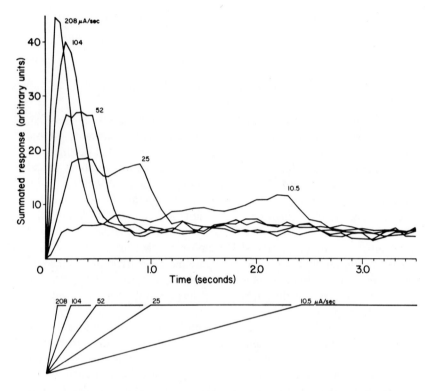

Fig. 1. Mean summated responses of the rat chorda tympani nerve to 25 μA anodal current presented at different rates of rise.

sensitive to onset rate. When NaCl was presented to the tongue at different rates of flow, the chorda tympani response reflected a sensitivity to stimulus onset rate paralleling that shown to linearly rising current (Fig. 3).

The initial transient phase of the chorda tympani discharge reflects the rate of stimulus onset. The sensitivity of taste units to changing stimuli would make them particularly responsive to rapidly occurring differences in gustatory input as would be likely during the ingestion of food. It is not clear whether taste receptor cells demonstrate a sensitivity to the rate of stimulus onset or whether it is a property of the first-order neuron, since studies of vertebrate taste receptor potentials have employed stimulus flow

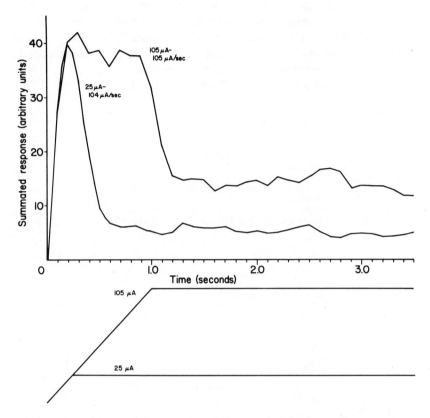

Fig. 2. Mean summated responses of the rat chorda tympani nerve to two stimuli differing in intensity but presented at the same rate of onset.

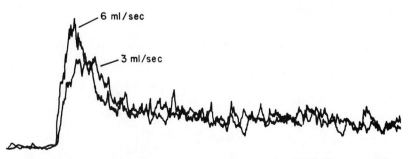

Fig. 3. Integrated chorda tympani responses to 0.1 M NaCl presented at 3 and 6 ml/sec. Total time = 7 sec.

rates slow enough to eliminate any phasic response that might have otherwise been present. The dependence of the transient response on the rate of stimulus change has important implications for the study of taste receptor potentials, cross adaptation, taste modifiers, or any studies in which stimulus onset rate may be a critical variable.

This work was supported by NINDS Grant NS-10211.

References
1. Halpern, B.P. and Tapper, D.N. 1971. Science, 171, 1256.
2. Heck, G.L. and Erickson, R.P. 1973. Behav. Biol., 8, 687.
3. Pfaffmann, C. and Powers, J.B. 1964. Psychon. Sci., 1, 41.
4. Sato, T. 1971. Brain Res., 34, 385.
5. Smith, D.V. and Frank, M. 1972. Physiol. Behav., 8, 213.
6. Smith, D.V., Steadman, J.W. and Rhodine, C.N. Unpublished.

General Anaesthetic Excitation and Inhibition of Insect CO_2 — Receptors:

An Interpretation

Mark Diesendorf

Department of Applied Mathematics, S.G.S.,

Australian National University, Canberra, A.C.T.

There is a large body of experimental observations concerning the membrane actions of general anaesthetics[1,2]. The main conclusions upon which there appears to be general agreement are (i) that when applied to a nerve axon or muscle cell the anaesthetic in most cases reversibly blocks the action potential without appreciably affecting the resting membrane potential (within 4-5mV)[1], and (ii) that no chemical reaction takes place between anaesthetic molecules and molecules of the membrane[2]. The responses of CO_2 — receptors of the honeybee[3,4] and the blowfly[5] to general anaesthetics are not in accord with general observation (i), and also differ with respect to each other in the nature of their individual responses. We propose that in each case the basic anaesthetic action is similar, and the differences in response are an expression of different sites of action and hence can assist in the separation of different stages of chemosensory transduction. The anaesthetics other than CO_2, which affect CO_2 receptors, are referred to here as "effectors", since under various conditions they can excite, or inhibit, or retard the response of a sensory receptor cell.

The similarities in response of the CO_2 — receptors of the honeybee and blowfly are:

(S1) Both respond in one or other of the observables (i.e. spike frequency or latency) to a similar spectrum of general anaesthetics including CO_2, N_2O, and the n-paraffins which exhibit a response increasing with chain length up to a cut-off (after pentane in the bee, butane in the fly)[6].

(S2) Both are considerably more sensitive to the CO_2 stimulus than to the effectors. Hence a CO_2 — specific acceptor exists for each.

The n-paraffin response (S1) is qualitatively similar to behavioural observations of the chemoreception of alcohols and aldehydes by the fly[7], and may be compared with the anaesthetic action of short-chain alcohols (C_2 to C_5) on the squid axon[8]. The cut-off, both in olfaction[9] and in general anaesthesia[2], has been interpreted as a molecular size effect.

However recent calculations of the chemical potential of thin liquid hydrocarbon films on water surfaces show a similar effect[10]: pentane, hexane and heptane form a stable wetting film, while octane and dodecane do not, in agreement with experiment. Therefore it is likely that the cut-off in both CO_2 receptors and axon anaesthesia experiments represents the point where adsorption of anaesthetic molecules at an interface (or on macromolecules) of the cell membrane, under the action of Van der Waals forces, ceases. If this interpretation in correct, then a change in dielectric constant at the interface, induced for example by feeding the insect with NaCl, should shift the cut-off point.

The differences between honeybee and blowfly receptors lie in the type of response elicited:

(D1) *Spike frequency:* In the bee, the effectors inhibit the spontaneous spike frequency, but have no significant observable effect on the phasic or tonic response to CO_2, apart from a phasic rebound[3]. In the fly, however, both CO_2 and the effectors, separately or together, produce an increase in spike frequency f which is an increasing function of a *total* stimulus concentration S: i.e.

$$f = f(S), \text{ with } S = [CO_2] + \alpha [E] \tag{1}$$

where E is effector and clearly $\alpha \simeq 0.01$ for Fig. 1 of Ref. 5.

(D2)*Latency:* In the bee the effectors produce a retardation in response (i.e. latency increase) which is proportional to the effector concentration at a given CO_2 concentration[3]. In the fly, all the effectors have the same latency for a given spike frequency, and this is shorter than the corresponding latency to CO_2[5]. (D3) An inhibitor of carbonic anhydrase activity, Acetazolamide, shifts the CO_2 − response curves to higher stimulus concentrations, but has no effect on effector responses, in the bee[4]. However the enzyme inhibitor has no significant effect on the fly receptor at all[5].

To reconcile these substantial differences within the known framework of anaesthetic actions we commence with observation (D2). We have shown previously that the measurements of latency τ in the honeybee CO_2 receptor obey an equation of the form

$$\tau - \tau_0 = a_1 [E] / [CO_2] + a_2 / [CO_2] \qquad (a_1, a_2, \tau_0 \text{ const}) \tag{2}$$

and that terms on the right-hand side of (2) represent two different stages of the transduction process. Further, it has been confirmed by the Acetazolamide experiments[4] (D3) that CO_2 specificity and spike frequency are determined at a primary stage by a specific molecular acceptor for CO_2, and that CO_2, together with effectors participates in non-specific action at a secondary stage which determines the latency changes (D2).

Therefore it seems a reasonable working hypothesis that the differences in anaesthetic response between the fly and bee receptors, and between these sensory receptors and the squid axon, arise from the different sites of action available to the anaesthetics. Then the honeybee latency changes could be accounted for by a reversible change of molecular ordering (see below) which shifts the membrane to a new equilibrium state. Thus the diffusion coefficient for a product of the CO_2 + acceptor interaction (e.g. a "transmitter" such as Ca^{++} ion[11]), which diffuses to another region of the membrane and there modulates ion conductance, can be modified. Alternatively, the *rate* of opening of molecular "gates" controlling the ion channels is modified directly. Neither the number of ion channels nor the rate of flow of ions through the channels need be affected, and so the spike frequency for CO_2 stimulus can remain unchanged[19].

By "molecular ordering" we mean collective changes in orientation, site, or internal conformation of the membrane macromolecules, leading to an entropy change of the system. Existence of such cooperative effects in nerve membranes is deduced from observations of changes in optical[12] and infrared[13] properties during excitation, and deviations of stimulus-response curves from the Michaelis-Menten hyperbolic shape. Theoretical models have been constructed which involve either cooperative specific adsorption of ions at protein sites[14,15], or ligand binding membrane "protomers" which have several conformational states[16]. For sequential binding of small anaesthetic molecules to long-chain macromolecules, exactly soluble models of statistical mechanics such as those applied to allosteric enzymes[17,18] will be appropriate.

In the fly, it is proposed that the effectors produce a change of molecular ordering in a domain which triggers directly the opening of molecular gates. Thus a modulation of ion conductance and hence spike frequency is produced. The small differences between the responses of individual effectors again originates in the difference in Van der Waals forces, but the much greater sensitivity for CO_2 probably arises from its hydrolysis. The CO_2 and cyclopropane data[5] can be fitted roughly by the equation.

$$f(S) \simeq a_3 S^{\frac{1}{2}} / (1 + a_4 S^{\frac{1}{2}}) \quad (a_3, a_4 \text{ constants}) \qquad (3)$$

which deviates from Michaelis-Menten form, so one is tempted to argue that the presence of H^+ or HCO_3^- leads to an enhanced cooperative interaction between molecular gates for CO_2 stimulus. That this is incorrect follows from eq. (1). The response curves[5] for $[CO_2] = O$ and $[cyclopropane] = O$, are almost identical once the cyclopropane concentration has been rescaled, and so the cooperatives must be the same[17,18]. A more likely mechanism for enhanced CO_2 sensitivity would involve the release of transmitter by acceptors specific to HCO_3^-.

The experiments on the squid axon show that anaesthetics interfere with the movement of both Na^+ and K^+ as they cross the membrane during an action potential. In this case the kinetic parameters involving rates of turning-on of ion conductance are not affected by anaesthetics[2].

Summarizing our interpretation: in the CO_2 — receptors of (a) the honeybee (b) the fly, and (c) in the squid axon, the anaesthetic responses are basically non-specific and involve molecular ordering which in receptor (a) modifies the *rate* of opening of ion gates, or the *rate* of diffusion of transmitter to the gates: in receptor (b) determines the number of gates opened; and in (c) controls the rate of ion flow through the molecular gates.

References
1. Seeman, P. 1972. Pharm. Revs. 24, 583-655.
2. Mullins, L.J. 1971. Handbook of Neurochemistry, Vol. 6 (A. Lajtha ed.) p. 395-421. New York: Plenum Press. Chem. Rev. 1954, 54, 283-323.
3. Stange, G. and Diesendorf, M. 1973. J. comp. Physiol. 86, 139-158.
4. Stange, G. 1974. J. comp. Physiol. 91, 147-159.
5. Stange, G. Olfaction and Taste V (this volume).
6. Stange, G. Olfaction and Taste V (this volume).
7. Dethier, V.G. 1954. J. Gen. Physiol. 37, 743-751.
8. Armstrong, C.M. and Binstock, L. 1964. J. Gen. Physiol. 48, 265-277.
9. Mullins, L.J. 1955. Ann. N.Y. Acad. Sci. 62, 247-276.
10. Richmond, P., Ninham, B.W. and Ottewill, R.H. 1973. J. Colloid Interface Sci. 45, 69-80.
11. Cone, R.A. 1973. Biochem. & Physiol. of Visual Pigments (H. Langer ed.) p. 275-282. Springer-Verlag.
12. Tasaki, I., Barry, W. and Carnay, L. 1970. In Physical Principles of Biological Membranes. (F. Snell et al. eds.) p. 17-31. New York: Gordon & Breach.
13. Sherebrin, M.H. 1972. Nature New Biol. 235, 122-124.
14. Ling, G.N. 1962. A Physical Theory of the Living State: The Association-Induction Hypothesis. New York: Blaisdell.
15. Karreman, G. 1973. Bull. Math. Biol. 35, 149-171.
16. Changeux, J.-P., Thiery, J., Tung, Y. and Kittel, C. 1967. Proc. Nat. Acad. Sci. 57, 335-341.
17. Thompson, C.J. 1968. Biopolymers 6, 1101-1118.
18. Diesendorf, M.O. Biopolymers (in press).
19. Molecular ordering can lead to an inhibition of spontaneous spikes by producing a slight increase in activation energy of the acceptors or of the molecular gates, or by re-orientating membrane dipoles to hyperpolarize the membrane slightly.

A Functional Arrangement of Labellar Taste Hairs in the Blowfly *Calliphora vicina*

F.W. Maes & C.J. den Otter

Department of Zoology, Groningen State University,

Haren, The Netherlands

Introduction

Peripheral coding of gustatory information has been extensively studied in blowfly taste hairs, where single-unit responses are rather easy to obtain. When considering the central nervous processing of this information, the question arises whether all taste hairs are functionally equal. At least for labellar hairs this question has received only indirect attention (1,6,8). We tried to find an answer by relating spike frequencies and amplitudes to the known arrangement of taste hairs (classification of den Otter (6)).

Material and Methods

Using the method of den Otter (7) responses were obtained from 110 labellar hairs in the aboral and adoral rows (Figure 1) of 9 female blowflies *Calliphora vicina* R.–D. on stimulation with 1 M KCl. Responses were quantified by counting the number of spikes in the first half-second of stimulation.

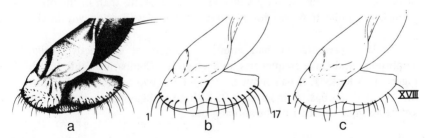

Fig. 1. a. Labellum of Calliphora, seem from the left. b. Aboral row, having 17 hairs, numbered with Arabic numerals; longer and shorter hairs alternate more or less neatly. c. Adoral row, 18 hairs, Roman numerals. (By courtesy of C.J. den Otter, Doct. Diss., Gronigen, 1971).

Results

The responses showed spikes from 1, 2, or sometimes 3 taste cells (Figure 2). The most frequently firing cells was a P-cell ("salt cell"; den Otter (5); other cells active were N-cells ("non-salt cells"). den Otter (5); other cells active were N-cells ("non-salt cells").

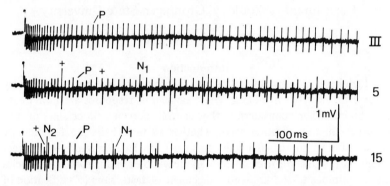

Fig. 2. Responses of labellar taste hairs to 1 M KCl. A frequently firing cell is always present (P). Often spikes from a second cell are seen (N_1); sometimes a third cell is active (N_2). Summations of P- and N_1-spikes occur repeatedly (+). Hair numbers are to the right of the recordings. Calibration bars apply to all recordings.

It appeared that the 110 sets of P- and N-spike amplitudes, which had been obtained for all aboral and most of the adoral hair numbers, can be SAB (short aboral), and AD (adoral), resp. The mean P-cell response turned out to be significantly larger in the LAB group than in the other groups. No indication was found for a further classification within the groups.

Table 1 presents the hair numbers, number of cells active, spike amplitudes, and spike frequencies for each group. Comparing Table 1 with Figure 1, it looks as if hair numbers 10 and 15 (two hairs each) fall into the "wrong" groups. Hair number 9, with an ambiguous length, can not yet be assigned to either of the aboral groups.

Discussion

Within each of the three groups of hairs the electrophysiological properties of the taste cells tend to be constant, although a certain amount of variability is present. Between the groups, however, differences occur. This strongly suggests that within each group the hairs are functionally equal, but that at least the LAB group differs functionally from the other two groups. The former suggestion implies a multiplicity in the afferent neural signals, which may serve to improve the "signal-to-noise ratio"; the latter implies that complete gustatory information is only present in the combination of the responses from taste hairs of different groups.

TABLE 1

Hair numbers, number of cells active, spike amplitudes, and spike frequencies for the LAB (long aboral), SAB (short aboral), and AD (adoral) groups of labellar hairs.

| | cell | group of hairs | | |
		LAB	SAB	AD
hair numbers		1-3-5-7-8- 10-11-13-17	2-4-6-12- 14-15-16	all adoral hairs
NUMBER OF	P*	36	21	53
cells active	N_1	36	3	17
mean amplitude (range) inμV	P	500 (300-700)	750 (600-900)	600 (450-750)
	N_1	800 (600-1200)	550 (500-600)	500 (350-600)
	N_2	~550	~1200	~1100
number of spikes ± s.d. in first half	P	57±12	30±7	31±8
	N_1	8±4	8±4	6±3
second	N_2	<1	<1	<1

* The number of P-cells equals the number of hairs tested.

References
1. Barton Browne, L. and Hodgson, E.S. 1962. J. cell. comp. Physiol. 59, 187-202.
2. Dethier, V.G. 1961. Biol. Bull. 121, 456-470.
3. Dethier, V.G. and Hanson, F.E. 1968. Proc. nat. Acad. Sci. (Wash.) 60, 1296-1303.
4. Goldrich, N.R. 1973. J. gen. Physiol. 61, 74-88.
5. den Otter, C.J. 1968. Neth. J. Zool. 18, 415-416.
6. den Otter, C.J. 1971. Neth. J. Zool. 21, 464-484.
7. den Otter, C.J. 1972. J. Insect Physiol. 18, 109-131.
8. Shiraishi, A. and Tanabe, Y. 1974. J. comp. Physiol. 92, 161-179.

The Pontine Taste Area: An Analysis of
Neural Responses in the Rat

Thomas R. Scott and Richard S. Perrotto
Department of Psychology, University of Delaware,
Newark, Delaware

The pontine taste area (PTA) is of particular interest to those concerned with the neural coding of taste information. Second-order (nucleus of the solitary tract, or NTS) and fourth-order (thalamic) afferents have been investigated, and significant basic discrepancies have been noted in neural responses (1, 3). The establishment of an intervening synapse raises the possibility of neural mediation between NTS and thalamic mechanisms. The purpose of this experiment was to record single and multiunit responses from PTA as evoked by a wide range of taste stimuli. This allowed comparison with NTS and thalamic data concerning the effectiveness of various stimuli in evoking responses, the relative similarity of those responses, spontaneous activity, breadth of neural sensitivity, and temporal response characteristics.

Methods

Subjects were male Wistar albino rats weighing 300-450 g. Surgical procedures, conducted under sodium pentobarbital anesthesia, are described elsewhere (4). Tungsten microelectrodes were used to record single and multiunit responses, which were amplified, filtered, and displayed by standard techniques. The stimuli employed are shown in Table 1.

Results and Discussion

Multiunit Responses:

Salty salts (NaCl, LiCl, Na_2SO_4) elicited a moderate phasic response which diminshed only slightly over ten seconds. Upon rinsing, the still robust tonic response decreased precipitously to spontaneous level, and often momentarily below it (Figure 1A). Bitter salts ($CaCl_2$, KC1, NH_4Cl) evoked a larger transient response which dropped more rapidly over time. Water rinse caused the remaining tonic activity to dimish only gradually back to base rate (Figure 1B). Inorganic acids (HCl, HNO_3) caused less of a transient response, but the tonic discharge rate, established within 0.5 sec. of stimulus onset, remained virtually undiminished for ten full

Stimulus	Effectiveness
0.1 M NH$_4$Cl	100.0%
0.3 M CaCl$_2$	99.8
0.1 M LiCl	98.0
0.1 M NaCl	97.4
0.3 M KCl	97.2
0.1 M Na$_2$SO$_4$	96.8
0.03M HNO$_3$	96.5
0.03M HCl	93.6
0.5 M Maltose	72.9
0.01M QHCl	70.6
0.5 M dl-Alanine	70.1
1.0 M Sucrose	66.8
0.1 M dl-Methionine	66.3
1.0 M Glycine	66.1
0.5 M dl-Valine	65.4
0.5 M Dextrose	63.4

Table I

Stimuli and concentrations, listed in order of effectiveness in exciting PTA cells. For multiunit data stimuli are shown as percentages of the most effective solution.

seconds. The distilled water rinse caused no significant change for several seconds (Figure 1C). Sugars evoked much less activity, but a rhythmicity was often apparent to sucrose, and, to a lesser extent, dextrose and glycine application (Figure 1D). Maltose, the remaining sugar, showed a time course and effectiveness more related to the salts, reflecting the non-sweet components which characterize its taste. Amino acids generally elicited unenthusiastic responses from PTA populations. Warm water (40°C) had no effect on these neurons, but cold water (10°C) always evoked strong activity. Touch and proprioceptive (jaw stretch) stimulation caused no response.

Single Neuron Responses:

The sensitivity of single neurons could be described as broad. Every salt and inorganic acid evoked discernible responses from all thirty neurons isolated. Sugars and amino acids were less effective, but quite often activated individual neurons. Discharge rates were generally low, with a mean spontaneous level of 1.8 spikes/sec., and evoked rates rarely exceeding 20.0 spikes/sec. Figure 2 shows a representative PTA neuron responding to stimuli representing all four "primary" tastes plus an amino acid.

Pontine taste neurons conform in many respects to the expectations generated for them from our knowledge of NTS and thalamic responses. Stimuli which are effective in NTS are effective in PTA. The time course of responses is similar, and the general trend for salty salts to evoke like responses, which have different characteristics from those elicited by bitter salts, was anticipated. Other PTA characteristics auger those seen at the thalamic relay. Pontine response rates decrease by a factor of about five from NTS levels to those of thalamic cells. There is a decrease in the mean

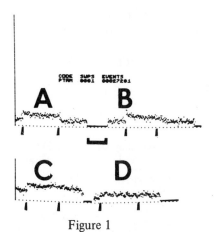

Figure 1

Post-stimulus time histograms for selected stimuli of one multiunit series. A. LiCl B. NH_4Cl C. HNO_3 D. Sucrose. Each dot represents neural activity summated over 128 ms. Arrows mark stimulus and water onsets. A five-sec time marker is shown.

ratio of phasic-to-tonic discharge rate. This ratio was 5:1 in chorda tympani (CT) recordings. NTS responses, while higher overall, showed an attenuation of the phasic response to 2.5:1 as compared to the steady-state. In PTA the ratio, calculated for the same stimuli, dropped to 1.5:1, and in the thalamus it was nearly 1:1.* Another higher-order characteristic displayed by PTA neurons involves inhibitory responses. NTS cells are almost never inhibited by chemical stimulation, but this phenomenon is rather common in the thalamus, and is not unusual in the pons. The clear impression is that pontine cells stand

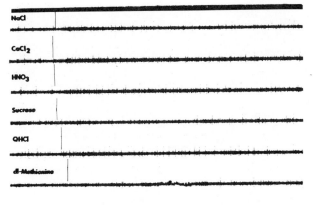

Figure 2

Single neuron response to a wide range of taste stimuli. The arrow indicates stimulus onset; a circle represents one second.

functionally as well as anatomically between second- and fourth-order taste neurons. They possess characteristics of both NTS and thalamic cells, and serve as a transitional stage for many of the neural changes which occur between the medulla and thalamus.

*This reflects the prevalence of inhibitory responses in the thalamus where an initial response decrement statistically offsets a phasic increase evoked by a different stimulus.

Supported by NIH grant 2 ROl NS 10405-03 CMS.

References
1. Doetsch, G.S. and Erickson, R.P. 1970, J. Neurophysiol. 33, 490-507.
2. Halpern, B.P., Bernard R.A. and Kare, M.R. 1962. J. gen. Physiol. 45, 681-702.
3. Scott, T.R. and Erickson, R.P. 1971. J. Neurophysiol. 34, 668-684.
4. Scott, T.R. and Perrotto, R.S. In preparation.

Linear Relation Between Stimulus Concentration and Primary Transduction Process in Insect CO_2 — Receptors

Gert Stange

Department of Neurobiology, Research School of Biological Sciences, Australian National University, Canberra, A.C.T.

Any sensory receptor can be regarded as a transducer which converts a specific stimulus energy into an electrical output, and for an understanding of primary mechanisms it is important to investigate which steps in the transduction system determine the response characteristics. In some mechanoreceptors[1], we find a linear transfer function, but in most sense organs nonlinear characteristics are observed which result in an extended dynamic range. From a theoretical study of several sensory systems Zwislocki[2] recently concluded that a first process in sensory transduction involves a linear integration of stimulus energies or a variable directly proportional to them, followed by further processes which provide a nonlinear transformation.

For the case of photoreceptors, direct experimental evidence for the linearity of the interaction between stimulus and primary acceptors is provided by the observation that even at the saturation level of receptor responses only a small fraction of the rhodopsin content is bleached, and by the finding that the miniature potential frequency and the early receptor potential, which presumably reflect primary events, are linearly related to the stimulus intensity over several orders of magnitude[3,4].

Beidler[5] showed that the responses of taste receptors can be related to the stimulus concentration by the mass-action law, and Kaissling[6] extended this formalism to insect olfactory receptors; he considered also the possibility that the saturation behaviour is determined by later steps in the transduction chain such as a nonlinear relationship between conductance changes and the receptor potential amplitude[7], implying that the number of acceptors is larger and that their stimulus affinities are lower than calculated on the basis of the mass-action law model.

For the antennal CO_2 — receptors of the honeybee, the blocking effect of the carbonic anhydrase inhibitor acetazolamide[8] indicates that a system with carbonic anhydrase specificity is involved in the primary reaction with the CO_2 — stimulus. Therefore it is interesting to compare known kinetic parameters of carbonic anhydrase isoenzymes with data obtained from insect CO_2 — receptors.

In several carbonic anhydrases[9], the Michaelis-constants of the CO_2 — hydration reaction are in the order of 10^{-2} M; an estimate based on the solubility of CO_2 in water shows that a CO_2 — content of 10^{19} molecules/ml in the gas phase is needed to yield an equilibrium CO_2 — concentration of 10 mM in an aqueous medium which might be assumed to prevail in the environment of a CO_2 — specific acceptor system. Under non-equilibrium conditions the CO_2 — concentration needed to half-saturate a system with carbonic anhydrase specificity would be even higher; in comparison, the half maximum response occurs below 3.10^{17} molecules/ml in honeybee CO_2 — receptors[8], and below 10^{17} molecules/ml in the CO_2 — receptors of the mosquito *Aedes aegypti*[10] and the blowfly *Lucilia cuprina* (Fig, 1).

This consideration suggests that within the dynamic range of those receptors any primary acceptor system with carbonic anhydrase specificity works in the lowest part of its adsorption isotherm, which must be linear. Independent evidence for linearity of a CO_2 — specific primary system has been obtained for the honeybee CO_2 — receptor[8] : N_2O, as well as other lipid-soluble anaesthetic gases such as Xe, the lower n-paraffins and cyclopropane, has an effect which is antagonistic to the CO_2 — response (suppression of spike activity followed by a phasic rebound)[11]. Acetazolamide reduces the sensitivity of the receptors to CO_2, but not to N_2O, and from experiments in which N_2O/CO_2 — stimuli were applied in the presence of acetazolamide it was concluded that N_2O interacts with an output variable derived from the CO_2 — and acetazolamide-specific system; this variable is linearly related to the CO_2 — concentration[8].

In extracellular recordings of single unit spike activity, a CO_2 — receptor on the antenna of female *Lucilia cuprina* was found to be several orders of magnitude less sensitive to acetazolamide than the honeybee CO_2 — receptor, and it also responds with excitation to stimuli which would cause an inhibitory response in the honeybee, so that the argument used for the honeybee receptor cannot be applied to the blowfly. Therefore it remains open whether the two species make use of different principles for the detection of CO_2, but the following two experiments suggest that the blowfly and honeybee receptors are in fact similar.

1. CO_2 — receptors of *Lucilia* were stimulated with mixtures of CO_2 and cyclopropane and the initial phasic peak frequency of the response was chosen as an assay for studying the interaction. Cyclopropane was used because it causes higher response magnitudes than N_2O, and because

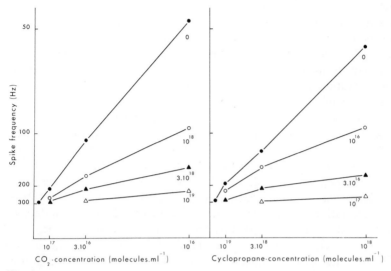

Fig 1.
Responses of a CO_2 − receptor to combined stimuli of CO_2 and cyclopropane. Figures in the plot give cyclopropane- concentrations (left) and CO_2 − concentrations (right) in molecules · ml^{-1}.

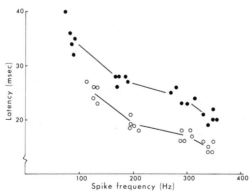

Fig 2.
Latency as a function of response magnitude for stimulation with CO_2 (●) and cyclopropane (O). Repeated measurements from a single receptor.

its low vapour pressure permits an unproblematic method of stimulation. As it is of particular relevance whether the maximum responses to both kinds of stimuli and to combinations are the same, the results (Fig. 1) are plotted in double inverse diagrams (cf also[6]). The dose-response curves converge to the same maximum values, and for combined stimuli there are no indications for major deviations from a simple additive interaction, suggesting that both kinds of stimuli interact at the same site.

2. For a given spike frequency the response to CO_2 appears later than the response to cyclopropane, and this difference in latencies is constant within the covered range of spike frequencies (Fig. 2). Other stimuli such as N_2O, C_2H_6 and C_3H_8 also cause faster responses than CO_2, and for a given response magnitude the latency difference is the same as for cyclopropane in an individual receptor, but between different receptors a considerable variability occurs: in 12 receptors studied latency differences between 2.5 and 12 msec were observed, with an average of 7 msec.

This observation is consistent with the hypothesis that in the blowfly receptor the CO_2 – stimulus interacts with a highly specific primary acceptor system which is bypassed by stimuli such as N_2O or cyclopropane. The observation that the receptor is much more sensitive to CO_2 than to the other stimuli indicates that this system has amplifying properties, and it is conceivable that such an amplification process takes time. If this interpretation is correct, then the concentration independence of the latency difference has the further implication that the concentration dependent formation time[6] of a complex between CO_2 – stimulus and acceptor is very short, and that therefore the concentration dependence of the latency is determined by later steps in the transduction chain.

The results presented in Fig. 1 permit the conclusion that the output variable of the CO_2 – specific system interacts additively with the effect of cyclopropane on a common rate limiting step, and therefore this variable is linearly related to the CO_2 – concentration.

The theory outlined is still hypothetical, but it demonstrates that stimuli such as anaesthetic gases or specific inhibitors are useful probes to obtain information on individual steps in the transduction process; it is desirable to test whether a similar approach is possible for other olfactory receptors.

References

1. Burgess, P.R., Perl, E.R.: Handbook of Sensory Physiology. vol. II (A. Iggo ed), p. 29-78. Berlin, Heidelberg, New York : Springer 1973.
2. Zwislocki, J.J. : Kybernetik 12, 169-183 (1973).
3. Cone, R.A.: Cold Spring Harbor Symp. Quant. Biol. 30, 483-491 (1965).
4. Scholes, J. : Cold Spring Harbor Symp. Quant. Biol. 30, 517-527 (1965).
5. Beidler, L.M. : J. gen. Physiol. 38, 133-139 (1954).
6. Kaissling, K.E. : Olfaction and Taste III (C. Pfaffmann ed.), p. 52-70. New York : Rockefeller University Press 1969.
7. Kaissling, K.E. : Handbook of Sensory Physiology, vol. IV, pt 1 (L.M. Beidler ed.), p. 351-431, Berlin, Heidelberg, New York : Springer 1971.
8. Stange, G. : J. comp. Physiol. 91, 147-159 (1974).
9. Carter, M.J. : Biol. Rev. 47, 465-513 (1972).
10. Kellog, F.E. : Insect Physiol. 16, 99-108 (1970).
11. Stange, G., Diesendorf, M. : J. comp. Physiol. 88, 139-158 (1973).

Gustatory Papillae in Advanced and Primitive Frogs:

Litoria vs. Leiopelma

John A. MacDonald

Department of Zoology, University of Auckland,

Private Bag, Auckland, New Zealand

New Zealand is inhabited by two very different kinds of frogs. The most common are several species of *Litoria (Hyla)* introduced from Australia in the 1800's. The native frog, *Leiopelma*, is rarely seen, and with *Ascaphus* of northwest America, is one of the most primitive of living frogs (Robb, 1973).

The tongue and fungiform papillae of *Litoria (Hyla) aurea* are very similar to those of other advanced anurans (Graziadei & DeHan, 1971; Raviola & Osculati, 1967; Stensaas, 1971). The tongue is elastic, free posteriorly, and can be everted to capture prey. The dorsal surface is covered with many small non-ciliated filiform papillae, and fewer large fungiform papillae, each of which bears a disc-shaped apical sense organ, 100-130μ in diameter, surrounded by a ring of ciliated cells. The disc is composed of two main cell types. Large interdigitating vesicular columnar cells, with nuclei in a discrete distal layer, form a polygonal surface array of microvilli. Putative receptor (rod) cells of a deeper nuclear layer send processes proximally toward the base of the disc, and distally between the columnar cells, where they end in protruding fibrillar cones. Desmosomes are common between distal columnar and rod cell membranes.

The tongue of *Leiopelma hochstetteri* differs from those of advanced ranid, bufonid and hylid anurans in several respects. It is inelastic and firmly attached to the buccal floor, with little free margin. The dorsal surface is covered with small ciliated papillae. Vase-shaped "taste-buds", 30-60μ in diameter, are housed in the apices of small papillae (Fig. 1), and are composed of cell types similar to those in the *Litoria* taste disc.

The distal portion of the taste bud is largely occupied by interdigitating vesicular columnar cells, with frequent desmosomes, pertaining to a distal nuclear layer. "Receptor" and sustentacular cells of the inner nuclear layer send processes both proximally and distally. Dense fibrillar pegs protrude from the surface and are probably continuous with receptor cells. Non-myelinated nerve fibres approach the base of the bud, but their contacts are unresolved. Lateral margins of all papillae are occupied by mixed ciliated cells and secretory cells, the latter packed with large (1μ) granules (Fig.2).

213

Fig. 1. *Leiopelma* taste bud, SEM. The surface of the taste bud lacks cilia and is covered with microvilla; cell borders are not apparent. Light dots are high points, probably rod cell pegs; dark cavities are evacuated goblet cells. Ciliated cells occur around the bud, and on adjacent papillae.

Leiopelma taste buds are intermediate between the bud-shape typical of urodeles (Fahrmann, 1967) and the disc of advanced frogs. Whether the *Leiopelma* taste bud is considered primitive or of secondary origin depends on further characterization of taste bud morphology and function in other anurans. Sections from a museum specimen of *Ascaphus truei* show tongue and taste buds similar to those of *Leiopelma,* therefore the *Leiopelma* features are probably primitive. Since *Leiopelma,* like many urodeles, does not evert the tongue in capturing prey, it is suggested that the disc-form of "typical" frog/toad taste buds is adapted to such use, perhaps mechanically isolating the sense organ from consequences of epithelial stretching.

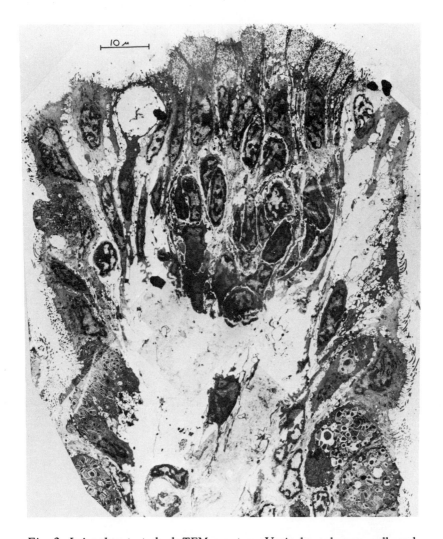

Fig. 2. *Leiopelma* taste bud, TEM montage. Vesicular columnar cells and distal nuclear layers are clearly visible. Rod cells are found in lower nuclear layer. The open hole to the left of the taste bud represents an evacuated goblet cell. Lateral margins of the papilla bear ciliated and secretory cells.

J. MacDONALD

References

Färhmann, W. Z.mikr.-anat. Forsch. 77:117-151 (1967).
Graziadei, P.P.C. & R.S. DeHan. Acta anat. 80:563-603 (1971).
Raviola, E. & F. Osculati. Ist. lomb. (Rend. Sci.) B101:599-627 (1967).
Robb, J. Reptiles and Amphibians, p.285-303. In: The Natural History of New Zealand, G.R. Williams (ed.). Wellington: A.H. & A.W. Reed 1973.
Stensaas, L.J. Am. J. Anat. 131:443-462 (1971).

The author is indebted to Don Ensor for supplying post-operative specimens of *Leiopelma*, and to Heather Roberts and Paul Hicks for substantial assistance with scanning and transmission electron microscopy.

Mechanisms of Lateral Interactions in Rat Fungiform Taste Receptors

Inglis J. Miller, Jr., Dept. of Anatomy, Bowman Gray School of Medicine, Wake Forest University, Winston-Salem, North Carolina 27103, U.S.A.

Lateral interaction occurs among inputs from adjacent fungiform papillae in the rat which evoke responses in the same chorda tympani nerve fibers.[1-3] A single isolated papilla was stimulated with NaCl and when potassium benzoate (KBz) solution was applied to papillae surrounding the isolated one, there resulted a depression of the ongoing response to NaCl. This depression was attributed to a differential effect of cationic excitation and anionic inhibition in compliance with the theory of Beidler[4]. To test the generality of this explanation of lateral depression, a series of electrolyte salts were used as surround stimuli in order to compare their effectiveness in producing lateral depression. The experimental method for surround studies of single rat chorda tympani fibers was the same as in previous experiments[2] and the response ratios for 10 components at 0.1 molar concentrations are shown in Table 1. A response ratio greater than 1.0 indicates enhancement, and a response ratio less than 1.0 indicates depression. All potassium compounds produced depression, which was somewhat surprising in the case of KCl, although the magnitude of the depression was greater for the

TABLE 1

Response Ratios for 0.1 M Salts

Compound	Response Ratio (S.E.)	N
NaCl	2.03±.46	7
KCl	.72±.04	12
KBz	.63±.04	11
KFo	.57±.05	12
KAc	.59±.04	10
KPr	.55±.04	10
TRIS Cl	.98±.07	10
TRIS Bz	.78±.02	7
TEA Cl	.92± 05	10
TEA Bz	.85±.05	12

Bz: benzoate, Fo: formate, Ac: acetate, Pr: propionate, TRIS: tris-hydroxymethylaminomethane, TEA: tetraethylammonium, Na: sodium, K: potassium, Cl: chloride.

compounds with large organic anions of benzoic and the carboxylic acids. NaCl was always excitatory and produced enhancement at the 0.1 M concentration.

When KCl is applied liberally over the tongue both the summated whole nerve response and the response of individual chorda tympani fibers are facilitatory at the 0.1 M concentration. KBz depresses the background activity of both whole nerve and single fiber responses at this concentration for whole tongue stimulation, yet both compounds produce a depression of the response elicited from a single fiber when they are applied to the surround of an isolated papilla which is simultaneously stimulated with NaCl. The potassium salts of formic, acetic, and propionic acids are similar to KBz in that they depress the whole nerve response and also produce a surround depression. Thus, the potassium ion appears to play a role in the lateral depression response. Salts with organic anions elicited greater depression responses than the chloride salts. With the organic cations TRIS and TEA little depression resulted with the chloride salt, but a greater depression was produced by the benzoate salts. The greatest depression was elicited when the surround stimulus was a potassium salt of an organic acid which might bode of synergy between potassium and the organic anion.

Electrical stimulation of a papilla adjacent to one which is isolated and stimulated with NaCl can produce lateral interaction with the response of the isolated papilla. A gold ball electrode of about 0.3 mm diameter and a constant-current stimulator were used to deliver the d.c. surround stimulus. A single isolated papilla was stimulated with 0.1 M NaCl for 30 sec and a d.c. electrical stimulus was delivered to an adjacent papilla during the middle 10 sec. A response ratio was calculated by dividing the number of single fiber impulses during a four sec interval of simultaneous stimulation of the isolated and surround papillae by the number of spikes in a four sec control period in which only the isolated papilla was stimulated. Figure 1 shows the response of one fiber explored in

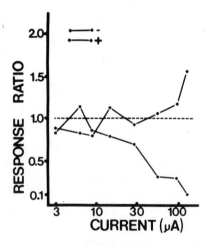

Fig. 1. A single fiber response to electrical surround stimulation.

this manner. Stimulation of the surround papilla with cathodal current produced depression at a current intensity of about 10 microamperes which increased with currents up to 145 microamperes. Anodal current produced enhancement at a current of 55 microamperes. Transient responses occurred with the onset and offset of the cathodal stimulus. These observations are consistent with the electrophysiological findings of Frank, Pfaffmann, and Bujas[5] in that the tonic phase of anodal stimulation was inhibitory. Bujas[6] pointed out that the sensation produced by cathodal stimulation in man lasts as long as the current is applied, which implies either a species difference between man and rat, as he has suggested, or that inhibition of ongoing neural activity may be a meaningful code in the elicitation of taste sensation.

Cross-innervation of rat fungiform taste buds have been examined recently in a light microscopic study of peripheral chorda tympani neurons[7]. Lingual nerve fibers were degenerated unilaterally leaving only chorda tympani fibers in the operated half, and tongues of 9 animals were reconstructed from serial sections. Figure 2 summarizes the findings from 407 papillae on the operated side and a representative number of papillae from the unoperated side. Bundles of chorda tympani fibers (solid lines) branched sending fibers to adjacent papillae. These fibers were found to terminate exclusively in the taste bud region of the papillae, and fiber branches could be traced to innervate adjacent taste buds. Represented on the right side are 21% of the papillae which received a straight bundle of fibers although some of these fibers may have branched proximally in the parent fiber bundle. Deep branch points were observed below 43% of the papillae and superficial branch points located in the sub-papillary dermis were found below 36% of the papillae. Nerve fibers illustrated by dashed lines (left) represent lingual trigeminal fibers as observed on the unoperated side of the tongue. Trigeminal fibers innervated the superior and

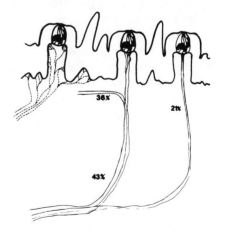

Fig. 2. Schematic reconstruction of fungiform papilla innervation. Solid lines: chorda tympani fibers; dashed lines: trigeminal fibers.

lateral walls of fungiform papillae as well as the common epithelium between papillae, but it is not known if trigeminal fibers penetrate the taste bud. The interconnection of adjacent taste buds by branching chorda tympani fibers and the absence of "free" nerve endings in this fiber population leads credance to the concept that lateral interactions in chorda tympani fibers are mediated through taste bud inputs.

Since interactions occur among adjacent fungiform taste buds, a study was made of the spatial distribution of fungiform papillae on the rat tongue[8] using the technique of Fish, Malone, and Richter[9]. The current study identified a mean ($\pm$S.D.) of 187.1 $\pm$9.2 (N=10) papillae per tongue which agrees with the earlier mean finding of 178.8 papillae/tongue. There was an approximately equal number of papillae on each half of the tongue and the distribution of papillae on the tongue surface was very similar among animals. The highest density of papillae was on the tongue tip with 3.4 papillae/mm^2 (anterior 4 mm) which accounted for 51.1% of the total number. The remaining 48.9% of the papillae were located on the dorsal surface which extended a length of about 20 mm to the intermolar eminence at the level of the molar teeth. The density on the dorsal surface was 1.3 papillae/mm^2. Each rat fungiform papilla contains a single taste bud, so that the distribution of fungiform papillae is equivalent to the location of taste buds on the anterior 2 cm of the rat tongue.

To test whether the distribution of papillae bears a relationship to the response elicited from the chorda tympani nerve, a diaphragm was applied to the tongue to divide the anterior 4 mm of surface from the mid-region, 5-20 mm posterior. The summated response was recorded from the entire chorda tympani nerve while approximately equal halves of the papilla population were stimulated on the tip and mid-region. NaCl, sucrose, and HCl were used to stimulate both regions and the steady-state responses were compared to the summated nerve response obtained from stimulation of the entire tongue (Fig. 3). Both regions (tip and mid-region) contained about

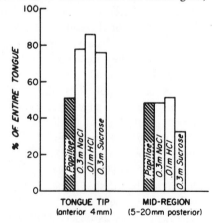

Fig. 3. Relative responses to stimulation of tongue tip and mid-region.

an equal number of the fungiform papillae but the responses elicited from the tip were 70-80% of the entire tongue response for several classes of chemical stimuli and for several concentrations of the same stimulus. Responses of the mid-region ranged from 33% of the whole tongue for sucrose to 54% for NaCl at these concentrations. Pfaffmann[10] has reported that single chorda tympani fibers receive input from more papillae on the tip than on the mid-region of the tongue. The greater response of the tip elicited in this experiment might be interpreted as resulting from a greater nerve overlap and more lateral enhancement on the tip than on the mid-region.

Lateral taste depression which was demonstrated in the rat with potassium benzoate[2] has been shown to result from other electrolyte stimuli. The potassium cation and the organic anions both seem to contribute to this effect. Branching in the primary neuron has been suggested as the mediating mechanism[2] and interconnecting branches have been demonstrated[7] between adjacent taste buds. A disparity in the distribution of fungiform taste buds on the tongue tip compared with the mid-region of the dorsal surface of the tongue allows the comparison of the responses of regions of high and low taste bud densities. Though both regions have approximately equal numbers of taste buds, the responses of the higher density tip are greater than the responses of the mid-region in comparable proportions for several classes of stimuli. Augmented responses from the region of high receptor density may indicate the operation of lateral interactions among adjacent taste buds as a fundamental process in peripheral gustatory neural encoding. A continuing investigation of the mechanisms of lateral interaction among taste receptors is being pursued with morphological evaluation of electrophysiological preparations.

References

1. Beidler, L. 1969. In Olfaction and Taste III, pp. 352–369.
2. Miller, I. 1971. J. Gen. Physiol. 57:1–25.
3. Miller, I. 1972. In Olfaction and Taste IV, pp.316–322.
4. Beidler, L. 1967. In Olfaction and Taste II, pp. 509–534.
5. Frank, Pfaffmann, and Bujas. Unpublished observations cited by Bujas[6].
6. Bujas, Z. 1971. In Handbk. Sens. Physiol., Chemical Senses 2-Taste, pp.180–199.
7. Miller, I. 1974. J. Comp. Neurol. 158:(No. 2).
8. Miller, I. and A. Preslar. (In press, 1975). Anat. Rec.
9. Fish, H., P. Malone, C. Richter. 1944. Anat. Rec. 89:429–440.
10. Pfaffmann, C. 1970. In Ciba Symp. on Taste and Smell, pp.31–50.

(This work was supported by the U.S. Public Health Service through research grant NS 10389 from National Institute of Neurological Diseases and Stroke.)

Molecular Structure and Stimulatory Effectiveness of Amino Acids in Fish Olfaction

Toshiaki J. Hara

Department of the Environment, Freshwater Institute

Winnipeg, Manitoba, Canada R3T 2N6

One of the most characteristic features in fish olfaction is that it takes place entirely in the aquatic environment. The carrier of stimulant molecules is not air but water, therefore, chemicals that are detected olfactorily by fish need not be volatile, but must be soluble in water. The spectra of chemicals detected by fish could be entirely different from those detected by terrestrial animals. In fact, recent electrophysiological studies show that certain amino acids, which are normally non-odorous to humans, are extremely effective olfactory stimuli and may play an important role, such as acting as chemical signals, in olfactory communication in fishes [1-4]. Further investigation of the specificity of olfactory stimulation of amino acids and analogues has led to the establishment of definite structure-activity relationships.

Olfactory Responses to Amino Acids

Olfactory electrical responses were measured in the olfactory bulb of rainbow trout (*Salmo gairdneri*) when the paired nares were stimulated with chemical solutions according to the method described previously [3,5]. The integrated bulbar response (Fig. 1a) increases exponentially with a logarithmic increase in stimulus intensity until the response reaches saturation. The range from threshold to saturation depends on the stimulatory effectiveness of the amino acid and rarely exceeds more than about 4 log units. General findings so far indicate that only α-amino acids are stimulatory, that L-isomers are always more stimulatory than D-isomers, and that the stimulatory effectiveness is not directly related to the essential amino acids.

Structure-Activity Relationships

Specificity of stimulation was studied by testing a number of amino acids and analogues for their stimulatory effectiveness as represented by the magnitude of the integrated bulbar response. All amino acids investigated were the L-forms and were tested at 10^{-4} or 10^{-5}M.

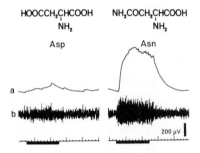

Fig. 1. Typical olfactory bulbar responses to aspartic acid (Asp) and asparagine (Asn). The upper traces (a) show the integration of the lower (b).

α -*Amino group.* Substitution of the α-amino group with -H (glycine vs. acetic acid, aspartic acid vs. succinic acid, alanine vs. propionic acid), -CH₃ (aspartic acid vs. methylsuccinic acid), -OH (aspartic acid vs. malic acid, serine vs. glyceric acid, alanine vs. lactic acid), or structural alteration of the α-amino group (acylation and methylation) resulted in markedly decreased responses or in entire elimination of the response. Amino acids whose amino group is located at a position other than the alpha were far less effective (α-alanine vs. β-alanine, α-aminobutyric acid vs. γ-aminobutyric acid).

α -*Carboxyl group.* Replacement of the α-carboxyl group with -H (aspartic acid vs. β-alanine, glutamic acid vs. γaminobutyric acid, serine vs. 2-amino-ethanol), -CH₃ (serine vs. 2-amino-1-propanol), or esterification of the α-carboxyl group (as in serine methylester), raises the threshold for response markedly.

α -*and β-Hydrogens.* α-Methylserine was far less effective than serine, suggesting a requirement for a free α-hydrogen. Replacement of a β-hydrogen with a functional group resulted in decreased activity (valine, threonine, and isoleucine).

Fourth α-moiety. The amino acids which satisfy the above conditions have shown to be all stimulatory. However, the final effectiveness of each amino acid depends largely upon the size of the fourth α-moiety. In glycine, the simplest amino acid, the fourth α-moiety is a single hydrogen atom. Overall molecular length exceeding six carbon atoms reduces effectiveness. The fourth α-moiety of aromatic amino acids is bulky and they are all only slightly stimulatory.

Amidation of the terminal carboxyl group

Dicarboxylic amino acids tested were only slightly stimulatory. However, amidation of the terminal carboxyl group increased activity markedly (aspartic acid vs. asparagine, glutamic acid vs. glutamine) (Fig. 1).

The amino acids inducing olfactory stimulation are thus characterized by being simple, short and straight-chained, with only certain attached groups. Glycine is the smallest amino acid molecule that fulfils all the requirements described. No chemical has been found to be more olfactorily stimulatory, when determined electrophysiologically, than the amino acid.

Hypothetical Receptor Site

It is suggested from the above findings that stimulatory effectiveness is dependent upon interaction of the amino acid molecules with receptor membrane structures of definite shape, size and charge distribution. The "amino acid receptor" thus can be considered to consist of an arrangement of charged atoms or groups located within a specific region of the molecular framework of the membrane. A hypothetical receptor site can be considered which involves two charged *subsites* capable of interacting with ionized amino and carboxyl groups of the amino acids. These two subsites are arranged in a specific order, clockwise or counterclockwise, around a centre which accommodates the α-hydrogen atom of the amino acid molecule, thus making only L-isomers accessible to the receptor site.

The three subsite system described above is a prerequisite for a compound to be stimulatory. There must however, be another receptor region which accommodates and recognizes the profile of the fourth α-moiety of the amino acid. Discrimination of the quality of the amino acid could take place at this region. The latter appears to have an extreme affinity for the amide function and less affinity for charged carboxyl and bisulfite groups. The amino acid receptor described here is similar to the AH-B system proposed by Shallenberger and Acree[6] for human taste sense, although in their system the D-amino acids are more readily accepted than the L-forms. However, the three-subsite system described here for fish olfactory system would seem to make the receptor more discriminatory towards odorous molecules.

References
1. Sutterlin, A.M. and N. Sutterlin, 1971. J. Fish. Res. Bd. Can., 28, 565-572.
2. Suzuki, N. and D. Tucker, 1971. Comp. Biochem. Physiol., 40A, 399-404.
3. Hara, T.J., 1973. Comp. Biochem. Physiol., 44A, 407-416.
4. Hara, T.J., Y.M.C. Law and B.R. Hobden, 1973. Comp. Biochem. Physiol., 45A, 969-977.
5. Hara, T.J., Y.M.C. Law and E. Van der Veen, 1973. J. Fish. Res. Bd. Can., 30, 283-285.
6. Shallenberger, R.S. and T.E. Acree, 1967. Nature, 216, 480-482.

The Problems of Perception of Leaf-surface Chemicals
By Locust Contact Chemoreceptors

by

E.A. Bernays, W.M. Blaney and R.F. Chapman

Centre for Overseas Pest Research, College House, Wrights Lane,

London W8; and

Zoology Department, Birkbeck College, Malet Street,

London WC1

It would be a remarkable feat if, blindfold, we could differentiate milk chocolate from plain chocolate simply by touching it with out fingertips. But this, apparently, is just what some insects can do. Grasshoppers, for instance, can differentiate acceptable from unacceptable food plants by drumming briefly on the leaves with their palps (Bernays & Chapman, 1970). Of course the insect is able to do this because it has contact chemoreceptors on the tips of its palps, but the faculty is even more remarkable than it appears at first sight. The grasshopper rejecting food after palpation makes contact only with the outer surface of the leaf and experiments have shown that the surface waxes alone provide the information on which the choice is based (Bernays, Blaney, Chapman & Cook, in press).

In other words, selection may be based on information received from contact with *dry, non-polar* and *non-volatile* compounds. How do these reach the dendrite endings within the terminal pore of the sensillum from the leaf surface? Surely their transport must depend on the nature of the material which bridges the gap between the leaf surface and the dendrites?

If this is the case we might expect all the dendrites within a sensillum to be similarly affected by a stimulating chemical. This is not to say that they will necessarily all respond in the same way since they may have their individual specificities, but a tendency to do so would be suggestive. Unfortunately it is difficult to obtain relevant and reliable information since the usual electrophysiological techniques involving stimulation and recording via a capillary over the tip of the sensillum are not appropriate for experiments with dry non-polar materials and sidewall recording is beset with difficulties and drawbacks (Dethier, 1974). The sensilla on the tips of the palps of locusts and grasshoppers have proved particularly intractable in this respect so that we have, of necessity, resorted to stimulation from a fluid-filled capillary over the tip. Nevertheless some suggestive results have been obtained.

That the sensilla really do respond to plant-surface waxes has been proved by stimulating them with waxes dispersed in an electrolyte solution by sonication. Different response patterns are produced by different waxes. More important, it has emerged that the neurones of the crested sensilla on the maxillary palps tend to be unspecialised in their responses, but the sensilla themselves show specialisation (Blaney, in prep.). This is just what might be expected if the material interposed between the outside environment and the tips of the dendrites was playing a significant role in perception. Critical experiments have only been carried out using fructose and sodium chloride, but less complete work on plant surface waxes indicates a similar tendency to specificity at the level of the sensillum rather than the neurone.

If the material at the tips of the dendrites is important in stimulus perception we might also expect the rate of transfer of the stimulant to the dendrite to vary with its chemical nature, in other words the latent period between initial contact and response need not be constant. We have no relevant data on the terminal sensilla of the palps of acridids, but it may be significant that Barton Browne & Hodgson (1962) recorded marked differences in the latent periods for fructose and sodium chloride in the labellar sensilla of *Lucilla*. It does not follow, of course, that the properties of the palp-tip sensilla would be the same, but the suggestion is there.

An alternative to the electrophysiological approach is to consider the physico-chemical properties of the material at the tip of the sensillum. But what is it? On the palps of *Locusta* the presence of an acid mucopolysaccharide immediately within the pores of over 80% of the terminal sensilla was demonstrated by a positive staining reaction with periodic acid-diamine staining (method of Spicer, 1965). The reaction occurred in less than 20% of the sensilla, however, after treatment with ovine testicular hyaluronidase (method of Ashhurst & Costin, 1971) indicating the presence of hyaluronic acid. It cannot be assumed that sensilla not staining with periodic acid-diamine or still staining after treatment with hyaluronidase were necessarily different from the rest, because the rates of penetration of chemicals into these sensilla is highly variable.

This is all we know of the material surrounding the tips of the dendrites, and no more evidence is available from other sensilla. In terms of the transport of non-polar chemicals this helps us not at all, but it does not follow that hyaluronic acid is the only, or even the most important, substance at the tips of the sensilla.

Here, then, we can do little more than draw attention to the problem, but what can be done to advance our understanding? First, there is no reason to suppose that the receptors of locusts and grasshoppers are peculiar in their ability to respond to dry non-polar materials. Critical experiments on host-recognition and mating in other insects would show whether or not this was a common faculty. Second, where electrophysiological studies do not necessitate the use of a stimulating/recording electrode over the tip of the sensillum, more attention to non-aqueous stimulants and to chemicals which normally occur in the ambit of the insect would be revealing. And third, microtechniques could be used to establish the chemical and physical nature of the material around the dendrites.

Sturckow (1970) has previously emphasised the potential importance of the 'viscous substance' in contact chemoreception although Dethier (1972) has questioned its role as a significant factor. His argument is not valid in the context of non-polar, non-volatile chemicals. In this situation the nature of the chemical bridge between environment and dendrite seems critical.

References

Ashhurst, D.E. & Costin, N.M., 1971. Histochem. J. 3:279-295.
Barton Browne, L. & Hodgson, E.C., 1962. J. Cell. Comp. Physiol. 59:187-202.
Bernays, E.A., Blaney, W.M., Chapman, R.F. & Cook, A.G., in press. Insect/Host plant Symposium, Budapest.
Bernays, E.A. & Chapman, R.F., 1970. J. Anim. Ecol. 39:761-776.
Dethier, V.G., 1972. Proc. Nat. Acad. Sci. USA. 69:2189-2192.
Dethier, V.G., 1974. in Experimental analysis of insect behaviour. L. Barton Browne (ed).
Spicer, S.S., 1965. J. Histochem. Cytochem. 13:211.
Sturckow, B., 1970. Adv. Chemoreception. 1:107-159.

Gunnar Bertmar

Section of Ecological Zoology

Dept. of Biology, Umeå University 901 87 Umeå Sweden

Reindeer have relatively large vomeronasal organs connected with both mouth and olfactory cavities. A technique for temporary elimination of the organs in living animals has been developed. The organs may be involved in drinking, grazing, social and sexual behaviour. Preorbital, tarsal and interdigital skin glands were found, and by analyzing and testing their structure and secretions they were shown to release scents functioning as pheromones. Two types of experimental techniques have been tested. Frequent sniffing the samples may be correlated with low rank, youth and female sex. Tarsal scents created more chemosensory reactions than interdigital scents. Reindeer may track skin gland scents of the same species.

Ecochemical studies on reindeer

A cross-scientific project of biologists and chemists has been started at Umeå University. The biological part of it is to study the role of the chemosensory organs in the activation of behaviour patterns of reindeer.*(Rangifer tarandus)*. In 1969 we started the project with a structural study of the reindeer nose. It showed that these animals have vomeronasal organs (VO), and that each organ is connected to both mouth and olfactory cavities by means of a nasopalatine (incisive) duct. The VO are relatively large and well differentiated compared to those of boxer and sheppard, two dog races which were studied at the same time. It could therefore be expected that the VO play an important role in the reindeer.

To study this we tested a technique for temporary elimination of the organ in living animals. A rubber-like mixture has been tested and shown to be inert, and caused only slight initial irritation. It was injected into the nasopalatine ducts and could be removed by a nylon thread fastened in the plug. When the tampon was removed after 5 days a mucus was forced out of the nasopalatine opening. Therefore, in animals where the plug was left for 3 months the secretion should have ceased and/or the mucus may have been removed by the highly vascular tissue of the submucosa. Preliminary results indicate that elimination of the VO may influence grazing, drinking and social behaviour. So far only yearlings of both sexes have been tested but observations this autumn on a 3 year old bull in estrus suggest that the VO are also involved in facial grimace called Flehmen, and that they are receptor organs for urine pheromones. The VO may therefore be important for sexual behaviour. Experiments now proceed in order to study effects in animals of different age, sex and social status.

We have also become interested in other possible natural inputs to the chemosensory organs, biological substances that are species-specific and well known to the animals for thousands of years and therefore should be interesting also for reindeer management and other practical purposes. Such an input might come from the big skin glands. We have found that reindeer have preorbital, interdigital and tarsal glands (Fig. 1)

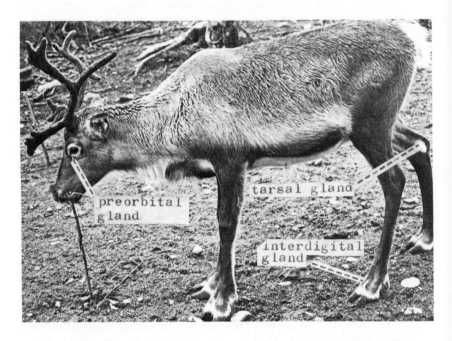

Fig. 1. Female yearling of a forest reindeer in Sept. 1974 showing position of skin glands.

In 1970 we therefore made a histological and cytological study of skin glands from adult forest reindeer slaughtered at winter. This suggests that the preorbital and interdigital glands (the tarsal gland has not been studied) have different types of morphology, structure and secretions, and that they also probably have different functions.

The behavioural effects of secretions from interdigital and tarsal glands of both mountain and forest reindeer have therefore been tested on forest reindeer. 9 animals were studies at Lycksele zoological garden in southern Lapland during part of the summer, autumn and winter in 1972, and since January 1974, 9 animals have now been studied in our enclosure in Umeå. In this work we co-operated with chemists trying to identify, isolate and synthesize the components of the glands. Volatile compounds from the tarsal gland have recently been isolated by a precolumn technique and analysed on a combined gas chromatograph-mass spectrometer. Saturated aldehydes and alcohols were found in both male and female glands (Andersson et al., 1974).

We are now testing natural and synthesized compounds of the glands. Since 1972 we have used a technique with test stations of varying number and position, each with a sample and a control 1/2-1 m apart. The samples are put on a clean glass bowl, and contain a total distillate, a fraction or a gland smear. The tests were spaced 15 min-1 hr apart, and some abiotic factors were also recorded. With this station technique we have found that the large skin glands are scent glands also in reindeer, and that some compounds of the secretions are pheromones. Calves and female yearlings with small antlers respond more often than other animals. As the age of the animals and the size of the antlers have importance for the heirarchy (Espmark, 1971), it therefore seems to be a positive correlation between low rank, youth and female sex on one hand, and frequent sniffing on the other hand. Licking of the samples occurs very seldom, and the tarsal scents created more chemosensory reactions than the interdigital scents.

In parallel with the station technique we are now using a scent track technique. Scents are put on the ground and/or the vegetation in a relatively intact part of the enclosure. One animal at a time is tested. Preliminary results indicate that reindeer are able to track skin gland secretions (pheromones) of other reindeer.

The zoologists working on the project are G. Bertmar, Y. Espmark, M-L Källquist and B. Leijon.

References

Andersson, G., Andersson, K., Brundin, A., and Rappe, C. 1974. Jou. Chem. Ecol. 1. (in press).
Espmark, Y. 1971. Jou. Wldfe. Mgmt. 35, 175-177.

Olfactory Bulb and Integration of some Odorous Signals in the Rat — Behavioural and Electrophysiological Study

by M. Cattarelli, E. Vernet-Maury, J. Chanel,

P. Mac Leod and A.M. Brandon

Laboratory of Psychophysiology — National Center

of Scientific Research and Claude Bernard

University, Lyon, France

The rat is able to distinguish between odours of grouped, isolated, stressed, rewarded and non-rewarded rats, and between odours of predatory and non-predatory animals (1-13). This work is the study of the behaviour of rats sniffing different odours in an open-field and, on the other hand, of the unitary responses of mitral cells, in the olfactory bulb, to the same odours.

Methods

The different olfactory stimuli used were isoamyl acetate and citral, urine and feces of male rats, feces of cat, fox and lion. Urine and feces were obtained from groups of five rats living as usual in their cage, or from an isolated rat (in an unknown cage), a stressed rat (by electric shock), a rewarded (by drink from the good arm of a T-labyrinth) or non-rewarded rat.

For the open-field test 220 Wistar male rats were used. Air was blown into the open-field through a first flask of activated charcoal and through a second flask containing the stimulus. For each rat the entrance-time (into the open-field) and different activities (walking, standing on hind legs, toilet, defecation, micturition) were noted. From these behavioural responses an *emotional reaction index* (Mann-Whitney U test) was calculated.

For the electrophysiological study 57 Wistar male rats (i.e., 143 mitral cells) were used. Among them 24 were anaesthetized with nembutal (37,5 mg/kg, i.p.) and 15 were curarized with flaxedil (3mg/kg, i.v.). In the third group of 18 rats the nervous tracts between higher centers and olfactory bulb were cut off just behind the anterior olfactory nucleus, under nembutal anaesthesia. In the three groups, rats were tracheotomized and a catheter was introduced, by this way, in the nasal cavity. All the incisions of the skin are infiltrated with 1% procain solution. The temperature of the rat was controlled with a rectal probe connected with an infra-red lamp. During all the experiments, deodorized air (1 liter/min) was blown through this catheter except during stimulation. Then the air flow was reversed: odorized air blown on the rat nose through a little funnel, was exhausted through the catheter. So olfactory mucosa was continuously mechanically stimulated. An extracellular tungsten microelectrode, prepared by Hubel technique (14) was implanted stereotaxically in the olfactory bulb. The amplified electrical activity of one mitral cell, visualized on oscilloscope screen, was recorded on a magnetic tape. Mitral cells are continually firing. When olfactory mucosa was stimulated by odorized air, either *no change* of the frequency of spikes discharges, or an *activation* (increase of frequency), or an *inhibition* (decrease of frequency), or a *diphasic response* (increase and then decrease of frequency) was observed.

Results

Behaviour — That the reassured rat is very active (it walks, stands on its hind legs, makes its toilet, etc . . .) and the alarmed rat frozen (it stays motionless) is well known. As other behavioural and electrophysiological studies (15,16) assume that isoamyl acetate odour is neutral, it is taken as a zero point. The emotional reaction index is then negative when the rat is reassured and positive when it is alarmed. For the different odours the mean *emotional reaction index* is:

Li	(Lion)	: − 3.5	deodorized air:	: + 0.8
R	(rewarded Rat	: − 3.3	NR (non-rewarded Rat:	+ 1.2
Ct	(Cat)	: − 3.1	St (stressed Rat)	: + 1.2
Gr	(grouped Rats)	: − 1.9	Is (isolated Rat)	: + 1.2
AA	(isoamyl acetate)	: 0	Ci (citral)	: + 3.2
			Fx (fox)	: + 8.1

Electrophysiology — For the same odours the *percentages* of *no change* (0), *activation* (A), *inhibition* (I) and *diphasic responses* (di) are calculated; rats are. . .

. . . .	anaesthetized				curarized				deafferented			
	O	A	I	di	O	A	I	di	O	A	I	di
Li	13	41	37	9	31	31	31	7	−	−	−	−
R	6	36	43	15	44	39	17	0	37	38	25	0
Ct	18	38	32	12	40	27	27	6	−	−	−	−
Gr	22	31	35	12	13	75	0	12	35	27	31	7
AA	9	35	25	31	20	30	40	10	−	−	−	−
NR	26	16	45	13	11	11	78	0	−	−	−	−
St	18	24	43	15	55	15	30	0	−	−	−	−

	anaesthetized				curarized				deafferented			
	O	A	I	di	O	A	I	di	O	A	I	di
Is	16	19	52	13	50	0	50	0	28	28	36	8
Ci	19	25	50	6	25	17	50	8	32	32	36	0
Fx	21	18	49	12	35	6	53	6	25	34	37	4

Conclusions

In the second table, stimuli are classified in the same order as in the first one, results are not exactly in the same order, but in the two tables reassuring and alarming stimuli are easily distinguishable. If *no change* and *diphasic* responses distributions are not significant, the highest *activation* of mitral cells is observed in animals which would have been very active in the open-field, the highest *inhibition* in rats which would have been motionless. These results are more significant in curarized than in anaesthetized rats and the difference between *reassuring* and *alarming* odours disappears when the olfactory bulb is deafferented. Higher nervous centers control the firing of mitral cells as they control behavioural responses. Modulations of firing and behaviour seem connected with the meaning of olfactory information.

Another statistical study (Fisher exact probability test) of electrophysiological results suggests that the curarized rat is able to distinguish each odour except the cat and lion.

References

1. Courtney, R.J. & al. ·Psychon. Sci. 1968, 12,315.
2. Curti, M.W. – J. Comp. Psy. 1942, 34,51.
3. Krames, L. & al. – J. Comp. Phys. Psy. 1973, 82,444.
4. Ludvigson, H.W. & al. – Psychon. Sci. 1967, 9, 283.
5. Means, L.W. & al. – J. Comp. Phys. Psy. 1971, 76, 160.
6. Morrison, R.R. & al. – Science. 1970, 167, 904.
7. Reiff, M. – Acta tropica. 1956, 13, 289.
8. Stevens, D.A. & al. – Behav. Biol. 1972, 7, 519.
9. Stevens, D.A. & al. – Behav. Biol. 1973, 8, 75.
10. Valenta, J.G. & al. – Science. 1968, 161, 599.
11. Vernet-Maury, E. & al. – C.R. Ac. Sci. 1967, 265, 1408.
12. Vernet-Maury, E. & al. – C.R. Ac. Sci. 1968, 267, 331.
13. Wassermann, E.A. & al. – Science, 1969, 166, 1307.
14. Hubel, D.H. – Science, 1957, 125, 549.
15. Tapp, J.T. & al. – Canad. J. Psychol. Rev. Canad. Psychol. 1968, 22, 449.
16. Pager, J. & al. – Physiol. Behav. 1972, 9, 573.

Coding of Odor Quality in the Insect Olfactory Pathway

J. Boeckh, K.D. Ernst. H. Sass and U. Waldow

University of Regensburg, Regensburg, Germany, BRD

This paper proposes possibilities for modes of connections between antennal receptors (especially olfactory) and central neurons. See Fig. 1 for diagrammatic formulation. The neuroanatomical data are based on degeneration experiments, on EM fiber counts in the antennal nerve and central tracts, and on silver impregnations (Golgi, modified after Colonnier; Bodian) (1,2,3,4). The neurophysiological data are from single cell recordings in antennal receptors and neurons in the deutocerebrum (2,3,5,6,7).

A. Receptor spectra

Reactions from 300 receptor cells in basiconic sensilla of *Periplaneta americana* were recorded; 150 compounds were tested. The receptors could be grouped according to their spectra. Those of a given type reacted invariably with the same graded specificity; that is, the rank order of each odorant rated by its relative efficiency remained constant for all members of a type.

The types were named for their maximum reaction compounds: pentanol, hexanol, octanol, alcohol-terpene, formic acid, butyric acid, and amine types. On the one hand there was considerable overlap in their spectra; on the other there were always effective compounds which did not occur in the spectra of other types (Fig. 2). The degree of uniformity in the reaction of the cells of any given receptor type

---

Figure 1. a: Neuro-anatomical data with numbers of axons, glomeruli, and neurons indicated. Inset (upper left in a): frontal view of brain. D, deutocerebrum; S, plane of section. b, c, d: hypothetical connections between receptors and central neurons. B, B-cell; glom, glomerulus; hygr, hygrorecptor; mech, mechanoreceptor; olf, olfactory receptor; R, reaction amplitude; stim, stimulus; temp, temperature receptor; TOG, tractus olfactorio globularis; W, W-cell. The precise location of B- and W-cells has not yet been determined.

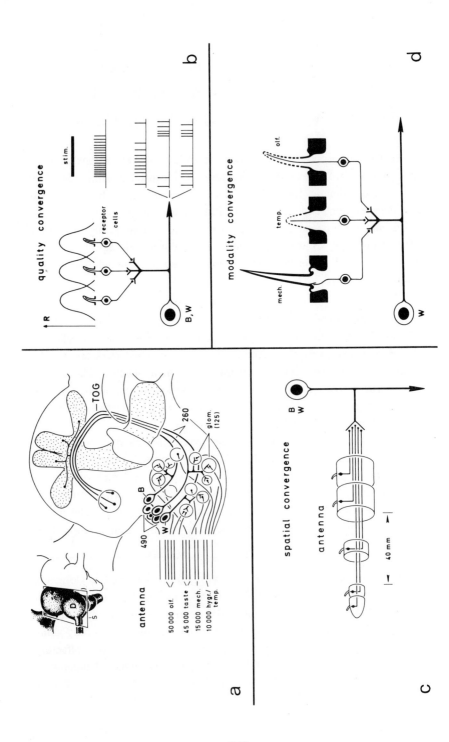

a

antenna

50 000 olf.
45 000 taste
15 000 mech.
10 000 hygr./temp.

490
W
B
260
glom. (125)
TOG

b

quality convergence

receptor cells

R

stim.

B, W

c

spatial convergence

antenna

40 mm

B
W

d

modality convergence

olf.
temp.
mech.

W

240

appears so high that probably odors can be distinguished only if each different odor stimulates different receptor types to a different degree. Naturally occurring foodstuffs for cockroaches such as fruit or meat contain many volatile compounds which occur in the spectra of different receptor types. Many of them are common both to different types of fruit and, in addition, to different receptor spectra. Therefore it is not surprising to find graded reactions of different receptor types to most odors of foodstuffs (Tab. 1).

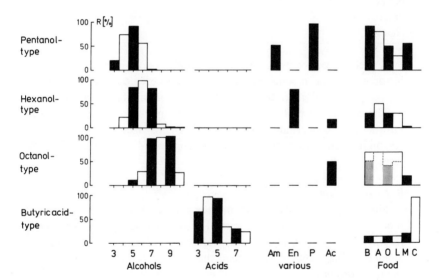

Figure 2. Sections from odor profiles of four types of receptors. Reactions (R) given in % of maximum impulse frequencies. Alcohols: homologous aliphatic alcohols from propanol (C3) to decanol (C 10). Acids: homologous fatty acids from propionic (C 3) to caprylic acid (C 8). Various: Am= amyl phenyl acetate, En= eugenol, P= pentyl acetate, Ac= ethyl acetate. Food: B= banana, A= apple, O= orange, L= Lemon, M= meat, C= cheese. Stippled lines in the profile of the octanol type indicate variations in receptor reactions believed to originate in variations of fruit odors.

Table 1.

Responses of different receptor types to complex odors. alc/terp. alcohol-terpene type; but acid, butyric acid type; O, no reaction; +, reaction less than 50% maximal; ++, reaction 50 to 100 % maximal.

stimulus	receptor type				
	pentanol	hexanol	octanol	alc/terp	but acid
lemon	+	+	+	+ +	o
orange	+	+	+ +	+ +	+
banana	+ +	+	+ +	+	+
apple	+ +	+ +	+ +	+	+
bread	+ +	+	+	+	+
meat	+ +	+	+	o	+
cheese	o	o	o	o	+ +

A preference, however, for certain complex odors seems to be indicated in some receptor in some receptor types. Thus each complex odor elicits a unique pattern of excitation from the receptor population in the antenna.

The conclusion is that it is precisely this unique pattern which conveys information concerning complex odors. No receptor was detected responding only to a single "key" — compound in the complex odors tested.

B. Neuroanatomy.

More than 120,000 axons from antennal receptors of different stimulus modalities terminate in the glomeruli of the deutocerebrum. 490 neurons (B-and W-cells, Fig. 1,a) send widely branching processes to these glomeruli. Only 260 of them send their axons via the TOG-tract on into the protocerebrum.

C. Central neurons.

A total number of 120 neurons was investigated. Two groups of neurons have been distinguished so far: "B-cells" and "W-cells". B-cells respond only by excitation and that only upon stimulation of the ipsilateral antenna with odors. No strict grouping of neurons according to odor specificity was apparent.

Most of them respond to lemon and/or orange odors, and to some alcohols and terpenes (Tab. 2). A few responses were found to meat and cheese. There is no conformity between the reaction spectrum of any receptor type and the spectrum of any B-cell. A comparison reveals a wider spectrum of central neurons for single compounds and a narrower spectrum for complex odors (Tab. 3). The first phenomenon can be explained in terms of a convergence of inputs from different types of antennal receptors (Fig. 1,b). The high specificity for fruit odors in B-cells can be explained only by complex speculations.

Table 2.

Reaction spectra from 6 out of 80 B-cells, +, excitation, O,no response.

B-cell Nr.	citral	spike	nerol	pentanol	hexanol	octanol	terpineol	cheese	meat	bread	apple	banana	orange	lemon
13		+	o	+			+	+	+	+	+	+	o	o
30	o	+	o	o	+	+	+	o	o	o	o	o	+	+
32	o	+	+	+	o	o	+	o	o	o	o	o	+	+
38	+	+	+	+	+	+	+	o	o	o	o	o	+	+
50	o	o	o	o	o	o	o	o	+	o	o	o	o	o
59				+	+	+	+	o	o	o	o	o	+	o

Each B-cell receives inputs from receptors of different antennal regions. Local stimulation at different antennal segments elicits similar reactions in a given central neuron even when the stimulated regions are separated by the whole length of the antenna (Fig. 3). Such a convergence of axons of many receptors upon single central neurons is well in agreement with the anatomical data (Fig. 1,a and c).

W-cells show a spontaneous activity which is changed in very complex ways by stimulation of the ipsilateral antenna (Fig. 4). Their odor spectra include fatty acids and amines rarely found in B-cells . W-cells react also to mechanical (bending, touching) and thermal stimulation of the antenna.

The response of a single W-cell to different stimuli can vary from excitation or inhibition to on-, off-, and on-off responses, and include all sorts of super-position of these modes. W-cells display both spatial and quality convergence. Further they receive a diversity of inputs from antennal receptors (via interneurons?) which differ even in stimulus modality (Fig. 1,d).

Table 3.

Comparison of spectra of receptors and B-cells. Symbols for receptor types as in Tab. 1, for reactions in Tab. 2.

stimulus		receptor type		
	11 B-cells	pentanol	octanol	alc/terp
lemon	+	+	+	+
orange	+	+	+	+
banana	o	+	+	+
apple	o	+	+	+
terpineol	+	o	o	+
3-methyl but.	+	+	o	+
octanol	+	+	+	o
pentanol	+	+	+	+

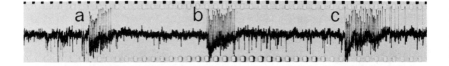

Figure 3. Reactions of a B-cell to stimulation of base (a), middle region (b), and tip (c) of the antenna with odor of bread.

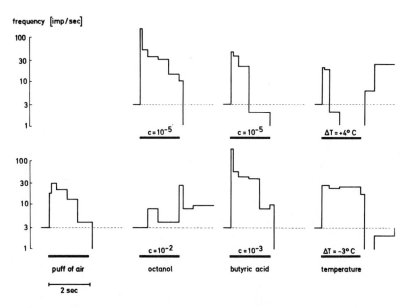

Figure 4. Reactions of single W-cell to stimulation of antenna with different modalities, qualities and intensities. Stippled line marks level of spontaneous activity. Touching antenna elicits reaction similar to that of a puff of air. c, concentration of odor in the source.

References

1. Boeckh, J., Sandri, C., Akert, K., Z. Zellf. 103, 429 (1970).
2. Boeckh, J., Verh. Dt. Zool. Ges. 1972, 66, 189 (1973).
3. Boeckh, J., J. comp. Physiol. 90, 183 (1974).
4. Boeckh, J., Ernst, K., Boeckh, V., in prep.
5. Sass, H., Verh. Dt. Zool. Ges. 1972, 66, 198 (1973).
6. Sass, H. Ph.D. − thesis, Regensburg, 1974.
7. Waldow, U., in prep.

Chemical Stimulation of the Cat's Tongue
Will Affect Cortical Neuronal Activity

A.R. Morrison and R. Tarnecki; Univ. of Penna.,

Philadelphia (U.S.A.) and Nencki Institute of

Experimental Biology, Warsaw (Poland)

Demonstration of the location of the gustatory cortex of the cat by the recording of neuronal responses to chemical stimulation of the tongue has yet to be shown convincingly, only 12 cells out of 181 examined having been classified as gustatory in the ventral coronal gyrus, the chorda tympani zone (Fig. 1) (1,3). With glass micropipettes we explored cortex lining the presylvian sulcus, shown anatomically (4) (Fig. 1) to be the projection zone of the thalamic taste area and a hidden portion of the chorda tympani zone. We provide direct proof that this cortex is a rich source of cells responsive to chemical stimulation of the tongue.

Recording in 6 cats began a few hours after administration of 35 mg/kg of pentobarbital sodium. The cats were paralyzed with Flaxedil, artificially respired and maintained at 36-38ºC. Vertical penetrations were made through the dorsal surface of the cerebrum at coordinates A24-26 and L4.5-5.3% agar in saline covered with melted bone wax, which hardened, prevented pulsations.

Stimulus solutions made with Ringer's and kept at 30ºC were 0.5 M NaC1, 0.1 M citric acid, 0.01 M quinine HC1, and distilled water. The procedure was to squirt 6 ml over the untouched tongue for 8 seconds. Recording sites were verified histologically.

Unit activity was monitored on an oscilloscope and audio monitor, which permitted detection of immediate changes in activity. Long-term changes in firing patterns and mean firing rates were determined by constructing interval histograms (IH) using a standard number of counts during resting and evoked activity on-line with an Anops laboratory computer fed by Schmidt-trigger pulses converted from spikes.

Responses to solutions, either excitation or inhibition, were obtained from 29 out of 55 neurons in the presylvian area in 6 penetrations in 5 cats. Electrical stimulation of the tongue dorsum with single 100µA, 0.5 msec square pulses produced field potentials in this region. Two other gustatory neurons were found in the precruciate gyrus (Fig. 1B). No silent cells were induced to fire. Mechanical effects were ruled out because changes were not noted during application of wash solutions. The other 26 neurons in the presylvian region could not be classified as definitely

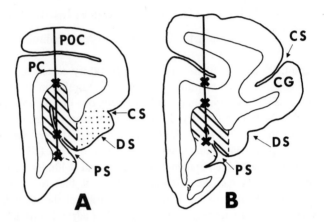

Fig. 1. Transverse sections from brains of 2 cats showing area explored. "A" is more rostral. Dots-chorda tympani zone explored earlier (1,3); hatch marks-gustatory area shown anatomically (4). A-dorsal X, potentials first evoked by electrical stimulation. Between the lower X's lay gustatory neurons analyzed. Track observed to midway through the hatched areas in A and B. B-Dorsal X, 2 additional gustatory cells lay here. Gustatory neurons were found between the lower X's. Gyri-CG, coronal; PC, precruciate; PoC, postcruciate. Sulci-CS, coronal; DS, diagonal; PS, presylvian.

gustatory for technical reasons, such as cell death or inconstant responses. In one other cat a more rostral penetration outside the thalamic taste area projection field (4) revealed 2 cells responding by inhibition to gentle stroking of the tongue, but no gustatory neurons were collected.

Each gustatory neuron could be classified qualitatively according to the change in firing rate heard immediately following application of a solution. The first eight were so categorized. Their responses were: Citric acid-6 excited, 1 inhibited, 1 no response; NaCl-3 excited, 3 inhibited, 2 no response; Distilled water-1 excited, 2 inhibited, 3 no response. For the remaining 21 the prolonged change in mean firing rate over a period of a minute or more was compared with that observed during three consecutive resting IH in order to provide more quantitative data. Neurons were considered to be influenced if their rate increased or decreased by 25%.

Most rate changes fell between 25 and 100%. Usually the firing rate and pattern then changed in the direction of the original resting rate and pattern during three successive IH. This procedure had the advantage of controlling for mechanical effects and providing a measure of the discharge patterns of gustatory neurons but the disadvantage of masking subtler early effects. Responses of these 21 cells were: Citric acid, N=19, 11 excited, 3 inhibited, 5 no response; NaC1, N=21, 8 excited, 3 inhibited, 10 no response; Quinine, N=8, 7 excited, 1 inhibited, 0 no response; Distilled water, N=18, 5 excited, 2 inhibited, 11 no response.

Cells could be excited by one or more stimuli and inhibited by others. Nineteen responded to two or more solutions. With excitation the peak firing frequency increased and with inhibition, decreased. Chemotopical organization was not observed due to the electrode approach used.

Thus, a cortical gustatory area, undoubtedly the major one, lies buried within the presylvian sulcus, confirming earlier anatomical findings (4). Prior difficulties in eliciting cortical responses in the cat with chemical stimulation resulted primarily from a lack of anatomical information and not for other reasons, such as anesthetic effects (See (4) for a complete discussion.) Chemical stimuli were found to excite some neurons and inhibit others. This observation supports results obtained in the thalamus of the rat (5) and in the cortex of unanesthetized dogs and rats (2). Although the only electrode track containing neurons responding clearly to mechanical stimulation of the tongue lay rostrally to the gustatory cortex, cells responsive to other stimulation, including thermal and mechanical, might be found in the gustatory region. Our primary aim in this initial study was to identify definitely gustatory neurons and to demonstrate that the presylvian cortex contains them in relative abundance.

Supported by NIH Grants NS06716 and NS08377. Experiments were done in Warsaw. We thank C. Borkowska, J. Lagowska and J. Rajkowski for assistance.

References

1. Cohen, M.J., Landgren, S., Strom, L. and Zotterman, Y. (1957). Acta Physiol. Scand., 40, Suppl. 135,1-50.
2. Funakoshi, M., Kasahara, Y., Yamamoto, T., and Kawamura, Y. (1972). In Olfaction and Taste IV, Schneider, D. (ed.) Wissenschaftliche Verlagsgesellschaft MBH: Stuttgart, pp. 336-343.
3. Landgren,S. (1957). Acta Physio. Scand., 40, 210-221.
4. Ruderman, M.I., Morrison. A.R., and Hand. P.J. (1972). Exp. Neurol., 37, 522-537.
5. Scott, T.R., Jr., and Erickson, R.P. (1971). J. Neurophysiol., 34, 868-884.

PHYLOGENETIC EMERGENCE OF SALT APPETITE

Chairman: C. Pfaffman
Moderator: D. Denton

Phylogenetic Emergence of Salt Taste and Appetite

S.F. Abraham, E.H. Blaine, D.A. Denton, M.J. McKinley,

J.F. Nelson, A. Shulkes, R.S. Weisinger and G.T. Whipp.

Howard Florey Institute of Experimental Physiology and

Medicine, University of Melbourne,

Parkville, Vic. 3052, Australia.

Substantial areas of the earth, alpine areas and the interior of continents are deficient in sodium as reflected in the sodium content of soil and vegetation. This follows from the fact that the Na content of rain diminishes with distance from the ocean, and in alpine areas freezing and thawing will accentuate leaching processes. So may seasonal monsoonal downpour in continental interiors.

In the absence of geological sources of salt, for example volcanic, the herbivorous animal species in such areas will suffer a varying degree of Na deprivation. This has been comprehensively investigated and validated in the alpine, and the monsoonal watered interior of Australia (Denton et al., 1961; Blair-West et al., 1968). Substantial Na deficit of native and introduced species of animals was shown by blood and urine electrolytes, increased blood aldosterone and renin levels, structural changes of adrenal glomerulosa and the salivary duct system, and, as we will recount, by salt appetite. Similar data has now been reported from Western and Eastern mountains and the interior of North America, Africa, and Central Asia (Botkin et al., 1973; Weir, 1972; Bazhenova and Kolpakov, 1969; Denton, 1967). There seems no reason to suppose that the conditions determining the present circumstances have not operated over much of geological time. Thus environmental Na deficiency has represented an important selection pressure during phylogenetic emergence of a wide variety of species of land animals (Denton, 1974).

Deficiency of Na reduces blood and extracellular fluid volume with impairment of functional capacity of circulatory adjustment. This can diminish speed and endurance during pursuit or flight, or, the distance ranged and thus grazing capacity. Also jeopardized is the all important capacity to continue these processes under reproductive conditions, with the obligatory additional sodium to be sequestrated in the products of conception, and during lactation. The effects may be further aggravated. The evidence of a substantially tropical − subtropical mode of existence of feral man and his evolutionary ancestors involves the perennial problem of temperature control with sweating and salt loss — sometimes aggravated by disease with pyrexia. Infection with diarrhoea will be most likely in gregarious species e.g. in bovidae with salmonella bacterial or viral diseases such as infectious bovine diarrhoea — mucosal disease. Environmental conditions involving trace metal imbalance, Cu^{++} deficiency or Mo^{++} intoxication, can also case diarrhoea and sodium loss. Any circumstances aggravating ecological deficiency will cause particular stress to ruminants which are dependent on circulation of large volumes of high salt saliva for normal digestion (Denton, 1957).

We will detail the effect on appetite behaviour of relatively small sodium deficit of 50-100 mEq further on, but the response of complex systems to minor deficit is well shown physiologically by the natriuretic response to 3rd ventricular injection of hypertonic saline (Andersson and Olson, 1973). As Fig. 1 shows, the single injection usually gives a clear-cut naturiuresis whereas, if the animal is depleted of some 60 — 150 mEq by parotid cannulation, the natriuretic response is greatly reduced or abolished, whereas the antidiuretic response still occurs.

A fact of general implication is that the impact of environmental Na deficiency falls directly and most severely on the herbivores. They eat the deficient grass and foliage. On the contrary carnivores, even in the most salt deficient regions, probably never experience deficiency because of the Na content, obligatory for life, in viscera and muscle of their herbivorous prey. And accounts by naturalists of salt appetite in the wild do not appear to directly involve carnivores (Denton, 1969). With primates and man, the major emphasis of anthropological evidence, as we have reviewed in detail elsewhere (Denton, 1974), is that over the Miocene to Pleistocene geological eras prehominids have been predominantly vegetarian or graminiferous. The hunting-meat eating development, as Leakey (1959) points out, has been relatively late — in the last two million years — and Clarke (1970) observes that even with some contemporary primitive hunting societies e.g. the Kalahari bushmen, the major components of intake may be vegetarian. With feral man, as for example, represented by the primitive highlanders of New Guinea, intake may be almost entirely vegetarian with a K/Na ratio of diet of 200-400:1; They have blood aldosterone levels four times western man and other chemical evidences of a relatively deficient sodium status (Scoggins et al., 1970). Historical records of various societies show a substantial human preoccupation with salt. A possibility to be considered is that this is a basis of cannabalism in some circumstances (Denton, 1974).

Salt responsive taste fibres occur in all vertebrates examined. The ubiquitous distribution of this modality of chemoreception probably reflects considerable survival advantage accruing with the ability to detect and ingest this substance.

In systematic field investigation we have shown wild kangaroos in the sodium deficient Snowy Mountains of Australia when encountering for the first time a cafeteria of filter paper blocks impregnated with strong solutions of NaCl, KCl, $MgCl_2$, $CaCl_2$, or water alone, show a fairly immediate choice of the Na blocks and rapid ingestion of them, with particular interest by lactating females.

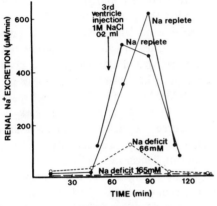

Fig. 1. The effect on renal Na excretion in a conscious sheep of 3rd ventricular injection of 0.2 ml of 1 M NaCl when Na replete (● – ●) and after depletion of 66 (○ – ○) and 165 mM (△ – △) of Na following parotid cannulation.

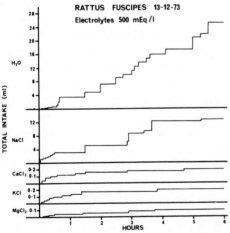

Fig. 2. The voluntary intake on first access to a cafeteria of water, and 500 mEq/1 solutions of NaCl, $CaCl_2$, KCl, and $MgCl_2$ of a native wild rat captured in the Na deficient Snowy Mountains region of Australia. Note the ordinate scale for NaCl is 10-fold the other electrolytes.

Native wild rats trapped in this region when placed in a drinkometer with a cafeteria of 500 mEq/1 solutions of NaCl, KCl, $MgCl_2$, $CaCl_2$ and water usually taste all solutions within a few minutes and show a specific and large intake of NaCl (Fig. 2). This large specific and immediate intake by animals most likely but not certainly, naive as far as such salt ingestion experience is concerned, cannot be attributed to retroactive appreciation of benefit of one of the four substances tasted. Even if there were time, it would be difficult to see how the animal would recognize which of the four had conferred benefit. The field data is consistent with laboratory data of our own on sheep, and data of Wolf (1969), Rodgers (1967) and earlier work of Pfaffmann and Nachman (1963) indicative that salt appetite and ingestion in response to initial experience of deficit is innate in the rat.

Botkin and colleagues' study of the moose population of Isle Royal suggested that Na availability controlled the population size. Calculated total Na content of all browse available on land indicated that it was insufficient to account for the Na of tissues produced in reproduction. Selective grazing in summer of submerged and floating aquatic plants having up to 500 times the Na content of land plants was the basis of a thriving population (Botkin et al., 1973). Weir's (1972) study of spatial distribution and density of elephants in an African national park has shown Na availability in water source or lick determined animal concentration. The effect predominated over browse availability.

The studies on wild sodium deficient kangaroos, and wombats in the Australian Alps, takes the phylogenetic emergence of the salt appetite system back at least to the Cretaceous geological era — 70-90 million years ago. There is little information on more primitive animals. Konishi and Zotterman (1963) have shown that carp have taste fibers responding to NaCl, and it is perhaps feasible that the salinity preference in some migratory fish (Baggerman, 1959; Hoar, 1959) represents the primitive emergence of neural mechanisms analagous to those which were eventually elaborated in more complex form to determine salt appetite in mammals.

In contrast to the paucity of data on salt appetite, aldosterone has been identified in reptiles, amphibians and teleosts including the South American lungfish, as well as in monotremes and marsupials. In ontogenetic studies, its synthesis, and increased secretion in response to ACTH, has been shown in the first 40-60 days of the 145 days of gestation in the sheep (Wintour et al., 1974).

Coming to laboratory analysis, we have studied sheep with parotid fistula. The animals lose 200-700 mEq of Na/day and large changes of balance can be contrived at will. If offered a sodium solution, they drink adequate and live in good health indefinitely. If the supplement is withdrawn for 12-72 hours, variable body deficits are produced. Fig. 3 shows representative data on an individual animal over months and there is a significant relation between deficit, and amount of 300 mEq/1 $NaHCO_3$

solution drunk, when the solution was presented for 15 min. The same relation is seen in adrenalectomized sheep with a parotid fistula (Denton et al., 1969). Similar studies on adrenalectomized rats shows them to overcompensate (Jalowiec and Stricker, 1973).

Sheep can also learn to press a lever to deliver a small volume of fluid into a bin in their cage. If each delivery is, for example, 50 ml of 300 mEq/1 NaHCO₃ = 15 mEq, then the animal will have to make and drink about 40 deliveries per day to compensate a salivary loss of 500 mEq of Na. Fig. 4 is constructed so that a longitudinal period is 24 hours and each

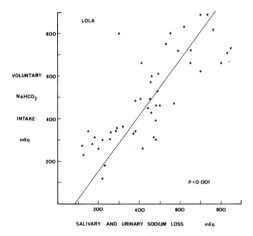

Fig. 3. The relation between body Na deficit and voluntary intake of 300 mEq/1 NaHCO₃ solution exhibited by a sheep, Lola, with a permanent unilateral parotid fistula. The sheep had access to a NaHCO₃ solution for 15 min following periods of 24-72 hr depletion of Na by fistula loss.

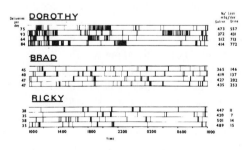

Fig. 4. Twenty-four records for 4 days of operant behaviour in 3 sheep (Dorothy, Ricky and Brad) with a permanent unilateral parotid fistula. Each vertical stroke represents delivery of 50 ml of 300 mEq/1 NaHCO₃ solution (15 mEq) to a drinking bin. Daily loss of Na in saliva and urine are shown also.

257

verticle stroke an operant act giving delivery. There are considerable differences in the behaviour of individual animals with a parotid fistula. Some, like Ricky (Fig. 4), appear to maintain balance by reacting to loss from the neutral point. They go deficient and correct themselves and always have little or no Na in the urine. Others, like Dorothy (Fig. 4), react as if to a set point far on the positive side of balance and maintain a large amount of Na in the urine. The record on Zeta (Fig. 5) shows the episodic nature of operant behaviour which may be seen in the face of essentially continuous salivary loss. With salivary loss at the rate of 10-25 mEq/hr, the behaviour suggests reaction to a body error of 40-100 mEq of Na. If the lever is taken out of the cage for periods of 6-72 hours with production of variable Na deficit, the rate of pressing over the two hours after the lever was replaced in the cage was commensurately related to sodium loss.

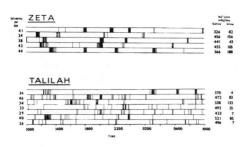

Fig. 5. Twenty-four operant behaviour records for two sheep Zeta and Talilah under the same experimental conditions as Fig. 4. (Figs. 4 & 5 reproduced by kind permission of Austr. J. Exp. Biol. and Med. Sci.).

This brief resume of results highlights the evident but variable hedonic component in sodium appetite superimposed on the basic drive activated and revealed by body deficit. Apart from bar-pressing itself, learned elements may also enter into the behaviour in such experimental circumstances (Abraham et al., 1973).

At this stage we would propose that there is a substantial neurophysiological organization subserving salt appetite. The activation, and capacity for specific satiation of the drive arising with deficiency is innate. Further, independently of deficiency, the appetite is activated as we have recently shown in wild rabbits (Covelli et al., 1973; Shulkes et al., 1972; Denton et al., 1971) by several steroid and peptide hormones involved in pregnancy and lactation.

This stimulation, despite normal body Na status, probably also reflects a naturally selected mechanism conferring survival advantage in marginal and deficient areas, in the face of the paramount needs of Na for the foetal tissues and for lactation.

Major questions for investigation in this field and some articles which have reviewed data and hypothesis are set out below.

1) What is the nature of the physiological process which generates an appetitive drive proportional to body deficit (Denton, 1966; Wolf, 1974; Richter, 1956; Denton, 1969; and Michell, 1975)?

2) What role, if any, does alteration in activity of specific salt taste receptors and fibres play in the genesis of appetite, and ingestion behaviour in salt deficient animals (Pfaffmann, 1963)?

3) What are the mechanisms whereby hormones of steroid and peptide types are able to generate appetitive behaviour (Fregly, 1967; Wolf and Stricker, 1967; Denton and Nelson, 1970)? Does the taste receptor mechanism change or is this effect seen only with deficiency of glucocorticoids in adrenal insufficiency (Richter, 1956; Henkin, 1967)?

4) How is an animal, a ruminant for example, able to rapidly satiate severe Na deficit accurately over 5 minutes of drinking and then stop with evident loss of motivation long before the material drunk can be absorbed from the gut and correct any putative chemical mechanisms generating appetite (Denton, 1966; Jalowiec and Stricker, 1973)?

5) The homeostatic systems of salt taste and appetite, and aldosterone secretion have emerged phylogenetically under the same selection pressure of environmental Na deficiency. To what extent are they functionally inter-linked and mutually influential in physiological organization – or are their operations essentially unrelated, though serving a common end (Denton, 1965, 1973)?

On the last question (5) the stimulating influence of aldosterone and DOC on salt appetite is well known though any physiological role is not clear. However, in Na deficient sheep with an adrenal autotransplant we have observed a remarkably large evanescent inhibition of aldosterone hypersecretion to follow 15-20 min after rapid satiation of salt appetite in a considerable number of animals. Aldosterone hypersecretion recovers by 100 min and then falls again as Na absorbed from the gut corrects deficiency. The phenomenon has particular interest as it still occurs if both angiotensin and KCl are infused at the time so that no change in these variables, or the other two factors known to influence aldosterone, occurs. The data is indicative of an unidentified factor in control of aldosterone secretion (Denton, 1973).

This evidence indicative of interrelation between salt appetite and, aldosterone control may be an instance of the more general phenomenon of gustatory metabolic reflexes – reviewed by Dr. Nicolaidis in this volume.

References

Abraham, S.F., Baker, R., Denton, D.A., Kraintz, F., Kraintz, L. and Purser, L. 1973. Austr. J. exp. Biol. med. Sci. 51, 65-81.

Andersson, B. and Olsson, K. 1973. Conditional Reflex 8, 147-159.

Baggerman, B. 1959. In Comparative Endocrinology, A. Gorbman (ed.), John Wiley & Sons Inc., New York, p.24.

Bazhenova, A.D. and Kolpakov, M.G. 1969. Proc. U.S.S.R. Acad. Sci. IV Physiology Conference of Central Asia and Kazakhstan, Novosibirsk, 9.

Blair-West, J.R., Coghlan, J.P., Denton, D.A., Nelson, J.F., Orchard, E., Scoggins, B.A., Wright, R.D., Myers, K. and Junqueira, C.L. 1968. Nature 21, 922-928.

Botkin, D.B., Jordan, P.A., Dominski, A.S., Lowendorf, H.S. and Hutchinson, G.E. 1973. Proc. Nat. Acad. Sci. USA 70, 2745-2748.

Clarke, J.D. 1970. The Prehistory of Africa. Praeger Publishers, New York.

Covelli, M.D., Denton, D.A., Nelson, J.F. and Shulkes, A.A. 1973. Endocrinology 93, 423-429.

Denton, D.A. 1967. Handbook of Physiology, 'Alimentary Canal Vol. 1', 433-459.

Denton, D.A. 1969. Nutrition Abstracts and Reviews 39, 1043-1049.

Denton, D.A. 1973. In M.P. Sambhi (ed.) Mechanisms of Hypertension, ICS 302, Proc. Int. Workshop Conf., 46-54, Excerpta Medica, Amsterdam.

Denton, D.A. 1966. Conditional Reflex 1, 144-170.

Denton, D.A. 1973. Conditional Reflex 8, 125-147.

Denton, D.A. 1957. Quart, J. Exp. Physiol. 42, 72-95.

Denton, D.A., Goding, J.R., Sabine, J.R. and Wright, R.D. 1961. In 'Salinity Problems in the Arid Zones', Proc. Teheran Symposium, UNESCO 193-198.

Denton, D.A. and Nelson, J.F. 1971. Endocrinology 88, 31-40.

Denton, D.A. and Nelson, J.F. 1970. Endocrinology 87, 970.

Denton, D.A., Nelson, J.F., Orchard, E. and Weller, S. 1969. In Olfaction and Taste III, C. Pfaffmann (Ed.) 535-547, Rockefeller University Press, New York.

Fregly, M.J. 1967. In Chemical Senses and Nutrition, M. Kare and O. Maller (Eds.) 115-138, The Johns Hopkins Press, Baltimore.

Henkin, R.I. 1967. In Chemical Senses and Nutrition, M. Kare and O. Maller (Eds.) 95-113, The Johns Hopkins Press, Baltimore.

Hoar, W. 1959. In Comparative Endocrinology, A. Gorbman (Ed.), John Wiley & Sons Inc., New York, p.1.

Jalowiec, J.E. and Stricker, E.M. 1973. J. Comp. Physiol. Psychol. 82, 66.

Konishi, J. and Zotterman, Y. 1963. In Olfaction and Taste I, Y. Zotterman (Ed.), 215-223, Pergamon Press, Oxford.

Leakey, L.S.B. 1959. Antiquity 33, 285.

Michell, A. 1975. In Olfaction and Taste V, D.A. Denton and J.P. Coghlan (Eds.), Academic Press, New York.

Nachman, M. and Pfaffmann, C. 1963. J. Comp. Physiol. Psychol. 56, 1007-1011.

Pfaffmann, C. 1963. In Olfaction and Taste I. Y. Zotterman (Ed.), 257-273, Pergamon Press, Oxford.

Richter, C.P. 1956. In L'instinct dans le Comportement des Animaux et de l'homme, Masson et Cie, Paris, 577.

Rodgers, W.L. 1967. J. Comp. Physiol. Psychol. 64, 49-58.

Scoggins, B.A., Blair-West, J.R., Coghlan, J.P., Denton, D.A., Myers, K., Nelson, J.F., Orchard, E. and Wright, R.D. 1970. Memoirs End. Soc. (U.K.), 'Hormones and the Environment', G. Benson and J. Phillips (Eds). 577-600, Cambridge University Press.

Shulkes, A.A., Covelli, M.D., Denton, D.A. and Nelson, J.F. 1972. Austr. J. exp. Biol. med. Sci. 50, 819-826.

Weir, J.S. 1972. Oikos 23, 1-13.

Wintour, E.M., Brown, E.H., Hardy, K., McDougall, J.G., Oddie, C.J. and Whipp, G.T. 1974. J. Steroid Biochem. 5, 366.

Wolf, G. 1969. In Olfaction and Taste III, C. Pfaffmann (Ed.), 548-553, Rockefeller University Press, New York.

Wolf, G., McGovern, J.F. and Dicara, L.V. 1974. Behavioural Biology 10, 27-42.

Wolf, G. and Stricker, E.M. 1967. J. Comp. Physiol. Psychol. 63, 252-257.

Salt Appetite, Its Physiological Basis and Significance

A.R. Michell

(Wellcome Fellow, Department of Medicine,

Royal Veterinary College, London, N.W.I.)

It is of more than academic interest to know how sodium intake is governed. Salt has been implicated in various clinical syndromes and although Western man probably takes too much, there is little idea of why, individually, he chooses to take any. Nevertheless "nutritionists are remarkably uninterested in salt"[8] and reviewers of regulation of body sodium frequently omit salt appetite entirely[19]. This reflects the difficulty in finding a convincing homeostatic rôle for a mechanism responding so sluggishly to sodium deprivation[15] when the intact renal-adrenal response (re-inforced in other tissues) is both swift and effective.

Salt appetite is the best example of a "specific appetite" and it clearly is enhanced by severe sodium depletion[9,29]. The line sometimes drawn between this and "need-free" salt ingestion is unfortunate while neither the physiological needs nor the cellular mechanisms underlying salt appetite are known. It is not inevitable, for example, that sodium deficits provoke a need for sodium in any biochemical mechanism, thanks to large reserves in bone. Links between deficits and intake in rats are fragile[28] and the crisis provoked by sodium depletion in ruminants is probably one of potassium metabolism[24] consistent with its dominant influence on their electrolyte physiology[32]. Indeed the problem of finding any aspect of extracellular sodium with a key role in governing salt appetite led to this rôle being ascribed to hypothetical sodium reservoirs[34]. Intuitively this is strange when only 10% of body sodium is permitted to remain inside cells and much chemical energy is devoted to excluding it. The possible relationship between sodium intake and potassium metabolism suggested that the actual process of sodium extrusion mediated by the enzyme Na-K ATP'ase offered a basis for salt appetite[22,23]. This "transport hypothesis" relates sodium appetite to a tangible, widespread and fundamental cellular system responsive to, and responsible for, changes in sodium and potassium distributions[13,17]. Interestingly, apart from its importance in excitable membranes[18], the enzyme has been implicated in the regulation of csf (cerebrospinal fluid) composition[6,16], salinity adaptation in fish[20], pigment control in prawns[12], production of aldosterone[5] and thirst[2].

The hypothesis is that sodium appetite is enhanced by accelerated sodium transport in some specific site(s) is compatible with some anomalous observations in the field[26] and possibly with the depression induced by anuria[14] (uraemia depresses Na-K ATP'ase activity[21,31]). More usefully, the hypothesis predicts that drugs affecting sodium transport should cause changes in sodium appetite inappropriate to their effect on sodium balance. Thus cardiac glycosides (inhibitors) should tend to cause renal salt loss[33] yet should depress salt appetite. Rises in extracellular potassium, intracellular sodium or mineralocorticoid activity should enhance sodium appetite by stimulating active transport of sodium and potassium[7,11] as should diphenylhydantoin (DPH)[35]. The concept that sodium appetite may reflect electrolyte transport rather than content has also been proposed by Ugolev and Roshchina[30].

Experimental Evidence

Digitoxin added to the drinking fluids (to give a daily dose of about 1mg/kg) significantly reduced the preference for 0.5% saline in Sprague Dawley rats (n=18, p=.001), not only during 16 days administration but 2-3 weeks beyond, consistent with its cumulative properties. Preference for 0.02% saccharin (n=11) was depressed only while fluids contained the drug; withdrawal restored normal preference. This suggested that the sustained effect on sodium preference did not involve "bait-shyness" and that there was probably no lasting influence on sweet preference. If, instead of digitoxin, the solutions contained 0.00015M quinine hydrochloride (Q HCl) sodium preference was replaced by aversion (n=11) but again withdrawal provoked recovery. Thus depression of fluid intake matching that caused by digitoxin (62%) did not cause persistent depression of sodium preference.

In order to avoid drugs in drinking fluids, digitoxin was given by stomach tube in a dilute alcohol base; control rats received base alone. Early results were encouraging, with controls unaffected, saccharin preference unaffected and persistent depression of sodium preference. By the end of a series of 12 rats (250g) given 1.5mg/kg daily, however, only trivial reductions in preference were seen, though they did persist. In 2 further experiments, each animal was its own control; preference was observed before intubation, during 8 days solvent intubation, 8 days drug intubation, and during the next 2 weeks. Some rats had digitoxin solution, but 2.5mg/kg; others had 3.7mg/kg suspended in 10% gum acacia. These doses seem high but rats are notoriously tolerant to cardiac glycosides[33]. The rationale for suspensions was to slow absorption as experiments with digitoxin in both fluids allowed drug intake throughout 24 hours. Taken overall, depression of sodium preference was statistically insignificant and did not persist.

All rats had been dosed early in their light period so it was decided to try giving the drug just before dark (solutions and suspensions, 1-2mg/kg). Again, saccharin preference was unaffected (n=10) add sodium preference (n=17) was undisturbed by intubation, but in this case digitoxin caused significant, persistent elevation of preference (control $\pm$ SEM = 66.5$\pm$5.3; intubation = 66.9$\pm$4.1; digitoxin = 78.9$\pm$2.2; subsequent week = 76.4$\pm$2.6). Urine output and total fluid intake greatly increased unlike the effect of cardiac glycosides intraperitoneally[2] or via the drinking fluids. The rise persisted whereas moderate reductions of food intake recovered rapidly once digitoxin was discontinued.

In view of the conflicting results by intubation, administration via the drinking fluids needed re-examination. This time, following control sodium preference, water alone (2 containers) was offered for 4 days, then digitoxin was added for 16 days. After withdrawal, water alone continued for 4 days. Sodium preference was then tested for 8 days; genuinely persistent depression should still be detectable. Neither with digitoxin nor Q HCl was it observed, in fact with both (n=6), preference was significantly higher than in the control period.

These experiments do not, therefore, yield adequate evidence to support the transport hypothesis. Denton & co-workers[10], however, showed that intracarotid ouabain diminished bar pressing in depleted sheep working for sodium. It also reduced voluntary drinking of sodium bicarbonate whereas water intake was affected much less. Ugolev & Roshchina[30] found that strophanthin-K enhanced preference for glucose-saline over glucose solutions in rats. Evidence with cardiac glycosides is thus indecisive and may remain so until receptor areas can be defined and drug actions accordingly restricted. Indeed one weakness of the hypothesis is that experimental interference with such a basic cell mechanism is likely to have numerous side effects. Thus similar problems arise with DPH. It does enhance sodium preference in rats (0.5% & 2% saline) despite a tendency to cause renal sodium retention[27]; whether it acts on the sodium pump, or through other effects[4] is open to question.

Since the sodium pump is stimulated by small rises in extracellular potassium, the observation that plasma potassium concentrations paralleled voluntary sodium intake in sheep[25] provoked interest. Nevertheless subsequent experiments failed to alter sodium preference during 2-hour periods when small increases (Ω 1mEq/L) were induced by infusions. On the other hand cellular potassium depletion, as predicted, does enhance salt appetite [1,3,36].

In conclusion, the sodium transport hypothesis of salt appetite, despite theoretical attractions, needs more evidence to sustain it. This may demand a prior definition of receptor areas, itself a field of conflicting data[34]. Two regions deserve special attention; firstly the "special permeability" areas of the brain where neurons may experience an environment more like extracellular fluid than the stability of csf. Secondly, tissues of branchial origin, which include areas responsive to taste, provide a plausible opportunity for evolution to adapt an enzyme system regulating sodium exchange with aquatic environments, into one governing sodium ingestion on land.

References

1. Adam, W.R. & Dawborn, J.K. (1972). J. comp. Physiol. Psychol. 78, 51.
2. Bergmann, F., Zerachia, A., Chaimovitz, M. & Gutman, Y. (1968). J. Pharm. Exp. Therap. 159, 222.
3. Blake, W.D. (1969). Ann. N.Y. Acad. Sci. 157, 567.
4. Bogoch, S. & Dreyfus, J. (1970). The Use of DPH. New York: Dreyfus Medical Foundation.
5. Boyd, J., Mulrow, P.J. Palmore, W.P. & Silva, P. (1973). Circuln Res. xxxii, 1-39.
6. Bradbury, M.W.B. & Stulcova, B. (1970). J. Physiol., 208, 415..
7. Charnock, J.S. & Opit, L.J. (1968). In "Biological Basis of Medicine" Ed. E.E. & N. Bittar. Vol. 1, p.69. New York: Academic Press.
8. Dahl, L.K. (1972). J. exp. Med., 136, 318.
9. Denton, D.A. (1967). In "Handbook of Physiology" Section 6, Vol. 1, p.433. Washington D.C.: American Physiological Society.
10. Denton, D.A., Kraintz, F.W. & Kraintz, L. (1970), Comm. Behav. Biol., 4, 183.
11. Dowben, R.M. (1969). Ed. Biological Membranes, 17, p.119. London. J. & A. Churchill.
12. Fingerman, M. (1969). Amer. Zool., 9, 443.
13. Flear, C.T.G. (1971). J. clin. Path., 23, suppl. 4, p.16.
14. Fitzsimons, J.T. & Stricker, E.M. (1971). Nature New Biol., 231, 58.
15. Fregly, M.J., Harper, J.M. & Radford, E.P. (1965), Am. J. Physiol., 209, 281.
16. Johanson, C.C., Reed, D.J. & Woodbury, D.M. (1974), J. Physiol., 241, 359.
17. Katz, A.I. & Epstein, F.H. (1967). Israel. J. Med. Sci., 3, 155.
18. Kerkut, G.A. & York, B. (1971). "The Electrogenic Sodium Pump". Bristol: Scientechnica.
19. Kramer, K., Boylan, J.W. & Keck, W. (1969). Nephron, 6, 379.
20. Maetz, J. (1971). Phil. Trans. Roy. Soc., 262B, 209.
21. May, E.S., Reinick, H.J., Mekhjian, H., Stein, J. & Ferni, T. (1973). J. clin. Invest., 52, 55a.
22. Michell, A.R. (1970). Proc. 78th Ann. Conv. Amer. Psychol. Ass., 211.
23. Michell, A.R. (1971). Proc. Int. Congress Physiol. Sci. (Munich) IX, 390.
24. Michell, A.R. (1972). Br. vet. J., 128, lxxvi.

25. Michell, A.R. (1972). J. Physiol., 225, 50P.
26. Michell, A.R. (1974). Br. vet. J., 130, vi.
27. Michell, A.R. (1974). J. Physiol., 237, 53P.
28. Morrison, G.R. & Young, J.C. (1972). Physiol. & Behav., 8, 29.
29. Richter, C.P. (1956). In M. Autori (Ed.) "L'instinct dans le comportement des animaux et de l'homme". p.577. Paris: Masson et Cie.
30. Ugolev, A.M. & Roshchina, G.M. (1965). Dokl. Akad. Nauk. SSSR, 165, 832 (Engl.).
31. Villamil, M.F., Rettori, V. & Kleeman, C.R. (1968). J. Lab. clin. Med., 72, 308.
32. Warner, A.C.I. & Stacy, B.D. (1972). Q. J, exp. Physiol., 57, 89.
33. Wilbrandt, W. & Lindgren, P. (1963). Eds. "New Aspects of Cardiac Glycosides", Vol. 3, p.47, p.233. London: MacMillan.
34. Wolf, G., McGovern, J.F. & Di Cara, L.V. (1974). Behav. Biol., 10, 27.
35. Woodbury, D.M. & Kemp, J.W. (1971). Psychiat. Neurol. Neurochir., 74, 91.
36. Zucker, M (1965). Am. J. Physiol., 208.

This presentation was made possible by generous financial support from Pedigree Petfoods Ltd. (U.K.)

A study of the role of olfaction and taste in the behavioural activity of ruminants towards the acquisition of sodium salts

F.R. Bell and Jennifer Sly

Department of Medicine, Royal Veterinary College,
London N.W.1, England

The interrelationship between the chemical senses of olfaction and taste with metabolism and nutrition has often been emphasised and culminated in the general thesis of self-regulatory processes especially in deficiency states (Richter, 1942). It has been suggested that in phylogeny, consequent upon evolution from sea to land forms, animals have a specific hunger for salt so that the extra-cellular fluid composition can be maintained through homeostatic mechanisms. This phylogenetic pressure it has been argued, falls hardest on ruminant herbivores causing these forms of artiodactyls to develop a specific salt appetite with central nervous control and through the endocrine system (Denton, 1969). It could well be, however, that the greatest continuing physiological pressure of ruminant herbivores is the maintenance of the whole vast substrate contained in the rumen in order to preserve the symbiont cellulase producing bacteria on which ruminant herbivores have become dependent for the extration of energy from cellulose since the Eocene Period, and which has allowed these animals to successfully occupy extraordinarily varied ecosystems from arctic to desert conditions.

A group of sheep in positive sodium balance have been examined using a two-choice preference test for sodium salt appetite (Bell, 1959). These normal animals show some variability between themselves but there is a marked preference for sodium bicarbonate over sodium chloride (Table 1). This result contrasts with that of Denton & Sabine (1961) where tests using short exposure to sapid solutions showed a preference for sodium chloride over bicarbonate and other salts. In our experiments sheep which prefer bicarbonate also prefer chloride but those with a marginal preference for bicarbonate show an aversion for chloride. Denton & Sabine (1961) showed that when sheep are Na^+ deficient, they have a clear preference for sodium bicarbonate over chloride. In the field sheep showing signs of Na^+ deficiency are more readily restored to health when they imbibe bicarbonate water rather than chloride water (Pierce, 1968).

Table 1. Sheep: taste preference for Na+ (% taken)

Test		03	07	34	58	111	Mean+S.E.
Water	(	48.3	41.5	53.4	49.2	52.1	48.9±2.1
Control	(	54.2	53.6	51.4	47.0	49.7	51.2±1.3
1%	(	76.8	76.6	65.5	67.6	77.4	72.8±2.5
NaHCO3	(	79.9	86.9	59.5	56.9	86.1	73.9±6.5
1%	(	62.4	56.8	48.1	33.8	61.0	52.4±5.3
NaCl	(	55.3	60.3	33.3	22.0	65.6	47.3±8.4

After ablation of olfactory bulbs

	03 Intact	07	34	111	Mean+S.E.
Water	52.8	50.7	53.1	56.4	53.4±1.6
1% NaHCO3	52.8	89.1	62.4	75.4	75.6±7.7
1% NaCl	70.6	84.1	55.0	53.3	64.1±10.0

Exteriorization of a parotid duct causes wastage of saliva outside the animal with the development of a pathological Na+ deficiency which the animal can compensate against if sufficient fluid is available (Bell, 1963). If sufficient water and Na+ salt is not available then the animal reduces its food intake and salivary secretion. In their natural habitat ruminants often show transient signs of Na+ deficiency and react by varying fluid and food intake to maintain mineral balances (Wilson & Hindley, 1968).

Table 2. Calf: 1% NaHCO3 preference with age

Period	Bicarbonate (% Intake)				Water	
	C3	C4	C6	C9	C11	C12
Milk fed	42.6	45.0	53.3	68.4	46.6	60.8
Weaning	49.9	30.9	33.3	66.2	63.9	44.7
Weaned	39.9	16.0	37.8	25.3	50.7	48.8
Weaned	24.9	7.4	25.0	9.2	—	—

Richter (1956) showed that young rats have an active appetite for isotonic NaCl and that mammals may have a universal liking for salt which if it is an inherited behaviour pattern may have a survival advantage. Mature calves unlike mature sheep, when in positive sodium balance show no preference for sodium salts. Very young calves have also been tested for preference to sodium salts while being milk fed, during weaning, and after being weaned (Table 2). During milk feeding and weaning no preference for sodium bicarbonate is seen and later the animals show a clear aversion to these salts. On the other hand when calves are weaned normally but without exposure to sodium salts they show a marked preference for 1% sodium bicarbonate when tested soon after weaning. It is possible therefore that ruminants may require greater supplies of sodium salts at this stage when the rumen is developing to its adult capacity. Since in the multiple choice situation ruminants appear to select by smell before drinking, taste preference was re-examined in sheep after ablation of the olfactory bulbs. These animals still showed almost exactly the same degree of preference for sodium bicarbonate over sodium chloride (Table 1). This suggests that although olfactory stimuli may lead the animal to its goal, final choice for imbibition is made by gustatory receptors. Sodium bicarbonate ingestion however may be hedonistic in sheep and totally dissociated from any Na^+ requirement.

There is anecdotal evidence that ruminants can detect sodium salts at considerable distances presumably by olfaction. To test this hypothesis normal and Na^+ deficient calves have been examined for olfactory and taste sensitivity in a suitable T-maze situation. The animals were run with 1% $NaHCO_3$ solution as the goal and reward. Normal calves produced low scores during 30 min. trial periods and did not drink bicarbonate (Table 3). In sodium deficient calves two out of three found the goal and licked the solution, the failed animal was shown the goal and allowed to lick. In the next eight trials all calves attained quickly the correct side but not always directly. In a further 10 tests all animals showed marked improvement in time and accuracy. When the maze side arms were closed by a low obstacle it was quickly crossed to achieve the goal.

These tests indicate that since sodium deficient animals show only slight improvement over normal calves at 7.5 m in locating $NaHCO_3$, olfactory stimuli are not effective at this distance. Positive behavioural responses at closer proximity (1.5m) suggests that olfactory stimuli are very effective at short range. The most striking result of the experiments, however, are the marked increase in motivation in the deficient animals which is readily enhanced by experience.

These preliminary tests suggest that the drive to ingest salt is not

Table 3. Calf: Response to NaHCO$_3$ goal in T-maze.

		Goal found			Mean time (sec)		
		C6	C7	C9	C6	C7	C9
1) Normal	/10	2	6	0	130	58	–
2) Sodium deficient							
a) Naive	/2	2	2	0	442	253	–
b) Experienced	/8	8	8	8	69	24	102
c) Repeat	/10	10	10	10	*73	11	44
d) Jump obstacle	/10	10	10	7	24	13	67

Note: Each calf sometimes made, and corrected, entries to the wrong side. When normal, no calf licked the bicarbonate. When sodium deficient each calf always licked the bicarbonate when found.
*Includes one long period.

a special situation for normal ruminants and the variation in the Na$^+$ content of the large volume of alimentary fluid can be met by normal physiological processes. This probably holds for animals nearing a Na$^+$ deficiency when both cellulose and water intake is reduced. Under these latter circumstances, animals may be activated through some Na$^+$ receptor in the alimentary tract or brain to move about more, so that when they come within a short distance of a sodium source, olfactory and gustatory receptors aid ingestion. Our experiments show that sodium deficient ruminants quickly learn the position of a sodium source which may explain subsequent return to the reward.

References

Bell, F.R., 1959. J. Agric. Sci. 52, 125.
Bell, F.R., 1963. 1st Internat. Symp. Olf. & Taste, 299.
Denton, D.A., 1969. Nutr. Abst. Rev. 39, 1043.
Denton, D.A. & Sabine, J.R., 1961. J. Physiol. 157, 97.
Pierce, A.W., 1968. Aust. J. agric. Res. 19, 589.
Richter, C.P., 1942. Harvey Lecture, 38, 63.
Richter, C.P., 1956. L'Instinct dans le Comportement des Animaux et de l'Homme, Masson et Cie, Paris. p.577.
Wilson, A.D. & Hindley., 1968. Aust. J. agric. Res. 19, 597.

This work was supported by the Agricultural Research Council.

Sensory Neuroendocrine Relationships in the Hydromineral Balance

Stylianos Nicolaidis

Laboratoire de Neurophysiologie Sensorielle
et Comportementale, College de France, Paris.

Although the determining signals for behavioural and metabolic events arise from imbalances of the internal milieu, we know that olfactory and oro-digestive stimulations from ingested material also play a role. Such a statement would sound obvious in the early ages of the physiology of regulation, since most regulatory responses were supposed to arise from peripheral signals. More recently, when internal imbalances have been recognized as the only signals generating corrective reactions, peripheral cues have been neglected. In this report, I shall review some of the recently investigated regulatory responses dealing with hydro-ionic balance and generated by exteroceptive orodigestive stimulations. These responses are anticipatory in the sense that they precede and are in the same direction as the changes which would be produced by the stimulating substances as a result of their post-asborptive systemic influence. Moreover, in many instances, responses of this kind seem to be preparatory towards post-absorptive effect of ingested material. Thus, they play a potentializing or optimizing role for a more appropriate corrective response. For example, Le Magnen (1) has shown that water intake was higher when a charge of salt was given orally than intragastrically. For better understanding, we distinguish between behavioural responses of intake and physiological responses of output, but this distinction may be inappropriate since the integrative mechanism of both seems to be identical.

1. EFFECT OF SENSORY SIGNALS ON THE CONTROL INTAKE OF WATER AND SALT. a) *Primary, secondary and extra-regulatory water intake*. Since Claude Bernard, the extent and the limits of the role of oral or/and gastric cues in the onset and the termination of drinking have been carefully investigated (2). But only very recently (Rowland & Nicolaidis 1971, 1974), it was shown (3, 4) that in rats the act of water intake could occur without any participation of oro-gastric cues. Under conditions of exclusive and permanent intravenous "drinking", regulatory responses were often deficient and the hydrational set point was switched towards very low levels. Comparison between intravenous and oral "drinking" throw a better light on the role of the latter. In orally drinking animals,

extra or intracellular dehydration seldom happens. Initiation of water intake is usually secondary (without actual systemic imbalances) but still regulatory, since such an actual repletion is provisional. So is satiety, where inhibition of thirst occurs before any systemic correction of dehydration. Oro-gastric cues play an amplifying role of exaggerating ingestive responses. Hedonic cues from oral receptors often lead to non-regulatory excessive drinking, or simply, to a permanent overhydrated state with chronic polyuria as compared to oliguric intravenously drinking animals. Oral drinking seems also to be a fundamental need superimposed on but independent of homeostatic signals. As Rowland & Nicolaidis (5) have recently shown in the rat, oral drinking is not extinguished when the animal receives a long term programmed (continuous, or discontinuous or both) intravenous infusion, in spite of the total or excessive (600%) supply of daily needs. In this need free situation, the residual oral drinking was usually equal to the minimum a rat requires for survival. A further reduction of the residual drinking was observed when infusions were given into the stomach (fistula technique). Total inhibition of oral drinking was observed when large gastric infusions were given through a naso-pharyngeal tube (6) so providing some thermal sensations at the level of the oral cavity. b) *Sensory control of salt appetite.* As Richter (7) has demonstrated, oral incentives are necessary for the development of regulatory consumption of NaCl. The salt deficient rats have not to learn to prefer (8) and to increase their salt consumption (9). Sodium deficiency markedly increases central responsiveness to unchanged peripheral signals mediating excitation from Na containing solutions (10). Although sodium appetite is proportional to deficits, the "stop" mechanism of intake is secondary (not related to actual imbalances of the internal milieu) and anticipatory. In Denton's experiment (11), the Na deficient sheep drank the appropriate amount of $NaHCO_3$ solution in 2-3 min. This early satiety was followed by an increase of plasma concentration only 15 to 30 min. later. Conversely, when Na deficiency was corrected by means of intragastric, intravenous, or intracarotid (12) infusions previous to presentation of $NaHCO_3$ solution, consumption was not or poorly affected. The improvement of regulatory responses when sensory signals precede the systemic ones is also demonstrated by experiments on adrenalectomized rats where NaCl solutions were offered after oral versus gastric preload of saline (13). Oro-pharyngeal cues alone are partially effective in satiating Na appetite. Sheep with oesophageal fistulas will drink $NaHCO_3$ solution in excess of deficit, but not as much as they drink water when they are thirsty (14). As a conclusion, in specific salt appetite, oro-digestive stimuli affect central specific arousal of consummatory behaviour in a similar way to that of drinking water. They generate anticipatory and optimizing neuro-endocrine responses and are also responsible for the development of secondary and occasionally non-regulatory intake of salt.

2. PHYSIOLOGICAL EFFECTS OF HYDRO-MINERAL ORO-DIGESTIVE STIMULATIONS. As in the case of control of intake, sensory stimuli generate synergic regulatory changes at the level of functions of losses. Two anticipatory reflexes of this kind have been shown (15, 16, 17) to affect diuresis and sweating in rats and humans. a) *Poto-diuretic reflex.* In rats, which were first stomach loaded with water to give a constant level of aqueous diuresis, stimulation of the mouth (the oesophagus was ligated) with 1-3 ml of water gave a large increase in diuresis starting within the first minute; 1ml H_2O intubated in the stomach had similar effect. Conversely, stimulation with 5% $NaCl$, at either oral or gastric level, gave an equally immediate although smaller, antidiuretic response. The short latency and the oesophageal ligation exclude the possibility of post-absorptive effect of the stimulating substances. I.V. injection of the same amount of water had no effect on diuresis while saline triggered an immediate diuretic anti-regulatory response. b) *Poto-hidrotic reflex.* When dehydrated humans drink water or saline in a hot room, they develop an almost immediate (2-7 seconds, latency) abundant sweating lasting several minutes. This response disappears after rehydration of the subjects.

3. NEUROENDOCRINE RELEASE. Can endocrine factors eliciting sodium appetite in need free animals (Aldosterone, DOCA, oestradiol, ACTH) be released in response to oro-digestive stimulation? Is such manipulation able to explain anticipatory behavioural and functional adaptations? As far as aldosterone secretion is concerned, Denton (18) observed in the sheep a biphasic fall of its circulating level only when $NaHCO_3$ was ingested orally (1st peak: 18-36mn). The plasma (Na^+) changes do not account for this endocrine response. When the solution was administered intragastrically, aldosterone fell uniformily and slowly, reflecting the absorption of Na. These changes parallel behavioural events such as satiation of Na appetite. Denton also found reflexly elicited Angiotensin II decrease halving in 20 min. without concomitant changes of Na and K content at systemic level. Exteroceptive reflexes of ocytocin release are well known. The fact that the poto-antidiuretic reflex persists after cordotomy favors the hypothesis of a vasopressin-mediated response.

4. NEUROPHYSIOLOGICAL MECHANISM. How is central activity (specific arousal) generated in response to internal changes and/or external announcement of their occurence? This question was answered in acute experiments of unit recording (19-20). Osmoresponsive units were first

localized in the vicinity of classic osmosensitive areas by means of intracarotid (IC) hypertonic or water injections. These specific units (unresponsive to general arousal) were then tested with lingual water and saline stimulations. Summarizing the essential findings, IC injection of 1M NaCl produced a rapid increase in activity in some units of the supraoptic nucleus. The same population was also activated by lingual stimulation with 1M NaCl. In typical cases H_2O IC or lingual, stimuli generate opposite decreased activity. These convergent effects on IC and gustatory stimulation of H_2O and NaCl were in the opposite direction in most responsive units distributed as a mosaic in the MFB and the pallido-hypothalamic tract. Similar convergent responses exist at the pre-commissural level but they tend to be less discriminatory for water and NaCl. The essential findings of this research were widely confirmed and extended to investigation of pathways involved (2). In similar regions of unanesthetized monkeys (22), there were specific changes of unit activity during the pre-consummatory (thirsty animal) as well as post-consummatory period (satiated animal) of active drinking. These findings are more consistent with the anticipatory poto-hidrotic or diuretic phenomenon than with motivational peri-consummatory states.

Convergence on the same integrative population from both interoceptive and exteroceptive signals allows the systemically determined activity to be alleviated or aggravated by peripheral cues. The algebric summation of such messages causes the specific arousal canalized towards neuroendocrine release of a) transmitters generating specific motivations and b) endocrine factors controlling peripheral organs. So announcement of future repletion not only allows anticipatory responses but, in addition, brings about a preparatory activation of secretory responsiveness and, as a consequence, potentialization and optimization of the post-absorptive effect of ingested metabolites.

References

1. Le Magnen, J. J. Physiol (Paris), 1955, 47, 404-418.
2. Epstein, A.N. − Handbook of Physiol., sect. 6, 1967, 1, 197-218.
3. Rowland, N. & Nicolaidis, S. − C.R. Acad. Sci., 1972, 275, 991-994.
4. Rowland, N. & Nicolaidis, S. − Journal Compar. Physiol. Psychol., 1974, 87, A, 1-15. − 1974, 87, B, 16-25.
5. Rowland, N. & Nicolaidis, S. − Lugano, (in press).
6. Epstein, A.N. − Science, 1960, 131, 497-498.
7. Richter, C.P. & Grasse, M. − L. Instinst., 1966, Paris, Masson, 577-629.
8. Epstein, A.N. & Stellar, E. − J. Compar. Physiol. Psychol., 1955, 48, 167-172.
9. Nachman, M. & Valentino, D.A. − J. Compar.Physiol.Psychol. 1968,62,2352-2355.
10. Pfaffmann, C. & Bare, J.K. − J. Compar. Physiol. Psychol., 1950, 43, 320-324.
11. Denton, D.A. − Handbook of Physiol., Amer. Physiol. Soc., Washington, 1967, 1, 453-459.
12. Beilharz S., Bott, E. Denton, D.A. & Sabine, J.R. − J. Physiol., (Lond.), 1965, 178, 80-91.

13. Nachman, M. & Valentino, D.A. – J. comp. Physiol. Psychol, 1968, 62, 2352-2355.
14. Bott, E.A., Denton, D.A. & Weller, S. – Hayashi Olfaction & Taste 2, London Pergamon Press, 1967, 415-429.
15. Nicolaidis, S. – J. Physiol., Paris, 1963, 55, 309-310.
16. Nicolaidis, S. – C.R. Acad. Sci., (Paris), 1964, 259, 4370-4372.
17. Nicolaidis, S. – N.Y. Acad. Sci., 1969, 157, (2), 1176-1203.
18. Denton, D.A. – Cond. Refl., 1973, 8, 125-146.
19. Nicolaidis, S. – N.Y. Acad. Sci., 1969, 157, 1176-1203.
20. Nicolaidis, S. – C.R. Acad. Sci., Paris, 1968, 267, 2352-2355.
21. Emmers, R. – Brain Res., 1973, 49, 323-348.
22. Vincent, J.D., Arnaud, E., Bioulac, B. – Brain Res., 1972, 44, 371-384.

Phylogenetic Emergence of Salt Appetite

Discussion and Chairman's Comments

Carl Pfaffmann, Chairman, The Rockefeller University,

New York, New York 10021, U.S.A.

Dr. Beidler: How might the concept of survival value apply to the differences in taste sensitivity observed in the rodent where sodium sensitivity is high and potassium relatively low, as compared to the carnivores where there is a high potassium sensitivity coupled with a low sodium sensitivity?

Dr. Denton: Having a good sodium sensitivity appears consistent with the idea that any organism relying upon an herbivorous or vegetarian diet might be subjected to more selection pressures relative to changing sodium content of the foodstuff with seasonal or other ecological factors.

Dr. Beidler: Would you comment on Richter's original idea that changes in sodium need would be reflected by changes in sensitivity of the taste bud but which has not been found by investigators?

Dr. Denton: When hypertonic salts are given to a repleted sheep, it can detect them perfectly well and may even refuse them, but when made deficient it accepts them and takes them avidly so that their acceptability is altered. One may presume that this change is a central event rather than some change of receptor.

Dr. LeMagnen: What is the effect in your preparation on water balance?

Dr. Denton: When animals are rendered depleted by a parotid fistula, water intake usually increases since animals lose fluid from the fistula, but the intake does not make up for the water loss, so, in fact, they go into a negative water balance during the first 48 hours. When repleted, water intake increases.

Dr. Zotterman: Besides taste, I think you should look at the afferent fibers from the rumen which might be sensitive to pH and other changes. These would provide important inputs in relation to appetite.

Dr. Denton: We have looked at this to some extent by giving the depleted animal sodium solutions by rumen tube five minutes before they were given the drinking test. Presumably the intubated solution would affect nerve endings in the rumen, but we have not recorded from the vagus nerve. However, when given salt solution to drink, they take the usual amount of solution after rumen intubation.

Dr. Nicolaides: Let me report that Dr. Sharma has recorded from vagal and sympathetic fibers in animals in normal and deprivated states. He found a tremendous influence of the change in composition of the body

277

on the afferent responsiveness of these nerves.

Dr. Coghlan: Dr. Michell's hypothesis is attractive but I find some difficulties with it. "Bone" sodium is only a reservoir if it is physiologically available rapidly. On the other hand, in deficiency it is an advantage to the animal to conserve potassium rather than to get rid of it. By retaining potassium it can use the 500 to 1,000 milli-equivalents (mEq) of sodium it has in the rumen by switching from predominantly sodium rumen to a predominantly potassium rumen.

Dr. Michell: With regard to "bone" sodium: in a 70 kg. man, this constitutes 2,600 mEq. Man reduces sodium loss in urine to zero when he loses only 200-300 mEq. In man half the bone sodium is rapidly available. In ruminants a much higher proportion of the bone sodium is rapidly available.

Dr. Coghlan: One of my students has data on that.

Mr. McDougall: In sheep 24 hours after the injection of Na^{24}, 20 to 30% of the bone Na was exchangeable Total bone sodium in sheep is not exchangeable, not metabolically available. If you acutely deplete a sheep of up to 600 mEq in three days, the bone sodium concentration decreases by 15 mEq/kg of bone; therefore, mobilizing only 70 mEq over a 3-day period when they are deficient by 600 mEq.

Dr. Michell: I accept that, but the occurrence of that loss, 600 mEq in three days is hard to envisage under ordinary conditions, particularly when the adrenal gland is intact, because animals reduce their sodium losses almost to zero after they have lost 200-300 mEq. The sheep is a species that evolved on a relatively small sodium intake, 30 mEq per day, and it copes with that by putting out very little in the urine. Coming back to the potassium in saliva, this is an aldosterone response, and its effects on potassium are at least as important as on sodium metabolism. Under sustained dosages of aldosterone, the kidney escapes from the sodium conserving effect, but it never escapes from the potassium excreting effect. The salivary effect of aldosterone is not just retrieving sodium, but it is dumping the toxic ion potassium into the rumen. In spite of this, the plasma potassium still rises in the depleted animal.

Dr. Coghlan: The plasma potassium does rise a little in sodium depletion but does not reach toxic levels.

Dr. Michell: May I make an additional comment. Sodium appetite varies during the estrous cycle of the sheep. It is high in the luteal phase and falls at the estrous stage for three or four days and then recovers. Sodium balance in the sheep shows a cumulative negative sodium balance in estrus. Sodium appetite falls but starts to recover three to four days after estrus, thus during a negative sodium balance there is a falling appetite. On the other hand, during this period just preceding and through estrus, there is a clear urinary potassium retention and three or four days after estrus with a rise in plasma potassium when there is no change in plasma sodium. I think this implicates potassium in these control mechanisms.

Chairman (Dr. Pfaffmann): Unfortunately time is running short, and this matter can be continued by the participants more directly in personal discussion. I'd like to ask Dr. Denton if he can account for the individual differences among his animals. Some work to hold themselves on the positive side of sodium balance with urinary excretion; others work to hold themselves just at balance with little sodium excretion.

Dr. Denton: I have no particular explanation. It appears to be the kind of variability often found in many biological situations. In relation to the preceding discussion, as I understand the proposition since the adrenal gland is so very good at conserving sodium, and we certainly agree with that, there is thought to be no need for sodium appetite. I believe there are two systems, aldosterone and sodium appetite. When the cow defecates it loses a fair amount of sodium. If it gets diarrhea or other disturbed gastric condition, it loses sodium in larger quantity. If it is living on a sodium deficient grass, it can become deficient and remain so for months unless it can find another salt source, such as a salt lick. Thus two systems have evolved to handle all situations where salt can be lost from the body.

Dr. Schneider: I'd like to call attention to the fact that salt balance and salt appetite occurs widely and is not restricted to herbivorous mammals. The group at Cornell University reports that the yellow swallow tailed butterflies actively seek out and search for refined sodium sources, in addition to feeding on the sources usually available to them. I wonder about the nectar feeding birds, how they supply their need for salt. We should look at this problem more widely in a comparative way throughout the whole animal kingdom.

Chairman: I'm glad to have this specific example of salt appetite in insects, as well as to have the general point of its widespread occurrence throughout the animal kingdom.

Chairman's commentary prepared after the meeting.

The papers and discussion in this session made it abundantly clear that the salt appetite mediated by way of the sense of taste is an important and widespread mechanism throughout the animal kingdom. There is still much to be learned about the detailed physiological interactions, as the discussion with Dr. Michell shows, but equally important and intriguing is the question of specifically how the nervous system "turns on" the appropriate drive.

Evidence not presented here indicates that salt appetite is an unlearned response, specific for sodium, not a generalized trial and error search. But this is not to say the behavior of salt appetite cannot be influenced by learning or experience so that the differences among experienced animals reported by Dr. Denton and their response to sodium deficit, some at the positive side, some just at balance, might conceivably result from differences in experience. Correcting salt deficiency would be associated with beneficial metabolic effects and thus associated hedonic consequence such that the animal would tend to do more than to make up for need when it encounters salt. Salt appetite thus may have both an obligatory homeostatic regulatory component and a reinforced or learned (facilitated) aspect which is not obligatory but which may lead to balance on the positive side. Perhaps intake in excess of need (overshoot so to speak) reflects this learned component.

GENERAL OLFACTION

Chairman: W. Whitten

Four Primary Odor Modalities of Man:
Experimental Evidence and Possible Significance

John E. Amoore

Western Regional Research Laboratory, U.S. Department of

Agriculture, Berkeley, California 94710, U.S.A.

The search continues for a truly fundamental classification of odors. Most methods of direct electrophysiological or biochemical experimentation are not suitable for application to the human species. Hence we must resort to indirect methods and particularly exploit certain "experiments of Nature" that are relatively common but quite harmless. I refer to the phenomenon of specific anosmia, the "anosmies partielles" that Guillot (1) suggested in 1948 may correspond with the "odeurs fondamentales" of the human sense of smell.

The first specific anosmia to be analysed in this way was the failure to smell isobutyric acid and related medium chain fatty acids (2). About 2% of people have this defect, which is most pronounced toward isovaleric acid (Fig. 1). I therefore suggested that these persons are defective in their perception of one of the primary odors, "sweaty."

Recently we have completed the analysis of a second primary odor by an exactly analogous procedure. About 20% of people fail to smell the heterocyclic Schiff's base 1-pyrroline (3). This defect is remarkably selective, and was found only with its homologue 1-piperideine, among 32 other purified nitrogenous bases, which included some very closely related structurally to the type compound (Fig. 2). The odor of 1-pyrroline almost exactly resembles that of human semen, so the adjective "spermous" is suggested for this new primary odor.

$$CH_2 \text{———} CH_2$$
$$CH_2 \qquad CH$$
$$N$$

Another interesting anosmia is demonstrated by trimethylamine which has an extraordinarily low threshold, 3 parts in 10^{10} of water, as compared with related bases. Yet about 5% of people fail to smell it at 1000 times its normal threshold concentration (3). I suspect that this tertiary amine represents another primary odor, which could be called "fishy". We have not yet begun the detailed chemical mapping of this anosmia.

Also relevant in this context is the well-known anosmia to various "musky" compounds such as macrocyclic lactones and steroids (1). About

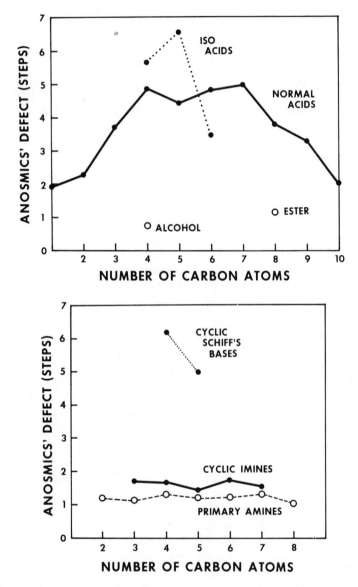

Fig. 1. (above). Specific anosmia to isovaleric acid and homologous carboxylic acids.
Fig. 2. (below). Specific anosmia to 1-pyrroline and other nitrogenous bases.

TABLE 1. *Additional reputed specific anosmias.*

Compound	References	Table II (5) line no.
Acetone	6, 7	3
Δ^{16}-Androsten-3-one	8, 9	44
4-(4'-*cis-tert.* Butylcyclohexyl)-4-methylpentan-2-one	8	44
l-Carvone	10	8
Diacetyl	W	35
2,4-Dimethyl-5-ethylthiazole	W	38
Formaldehyde	11, 12	51
Isoamyl alcohol	W	31
Lyral	13	15
Methyl ethyl ketone	7	3
N-Methylpyrrolidine	3	21
Pyridine	14	23
l-Pyrroline	3	29
Trimethylamine	3	21

W= Unpublished observation at Western Regional Research Laboratory.

7% of people cannot smell the synthetic musk pentadecalactone. Chemical mapping of this anosmia is still awaited. In collaboration with Dr. D. Whissell-Buechy it has been shown (4) that pentadecalactone anosmia is inheritable as a simple Mendelian recessive character (not sex-linked). This result lends considerable support to our assumption that odor-blindness is an inborn molecular-biological defect, probably in the odorant receptor protein.

I am still interested in hearing of any new chemical examples of human specific anosmia. Sixty-two compounds were tabulated in my paper at ISOT III in 1968. Meanwhile 14 additional compounds have been accumulated (Table 1). Tentative assignments are made to the appropriate line of Table II in the earlier compilation (5). This brings the total number of reported odor-blindness compounds to 76. Some of the new compounds seem to demand placement on previously unoccupied lines of the Table, bringing the likely number of primary odors up to 31 (plus formaldehyde which may represent a trigeminal sensation). The final number of human primary odors remains an open question that can only be settled by a comprehensive survey and analysis of all known forms of specific anosmia, together with confirmatory genetic and psychophysical studies. At present only line 25 (sweaty) and line 29 (spermous) represent reasonably confident assignments of primary odor status, epitomized by isovaleric acid and 1-pyrroline.

Each primary odor represents a distinct modality of the sense of smell. The corresponding receptor mechanisms are both selective and sensitive toward certain volatile compounds. If we assume that all structures and functions of the body have now, or once did have, some survival value, we are led to speculate just what may be the significance of the particular olfactory specificities that mankind possesses.

Isovaleric acid and trimethylamine are widely distributed in natural products. They are present in many foodstuffs, especially as a result of bacterial degradation, whereupon they are released from precursors such as leucine or choline. These facts suggest that the sweaty and fishy primary odors may be helpful in recognizing foods, and particularly in warning against consumption of spoiled foodstuffs which might contain bacterial toxins. 1-Pyrroline itself has rarely been identified in natural products, but one often encounters its odor, which is usually ascribed erroneously to its oxidation-prone precursors putrescine or pyrrolidine (3). It probably contributes to the odor of spoiled meat. Hence the spermous primary odor may also have some value as a warning against poisons. The pure musky primary odor is seldom found in natural foods, so we will have to look elsewhere for its significance.

It is an intriguing fact that all these four primary odors are actually produced by the human body. (This was a weighty factor in my selecting the corresponding specific anosmias for early examination.) The odors are generated in localized regions of the body, and/or in special circumstances, which in any other species would be unhesitatingly ascribed to pheromone systems (15). The data and their tentative interpretation are summarized in Table 2.

An odor resembling isovaleric acid is detectable in stale underarm sweat. The apocrine sweat glands release their secretion profusely in situations causing alarm or other emotional stress. Hence this could represent an alarm pheromone. Curiously enough, isovaleric acid itself, together with isobutyric and isocaproic acids, was identified by Michael in the vaginal pheromone (copulin) produced by the rhesus monkey at the time of oestrus (16). Analytical evidence implies that the same sex pheromone is produced by the human female (17).

The odor of 1-pyrroline is notable in human semen, and also can be detected in the sweat of the male pubic area (3). The semen contains a

TABLE 2. *Possible pheromonal significance of human primary odors.*

Probable primary odor	Established primary odorant	Presence in human secretions	Possible pheromonal significance
Sweaty	Isovaleric acid	axillary sweat[†] vaginal secretion	alarm pheromone ovulation indicator
Spermous	1-Pyrroline	male pubic sweat[†] semen[†]	aphrodisiac mating marker
Fishy	Trimethylamine[*]	menstrual sweat[†] menstrual blood	oestrus indicator menstruation indicator
Musky	Δ^{16}-Androsten-3-one[*]	smegma[†] male urine	response substance territorial marker

[*] Provisional assignment, pending complete anosmic mapping.

[†] Suspected from olfactory observation; not yet confirmed by chemical analysis.

high concentration of spermine, together with spermidine, putrescine, and diamine oxidase (18) which should be expected to generate the typical odor, even though 1-pyrroline itself has not yet been identified. I suspect that 1-pyrroline may once have been a male sex pheromone for our species.

Trimethylamine was isolated from human menstrual blood by Klaus, and is probably responsible for the characteristic herring-brine odor of the discharge and of stale menstrual sweat (19). The same author pointed out nearly half a century ago that the odor of a menstruating woman brings many animals into a state of sexual excitation. Hence trimethylamine may be a common oestrus pheromone for several mammalian species, but its value may be obsolete for the human female, where ovulation (fertility) occurs some 10 days after menstruation ends.

The musky primary odor is represented by macrocyclic ketones and lactones, and by steroids of the androstane series. They have the properties of sex pheromones and/or territorial markers for many mammalian species, and may have some residual significance in these respects for mankind (15).

Δ^{16}-Androsten-3 one appears to be responsible for the musky odor of stale human urine (20). This compound is included in Table 2 as a typical example but several other musky smelling compounds occur in these secretions so their relative importance is uncertain.

It would appear to be more than a coincidence that *Homo sapiens* would produce these four odorants in circumstances related to danger, reproduction and sovereignty and, furthermore, that he would have chemically specific and very sensitive receptor sites for these same odorants in his nose. Here is considerable presumptive evidence for at least four, albeit vestigial, pheromone systems in man. Although some of the chemical confirmation and behavioral evidence is lacking, the demonstration of these four primary odor receptor systems in the human sense of smell lends strong support to Comfort's assertion that human pheromones are a more than vestigial reality, thinly veiled by the taboos of civilization and the promotion of deodorants (15).

References

1. Guillot, M. 1948. C.R.H. Acad. Sci. 226, 1307-1309.
2. Amoore, J.E. 1967. Nature, London, 214, 1095-1098.
3. Amoore, J.E., Forrester, L.J. and Buttery, R.G. J. Chem. Ecol. (submitted).
4. Whissell-Buechy, D. and Amoore, J.E. 1973. Nature, London, 242, 271-273.
5. Amoore, J.E. 1969. Olfaction and Taste III, edit. C. Pfaffmann, Rockefeller University Press, New York, 158-171.
6. Blondheim, S.H. and Reznik, L. 1971. Experientia, 27, 1282-1283.
7. Forral, G., Szabados, T., Papp, E.S. and Bankovi, G. 1970. Humangenetik, 8, 348-353.
8. Beets, M.G.J. and Theimer, E.T. 1970. Taste and Smell in Vertebrates, edit. G.E.W. Wolstenholme and J. Knight, Churchill, London, 313-323.
9. Griffiths, N.M. and Patterson, R.L.S. 1970. J. Sci. Food Agric. 21, 4-6.
10. Friedman, L. and Miller, J.G. 1971. Sciences 172, 1044-1046.
11. Douek, E.E. 1967. J. Laryng. 81, 431-439.
12. Pierce, T.W. 1968. Letter of September 26.
13. Theimer, E.T. 1969. orally, August 7.
14. Diamente, D. 1971. letter of April 7.
15. Comfort, A. 1971. Nature, London, 230, 432-433.
16. Michael, R.P., Keverne, E.B. and Bonsall, R.W. 1971. Science, 172, 964-966.
17. Michael, R.P. 1972. Acta Endocrinol. 71, Suppl. No. 166, 322-363.
18. Williams-Ashman, H.G. and Lockwood, D.H. 1970. Ann. N.Y. Acad. Sci. 171, 882-894.
19. Klaus, K. 1927. Biochem. Zeit. 185, 3-10.
20. Prelog, V., Ruzicka, L. and Wieland, P. 1944. Helv. Chim. Acta, 27, 66-71.

Olfactory Adaptation in the Rat

D.G. Laing* and A. Mackay-Sim**

*CSIRO Division of Food Research, Sydney, Australia.

**Macquarie University, Sydney, Australia.

Elucidation of the mechanisms underlying olfactory self and cross-adaptation effects in vertebrates, has been hindered because studies have largely been restricted to psychophysical measures with humans. This has primarily been due to the absence of suitable quantitative behavioural techniques for use with other vertebrates, and the limitations imposed in electrophysiological studies where only responses from specific regions of the olfactory system have been recorded (1,2). There is therefore a need for studies with other vertebrates where both electrophysiological and psychophysical data can be obtained and correlated.

Recently, a technique was described (3) for delivering odorants at specific concentrations and precise periods into the nasal cavity of a freely moving rat. This technique was therefore used in the present behavioural study, where the aim was to investigate olfactory self and cross-adaptation in the rat.

Experimental

The subjects were 5 male Sprague Dawley rats aged 6-12 months during testing. A stainless steel cannula was chronically implanted (3) into the nasal cavity of each animal at the age of 4 months. The odorants used were propanol (B.D.H.) and heptanol (Fluka A.G.). Each compound was purified as described previously (3) and in each case the final purity was in excess of 99%. The odorants were delivered in an air stream (40 ml/min) into the nasal cavity by a modified air dilution olfactometer (3). This instrument was connected to the cannula by a Teflon tube.

The animals were trained and tested in a Perspex chamber (37 x 23 x 33 cm) with a grid floor. This chamber was fitted with 2 response levers on either side of a drinking cup mounted on its front wall, and with a single lever mounted on the rear wall. Animals were deprived of water for 22 hr each day during training and testing. They were trained to press the lever at the rear of the chamber to initiate the delivery of the adapting and test odorant, or an air blank. Then, at the sound of a low frequency buzzer given 12s later, to press the left front lever if the test odor was detected, or to press the right front lever if the test odor could not be detected. If a front lever was not pressed during the 10s period after the initial sounding of this buzzer, the trial was repeated. When the correct lever was pressed a reward of water was given. Selection of the incorrect bar resulted in no reward and was indicated by the sounding of a high frequency buzzer (0.5s). An adaptation test trial consisted of the presentation of the adapting odor or air for 3s, an inter-stimulus interval of 3s, followed by a 3s delivery of the test odor or air (blank). Thus the sounding of the low frequency buzzer coincided with the presence of the test odor or blank, since they took 2s to reach the nasal cavity from the olfactometer. These test conditions were similar to those used with humans (4). There were 4 types of test trials: (i) where the adapting odor preceded the test odor e.g. heptanol (H) − propanol (P), i.e. (HP), or propanol-propanol (PP); (ii) adapting odor-blank (A), i.e. (HA) or (PA); (iii) blank-test odor (AP); and (iv) blank-blank (AA). Equal numbers (15 or 20) of each type were randomly distributed throughout each session according to modified Gellerman sequences (5). At the beginning of each session, an example of each trial type was given to familiarise the rat with the test conditions.

Before the start of the adaptation experiments, the threshold of animals for one or both odorants was determined. The test procedure was similar to that described above except that only type (iii) and (iv) trials were given. Initial training was to propanol or heptanol. Animals were considered trained when their performance exceeded 85% correct responses (R+) to the odorant and air for 3 consecutive sessions. Once this learning criterion was attained, threshold testing trials were run at successively lower concentrations each day until the rat obtained an R+ score of 50% to the odorant at a given concentration, at 2 sessions. Following this, either the threshold for the second odorant was determined, or the concentration of the odorant was adjusted daily until the R+ level required for a particular adaptation experiment was established. Only propanol was used in self-adaptation experiments, whilst in cross-adaptation measures heptanol was the adapting odorant, and propanol the test odorant.

Results

As indicated above, the basic measure was the first lever press, whether correct or incorrect, following an interval of 12s after the rat had initiated a trial. All animals attained the learning criterion in 417±43 trials.

The threshold concentrations of propanol and heptanol determined for the rats are given in Table 1. The values for heptanol are similar to those reported previously, whilst those for propanol are generally lower (6,7).

In the adaptation experiments a wide range of test conditions were used, since it was difficult, for behavioural reasons, to accurately adjust

Table 1.

Rat No.	Propanol (mg/1)	Heptanol (mg/1)
10	1.3×10^{-1}	6.0×10^{-4}
6	4.5×10^{-2}	6.5×10^{-4}
736	6.6×10^{-5}	2.5×10^{-6}
734	7.0×10^{-4}	
732	7.0×10^{-4}	

Summary of threshold measurements.

the level of R+ of animals to a preconceived value. Adaptation results were analysed using a two-tailed t test. This statistic was used to determine if a significant difference existed between the R+ recorded in trial types PP v AP, AA v PA, HP v AP, and HA v AA. However in no instance was there a significant difference in R+ to AA v PA and AA v HA.

Figure 1a shows the results of the self-adaptation experiments. The major effects observed were that (i) high adapting stimulus concentrations i.e.>5x threshold reduced the detectability of the test stimulus (rat 732). This effect appears to operate even with high test stimulus concentrations (rat 736); (ii) low adapting stimulus concentrations facilitated the detection of threshold concentrations of test stimuli (rat 10 and 734), reduced the detectability of suprathreshold test stimuli (rat 10 and 734), or had no effect (rat 736).

Figure 1b shows the results of cross-adaptation experiments. The

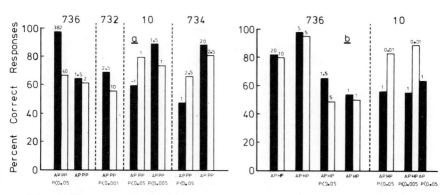

Fig. 1: Percent correct responses of rats in (a) self-adaptation and (b) cross-adaptation experiments. The numbers immediately above each bar pair indicate the test (left) and adapting (right) concentrations relative to the threshold value. Significant differences between AP v PP and AP v HP are shown.

results indicate that high concentrations of heptanol either reduce the detectability of propanol (rat 736), or have little effect. In contrast subthreshold adapting concentrations of heptanol facilitated the detection of threshold concentrations of propanol (rat 10). This effect was also observed using a second test method where the results of 3 experiments were compared. Trial types AA and AP were used in the first and third experiments, whilst only HP and HA were used in the second. Significant differences existed between the R+ for AP v HP (Fig. 1b), but not between AA v HA nor between the values of AP in Experiments 1 and 3.

Discussion

The results clearly show that behavioural methods can be used to demonstrate olfactory self and cross-adaptation in the rat. They indicate that sensitivity can be enhanced or reduced by either adaptation method. In addition, the effects at low concentrations do not appear to be dependent on any one physicochemical property of the stimulus. The lipid-water solubility of odorants for example, has been suggested (4) as a major factor in determining the adaptation effect in humans. However, stimulus-oriented hypotheses do not account for facilitation by self-adaptation or subthreshold effects. It is therefore suggested that adaptation effects at low adapting and test concentrations are determined by neural factors rather than stimulus properties. Indeed the present low concentration effects may provide evidence for the centrally-controlled facilitatory feedback system proposed recently (8) to operate in the rabbit olfactory bulb.

Some of the effects observed in the present work have not been reported in studies with humans. Self-adaptation for example, has been shown to produce masking (4), but never facilitation. In addition, the enhancing effect of subthreshold quantities of heptanol on the detection of propanol, has not been observed, although threshold adapting concentrations of hexanol can enhance detection of threshold concentrations of propanol or butanol (11). However, there is general agreement between the present cross-adaptation results and those from humans. High adapting concentrations of heptanol have been reported to mask butanol (9) and hexanol to mask propanol (10), whilst facilitation was the most common effect with low concentrations of each odorant (4).

Finally, it is suggested that facilitatory mechanisms may play an important role in chemical communication. The reception of low quantities of a pheromone or foreign odor for example, may not result initially in conscious recognition of the substance, but may prime or activate specific neurones or neural circuits (8) thereby enhancing the animal's sensitivity for the substance.

Acknowledgements

Thanks are extended to Dr K.D. Cairncross who provided part of the facilities. A. Mackay-Sim was supported by a Commonwealth Scholarship.

References

1. Ottoson, D., 1963. Olfaction and Taste I, edit. Y. Zotterman, Pergamon Press, London, 35-44.
2. Moulton, D.G., 1963. Olfaction and Taste I, edit. Y. Zotterman, Pergamon Press, London, 71-84.
3. Laing, D.G., Murray, K.E., King, M.G. and Cairncross, K.D. 1974. Chem. Sens. Flav., 1 (2), 197-212.
4. Corbit, T.E. and Engen, T. 1971. Perc. and Psychophys., 10, 433-6.
5. Gellerman, L.W. 1933. J. Genetic Psychol., 42, 207-8.
6. Laing, D.G., Chem. Sens. Flav., (submitted).
7. Moulton, D.G. and Eayrs, J.T. 1960. Quart. J. Expt. Psychol., 12, 99-109.
8. Nicoll, R.A. 1971. Science, 171, 824-6.
9. Engen, T., 1963. Amer. J. Psychol., 76, 96-102.
10. Rovee, C.K., 1972. J. Exp. Child Psychol., 13(2), 368-381.
11. Engen, T. and Bosack, T.N. 1969. J. Comp. Physiol. Psychol., 68, 320-6.

Multiple function of the Anterior Olfactory Nucleus (A.O.N.): lateral discrimination and centrifugal control.

G. Daval and J. Leveteau

Laboratoire de Neurophysiologie comparee.

Universite Paris VI France

In previous experiments (1) we have reported on the existence of interbulbar reciprocal lateral inhibition, the effect of which was maximum when the heterolateral stimulation was delivered 3 msec. before a homolateral one. We have also shown (2) that the inhibition does not involve the higher olfactory centers but centers located in the retrobulbar area; indeed, the evoked electric activity as well as the unitary responses of the A.O.N. disappeared after sectioning the anterior commissure. Thus we have reported physiological experiments showing that the spontaneous activity of the A.O.N. could be modified by homolateral, heterolateral or bilateral olfactory stimulation. The effect was maximum for a time difference of 2 msec. during bilateral stimulation suggesting that the early interbulbar inhibition is relayed through the A.O.N. More recently, in order to investigate the role of the A.O.N. in the interbulbar connections, systematic exploration of all possible pathways between both olfactory bulb (O.B.) and A.O.N., was undertaken (3). Strong evidence was found for the presence of direct connections from mitral as well as tufted cells in the A.O.N., and a synaptic relay of interbulbar fibers lying in the ventro-lateral part of the A.O.N. The present experiments were designed to analyze the functional aspects of the homolateral O.B.-A.O.N. connections and the coding of interstimulus intervals during bilateral odour stimulation.

Material and Methods

The experiments were performed on young rabbits weighing between 1 and 1.5 kg. Both olfactory bulbs and retrobulbar regions were exposed under light thiopentone anesthesia. The animal was then curarized and placed under artificial respiration. Two types of recordings were made:

1) **Field** potentials were picked up in the A.O.N., pars ventralis, in response to electrical stimulation (of variable duration and intensity) of the homolateral olfactory bulb through a bipolar electrode comprising two silver wires, 100μ diameter, insulated down to the tip and contained within a stainless steel needle of 0.4mm outside diameter.

2) Single unit activities were recorded through a conventional unipolar

glass micropipette descending in the A.O.N., in response to:

a) either electrical stimulation of the O.B.

b) or bilateral olfactory stimulation consisting of olfactory stimuli delivered through intranasal cannulae independently adjusted in time and concentration on each side; the time interval between both sides could be varied from −6 to +6 msec. In many experiments the recording microelectrodes were filled with a solution of potassium acetate saturated with methyl blue which was used to mark the recording site by the technique of Thomas and Wilson (4).

Results

1) A.O.N. field potentials. Stimulation of the O.B. with a brief electric shock (0.3 msec.) elicited in the ipsilateral A.O.N. a slow diphasic positive-negative variation of potential which resulted from the activation of the mitral cells (2). The initial positive response could be separated into two phases by a long duration stimulation (3 msec.) of higher intensity located in the external plexiform layer (3). The independence of the two waves was demonstrated by measurement of their latencies. The latency histogram obtained from 55 different recordings showed a bimodal distribution (Figure 1).

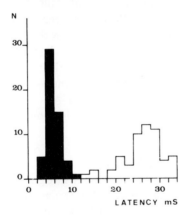

Fig. 1 Histogram of latencies of field potential elicited in the A.O.N. by electric stimulation of the O.B.

2) A.O.N. unit response to electric stimulation of the O.B. In another series of experiments, extracellular recordings were obtained from 283 A.O.N. units, in response to brief electric stimulation (0.3 msec.) of the ipsilateral O.B.

17 units responded with a short latency (2.15 ± 0.29 msec S.D.). The

latencies were constant at threshold stimulus strength and became briefer when the intensity of the stimulus was increased. There spikes followed high rates of stimulation. On this basis, it can be concluded that these units were antidromically activated. The other 266 units responded with a longer latency of between 4 and 21 msec. and could only follow very low stimulation frequencies (5 to 15 Hz), suggesting that these neurons are orthodromically activated through a synaptic relay. Classifying all the units according to their latencies, resulted in a histogram (Figure 2) with a multimodal distribution: four populations of units can be distinguished (4-6 msec.; 11-12 msec.; 11-17 msec. and 20-21 msec.). Student's test confirmed that all the differences were highly significant (p<0.001).

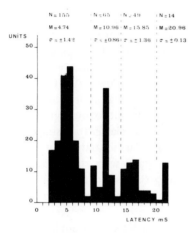

Fig. 2. Histogram of latencies of unit responses elicited in the A.O.N. by electric stimulation of the O.B.

3) A.O.N. unit response to bilateral olfactory stimulation. Extracellular recordings were obtained from 40 A.O.N. units in response to bilateral olfactory stimulation.

9 units were located in the A.O.N., pars dorsalis. Characteristically, they responded differently according to the time interval between the onset of the odour stimuli during bilaterial olfactory stimulation; they could, for instance, be activated for every interval less than 2 msec. and inhibited for every interval greater than this. Many other response pattern were observed.

31 units were located in the A.O.N., pars ventralis 4 of these units responded like the 9 units located in the A.O.N. pars dorsalis; the other 27 units were not sensitive to different time intervals but gave either a constant inhibitory or a constant excitatory response.

Discussion

The existence of the late wave superimposed on the positive wave of the field potential recorded in the A.O.N., in response to long duration shocks of the O.B., indicates that two distinct afferent pathways pass via the lateral olfactory tract (LOT) This is in complete accordance with the finding by Nicoll (5) of two groups of afferent fibres contained in the LOT. Moreover, our findings also suggest that both sorts of second order neurons of the O.B. send some terminals into the A.O.N.; the first positive wave, the latency of which is 5.18 msec., would certainly correspond to the afferent volley from mitral cells whereas the second positive component, with a latency of 25.8 msec., would correspond to the afferent volley from the tufted cells.

Only 6% of the units recorded responded with a very short latency and followed high stimulation rates. These units were probably antidromically activated, demonstrating the existence of a direct pathway between the A.O.N. and the ipsilateral O.B. This is in accordance with the anatomical data of Valverde (6,7), Scalia (8), Price and Powell (9, 10).

All the other units were activated through at least one synaptic relay since they had longer latencies, and could not follow high stimulation rate. These units can be classified by their latencies in four statistically distinct groups. Furthermore, there is a regular separation of about 5 msec. between each population. These data seem to indicate that some reverberating loop exists confirming Cajal's findings, but it is at present impossible to say whether it involves the O.B. or it is restricted to the A.O.N. interneurons.

Two kinds of units responding to bilateral stimulation were found. Some, located in the A.O.N., pars dorsalis, could distinguish different time interval; it is obvious that any given interval registered in the A.O.N. by a specific pattern of activation-inhibition, provides the animal with a powerful tool for precisely discriminating the first-stimulated side. The other group, located in the A.O.N., pars ventralis, insensitive to time interval variations must relay centrifugal afferents.

Conclusion

It is surprising that it is during the very first milliseconds of the bilateral olfactory stimulation that is decided, by the process of mutual inhibition, which olfactory bulb takes precedence over the other, and keeps it, during the whole response.

References

1. Leveteau, J., Daval, G. and MacLeod, P. Physiol. and Behav. 4, 479-482 (1969).
2. Leveteau, J., Daval, G. and MacLeod, P. In Schneider, D. (Ed.) Olfaction and Taste, IV, Seewiesen (1972), 135-141.
3. Daval, G. and Leveteau, J. Brain Res. 78, 395-410 (1974).
4. Thomas, R.C. and Wilson, V.J. Sci. 151, 1538-1539 (1966).
5. Nicoll, R.A. Nature (London), 227, 623-625 (1970).
6. Valverde, F., Anat. Rec. 148, 406-407 (1964).
7. Valverde, F., Cambridge Mass., Harvard University Press, 1965.
8. Scalia, F., J. Comp. Neurol., 126, 285-310 (1966).
9. Price, J.L. and Powell, T.P.S. J. Anat., 107, 215-237 (1970).
10. Price, J.L. and Powell, T.P.S. J. Anat., 107, 239-256 (1970).

Cortical Neuron Responses to Odours in the Rat

Giachetti, I. and Mac Leod, P.

Laboratoire de Physiologie Sensorielle, E.P.H.E.

College de France, Paris.

Since the demonstration by Powell (8) of olfactory connections between the prepyriform cortex and the ventromedial and dorsomedial nuclei of the thalamus, very few attempts have been made to find an electrophysiological support to this anatomical finding. Previous experiments had showed that the gustatory afferent pathway ends in the ventromedial nucleus and projects to the neocortex in an area close to the somatic I area of the tongue (1). These observations bringing together these two evidences suggest the possible occurence of an olfactory projection in the neocortex superposed to the gustatory projection.

Actually, Motokizawa has just shown that the thalamus receives an olfactory input (7) reflected by field potentials as well as by unit responses to odours, but surprisingly limited to the dorsomedial nucleus. This background led us to look for a possible olfactory response in the neocortical somatic and gustatory areas. The evidence of such a response would allow for a comparison between the olfactory discrimination abilities of neocortex and pyriform cortex.

Methods and Results

I–Pyriform units.

The animals, male Wistar rats, were lightly anesthetized with Nembutal (30mg/kg). The prepyriform cortex was located according to the technique of Biedenbach and Stevens (2): the glass recording micropipette was placed within the prepyriform area at the turnover point of the field to record potential elicited by stimulation of the mitral layer of the olfactory bulb.

The 5 stimuli previously used in our experiment of mitral cells (4), were delivered through a nasal cannula, in random order by an automatic programmed device.

We obtained the following percentages of responses:

27% of positive responses

7% of negative responses

66% of non responses

Haberly (6) did the same experiment on deeply anesthetized rats and obtained 80% negative responses and 20% positive responses. We

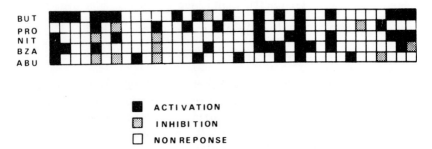

```
■  ACTIVATION
▒  INHIBITION
□  NON REPONSE
```

Fig. 1. Responses of 36 units to 5 stimuli: BUT: butanol, PRO: propanol, NIT: nitrobenzen, BZA: benzaldehyde, ABU: butyric acid.

interpreted this discrepancy according to Freeman's experiment (3) as due to a differential action of the anesthesia in relation to the cell dimension (9). The 10 paired correlation coefficients between the 5 responses profiles of the 36 recorded units are low, with a mean value of 0.105, not significantly different from the mean value of 0.011 found for the mitral cells.

The tridimensional representation of the results of the principal components analysis confirms that there is no evidence of convergence at the pyriform level. The representative points of the 5 stimuli are evenly distributed in the whole space and do not show any significant correlation.

II– Neocortical units.

The animals used were male SPF rats, curarized with Flaxedil, locally anesthetized with Xylocaine and maintained under artificial respiration. This was a necessary condition to record evoked potentials in the cerebral cortex. The first somatic area of the tongue was located by recording the surface potentials elicited by stimulation of the anterior contralateral part of the tongue (10). After location of the tongue somatic area, the homolateral olfactory bulb was stimulated with a bipolar concentric electrode, delivering square pulses of 0.5 msec duration, 0.5 Hz, and 10-15 V amplitude. Stimulation of the mitral cells layer resulted in typical evoked potential responses (Fig. 3) with a mean latency of 15-20 msec all over the gustatory cortical area defined by Yamamoto and Kawamura (Fig.4).

Unitary activity extracellularly recorded in the same area exhibited discriminatory responses to odours (Fig. 5).

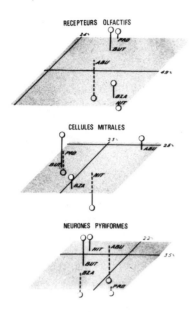

Fig. 2. Spatial distribution of 5 odour representative points given by the principal components analyses (5).

Discussion

These preliminary results bring out rather convincing evidence of an olfactory input to the neocortical area which receives a gustatory input. This area lies just caudal to the somatic I area of the tongue, quite in accordance with Powell's anatomical data. However, no olfactory response has yet been recorded in the ventromedial thalamus, where the gustatory relay occurs (1). This point deserves further investigation. Similarly, the physiological meaning of the olfactory input to the dorsomedial thalamus (7) is still unknown.

The minimum latency of the somatic evoked potential of the tongue is approximately 10 msec. in comparison to 15 msec. for the olfactory response. This difference can be explained by a smaller diameter of the olfactory fibers connecting the pyriform cortex to the cortex to the thalamus.

The possibility of a somesthesic response conveyed by trigeminal fibers excited by the intrabulbar electrode was ruled out by local anesthesia of the upper trigeminal branches.

Although the number of unit responses presently recorded is too small to be statistically evaluated, there is undoubtly a certain amount of odour discrimination in the neocortex. The possible convergence of gustatory and olfactory inputs on the same neurons is currently under investigation.

305

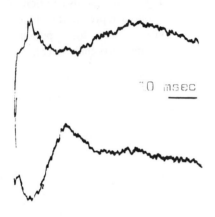

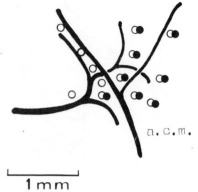

1 mm

Fig. 4.

Fig. 3. Upper trace: somesthesic tongue evoked potential. Lower trace: olfactory evoked potential.

o: loci of olfactory evoked potentials

●: loci of somatic evoked potentials

a.c.m.: A.cerebri media

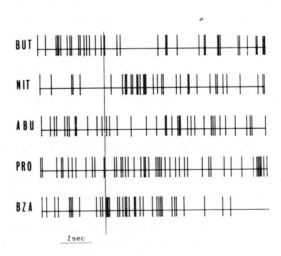

Fig. 5. Spikes from one unit in response to 5 odours: inhibitory response to BUT, excitatory response to NIT and BZA.

References

1. Benjamin, R.M. and Pfaffmann C. 1955 J. Neurophysiol. 18, 56-64.
2. Biedenbach, M.A. and Stevens, C.F. 1969. J. Neurophysiol. 32 193-203.
3. Freeman, W.J., 1972. J. Neurophysiol. 35 745-761.
4. Giachetti, I. and MacLeod P. 1973 J. Physiol. (Paris) 66 399-407.
5. Giachetti, I. and Mac Leod P. 1974. Chemical Senses and Flavor, in press.
6. Haberly, L.B. 1969. Brain Research 12 481-484.
7. Motokizawa, F. 1974. Brain Research 64 334-337.
8. Powell, T.P.S., Cowan, W.H., and Raisman, G. 1965. J. Anat. London 99 791-813.
9. Stevens, C,, 1969. J. Neurophysiol. 32 184-192.
10. Yamamoto, T. and Kawamura, Y. 1972. Physiol. and Behav. 9 789-793.

Neurophysiological Studies on the Prefrontal Olfactory Center in the Monkey

T. Tanabe, M. Iino, Y. Oshima and S.F. Takagi

Dept. Physiol., Sch. Med., Gunma Univ., Maebashi-city Japan

Research works on the olfactory areas in the brain have been extremely few and specific studies on an olfactory area in the neocortex has been made only by Allen. Using the olfactory conditioned reaction, and the electrophysiological technique, Allen suggested that there may be an olfactory projection area in the prefrontal lobe (1,2). His research, however, had not been confirmed until today. The present research was intended to confirm and where possible, to further develop his results (8).

(I) Electrophysiological studies

Under nembutal narcosis, single electric shocks were applied to the olfactory bulb (OB) or to the prepyriform cortex, and the evoked potentials were recorded in the ventro-latero-posterior portion of the orbitofrontal area (LPOF). It was further found that these evoked potentials were no longer elicited by the OB stimulation after the prepyriform cortex was destroyed.

(II) Behavioural studies

Normal monkeys have a habit of smelling something before they eat it. Two different odours were applied to two small pieces of bread respectively and in addition, a bitter taste was added to one of the two pieces. When these two breads were repeatedly given to the monkeys, they became conditioned to discriminate the two pieces of bread, because they dislike any bitter taste. When the LPOF was removed in these conditioned monkeys, the rate of successful discrimination decreased remarkably i.e. by about 50%. In contrast, when the other portion of the orbito-frontal area was removed, the rate decreased much less. From the results of these behavioral and electrophysiological experiments it was concluded that there exists an olfactory area in the orbito-frontal area.

(III) Cellular responses to odours

Eight kinds of odours, dl-camphor (CM), cineol (CL), borneol (BL), dichloroethane (DE), γ-undecalactone (UND), methyl cyclopentenolone (CLT), iso-valeric acid (VA) and iso-amylacetate (AA) were applied at a concentration of 10^{-2}.

Acute experiments

Monkeys were anaesthetized with nembutal. A tungsten microelectrode was inserted into the LPOF and spontaneous discharges of 270 cells were recorded.

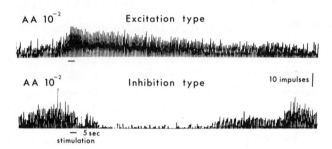

Among them only 29 cells (about 10%) responded to one or more odours. One of the examples is shown above.

Chronic experiments

By using the Evarts type micromanipulater, a tungsten microelectrode was inserted into the same portion of the chronic unanaesthetized monkey restrained in a monkey chair. Thus, spontaneously occuring spike discharges were recorded.

Out of the 119 single cells recorded, 44 cells reacted to one or more of the eight odours (37%). An example is shown below. Thus, nearly four times as many cells as in the acute experiments responded to one or more

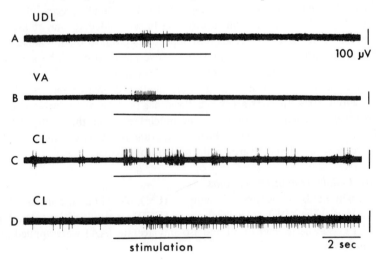

odours in the chronic experiments. This is one of the most important findings of our studies.

Comparison of the cellular responses in the acute and chronic experiments.

(a) When the response types were examined, half of the cells showed increases of the discharges (+type) and the other half revealed inhibition (−type) in the acute experiments. In contrast, in the chronic experiments, 70.5% of cells showed the +type, 26.9% the −type and 2.6% the mixed type (a decrease followed by an increase in the firing rate). Thus, the response types were more complicated in unanaesthetized monkeys.

(b) When the number of the cells which responded to each of the 8 odours was examined in the acute experiments, the cells which responded to AA, VA and DE were many (9, 7 and 6 cells respectively), while the cells to CM, BL, UND and CLT were few (2,2,2 and 1 cell). Thus, 8 odours were divided into two groups from the rates of responses. In contrast, in the chronic experiments, the 44 cells responded uniformly to any of the 8 odours. The results in (a) and (b) show very important differences, and indicate that when the responses of cells to odours are studied, unanaesthetized animals should be used.

(c) The lengths of the cellular responses: In the acute experiments, about 80 percent of the cells continued their spike discharges for 5 to 20 sec, once they were stimulated; 10 percent of the cells continued their discharges for longer than 60 sec, while only 10 percent of the cells discontinued their discharges within 5 sec. In contrast, in the chronic experiments, about 50 percent of the cells discontinued their discharges within 5 sec and about 80 percent of the cells did so in less than 10 sec. This is also an important finding to be remembered in olfactory research.

From these comparisons, it is clear that the cellular responses to odours become very different, depending upon whether the animals are anaesthetized or not.

In most of the olfactory experiments so far performed (3,4,5,6,7), the animals were anaesthetized. It could be said, therefore, that the results obtained in those experiments are incorrect, and that the responses of the cells in the olfactory nervous system should be studied in unanaesthetized animals in the future.

(IV) A characteristic of the cellular responses in the prefrontal olfactory area

In the chronic experiments, numbers of the cells which responded to one, two, three and four odours were counted. Out of the 44 cells, 22 cells (50%) responded to one odour, 13 cells (29.6%) to two odours, 6 cells

(13.6%) to three odours and 3 cells (6.8%) to four odours. Not a single cell responded to more than five odours. Consequently, it is apparent that most of the cells in the prefrontal olfactory area belong to the 'specialist type' and hence play a very important role in the discrimination of odours. This is in contrast with the findings of other investigators that most of the cells in the OB, the pyriform cortex and the amygdala responded with the 'generalist type' (3,4,5,6,7).

References

1. Allen, W.F., Amer. J. Physiol., 128 (1940) 754-771.
2. Allen, W.F., Amer. J. Physiol., 139 (1943) 553-555.
3. Cain, D.P. and Bindra, D., Exp. Neurol., 35 (1972) 98-110.
4. Haberly, L.B., Brain Res., 12 (1969) 481-484.
5. Mathews, D.F., J. gen. Physiol., 60 (1972) 166-180.
6. Mathews, D.F., Brain Res., 47 (1972) 389-400.
7. Shibuya, T. and Tucker, D., Olfaction and Taste II (1967) 219-233, T. Hayashi, the editor, pergamon press, Oxford.
8. Tanabe, T., Iino, M., Ooshima, Y. and Takagi, S.F. Brain Res., 80 (1974) 127-130.

Psychophysical Recognition of Functional Groups on Odorant Molecules

Rollie Schafer and K.R. Brower

Dept. Zoology, University of Michigan, Ann Arbor, MI
and Dept. Chemistry, NMIMT, Socorro, NM, USA

Many theories have been proposed to explain the physico-chemical basis of odorant-acceptor site interaction during olfactory excitation (1). Although the possible role of functional groups on the odorant molecule has been suggested (2), virtually no experimental study has directly examined the contribution of functional groups to odor quality. We set out to perform such a study by enlisting professional organic chemists in a program of organoleptic testing (3). Experienced organic chemists are reputed to be able to name organic unknowns by chemical class, functional groups, or certain heteroatoms solely on the basis of odor.

Each of the 73 subjects tested received a detailed protocol and a testing kit consisting of 36 odorant-containing vials. The participants smelled each of either 24 or 36 vials in random order in two or three sessions, allowing at least two minutes for consideration of each odorant. Subjects were asked to not consult with others or to make comparisons with chemicals on hand. Relatively uncommon reagents (mostly Eastman white label purity) were chosen to negate the effects of first-hand experience with specific odorants or visual cues such as viscosity, physical state or color. One vial contained water as a control for cross-contamination, and two vials held the same odorant (methyl isobutyl carbinol). After smelling each odorant, the subjects entered their impressions on a report sheet listing presumed functional group(s), heteroatom(s), chemical class, possible identity and general remarks. It should be stressed that no system of functional classification was suggested to the subjects, nor were any examples of functional classes given in the instructions.

The odorants used and their functional classification were as follows: *Alcohols*—2-ethylhexanol, 4-methylcyclohexanol, methyl isobutyl carbinol (in duplicate); *Amines*——N—methylmorpholine, triethylamine, 2-picoline; *Misc. N–compounds*——valeronitrile, N-vinyl-2-pyrrolidone; *Sulfur compounds*——cyclohexyl mercaptan, phenyl ethyl sulfide, tert. —butyl disulfide; *Esters*——isoamyl butyrate, methyl benzoate, ethyl cinnamate; *Hydrocarbons*—mesitylene, cyclohexene, decane; *Phenols*——m-cresol, *m-methoxyphenol*, 2-napthol; *Ketones* ——4-methyl-2-pentanone, cyclopentanone, mesityl oxide; *Carboxylic acids*——phenylacetic acid, isovaleric acid; *Ethers*--methyl p-cresyl ether, 1,2-dimethoxy-ethane, isopropyl ether; *Aldehydes*--heptaldehyde, furfuraldehyde, citronellal; *Halides*--1,4-dichloro-butane, β-bromostyrene; *Control*--H_2O.

The data presented in *Figure 1* indicate that all major functional types of organic compounds have a measurable degree of recognizability, with the exception of ethers and halides. Fewer than one percent of the responses were exact identifications, except in the case of water which was identified correctly by eleven subjects and specifically noted as having no odor by another fifteen. No consistent pattern of cross-contamination was present in the misidentifications of the vial containing water. The distribution of functional guesses was similar for the duplicate vials containing methyl isobutyl carbinol; alcohol, 14 & 13; ester, 8 & 7; ketone, 15 & 17; terpene, 9 & 11; and camphor, 3 & 4.

The most highly recognizable functional types were the *amines* (87% correct identifications), *sulfur compounds* (61%), *esters* (64%), *phenols* (62%), and *carboxylic acids* (53%). Recognizability was marginal for the *alcohols* (25%), *hydrocarbons* (36%), *ketones* (42%), and *aldehydes* (42%). *Ethers* (20%) and *halides* (16%) were nearly, if not totally, unrecognizable. The sulfur compounds were diluted with mineral oil in an attempt to prevent cross-contamination, and the unexpectedly low score of 61% correct identifications may be a reflection of excessive dilution.

The data also show that certain confusions of type conform to a pattern, especially among the oxygen-containing functions. For example, alcohols were often called ketones, and ketones were often mistaken for esters. However, none of the subjects called any of the phenols an alcohol, and only a small number of subjects called any of the alcohols a phenol. There were 18 reports on a camphor or menthol quality for 4-methyl-2-pentanol, isopropyl ether, and 4-methylcyclohexanol. There were no such reports for 1,4-dischlorobutane, cyclohexanone, or 2-ethylhexanol. This suggests that the camphoraceous quality is probably conferred by a certain degree of branching in oxygenated aliphatic or alicyclic compounds containing five to ten carbon atoms. Examples of other compounds which have a camphoraceous quality are hexamethylacetone (almost a perfect smell-alike), pinacolone, neopentyl alcohol, and dipivaloyl-methane.

We conclude that the members of several chemical classes such as the amines, sulfur compounds, and esters exert a common mode of action on the olfactory receptors. Obviously there is no one-to-one correspondence of functional types to odor types as evidenced by the poor recognizability of alcohols, ethers, and halides. Likely physical characteristics which are common to members of a functional class and cause them to be bound at

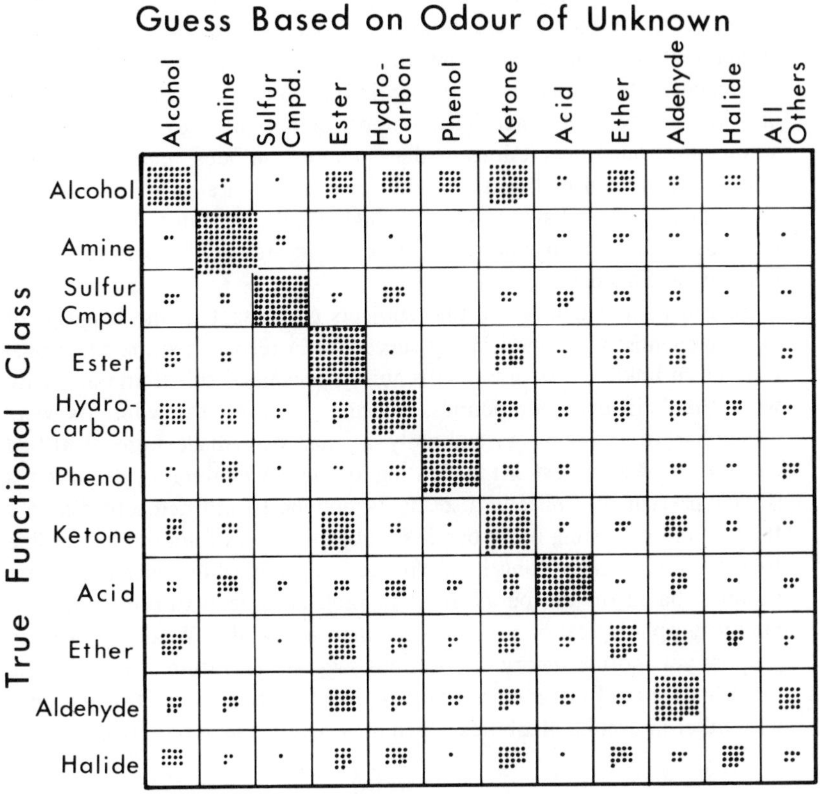

Guess Based on Odour of Unknown

Figure 1. Individual responses of organic chemists to odor unknowns. A dot on the diagonal represents a single correct response. A dot outside the diagonal represents a misidentification. The paucity of entries in the *All Others* column deserves attention, since no system of functional classification was suggested to the subjects. There are fewer dots in some rows because some subjects did not attempt identification of all of the odorants in their testing kit, or if they attempted an identification, they made no entry for the compound.

the same type of receptor site on olfactory neurons would include hydrogen-bonding capacity, acid-base reactivity, charge-transfer complexation and *softness* or electrical polarizability (4). The data indicate that aromatic compounds are easily distinguished from aliphatic compounds. Either softness or charge-transfer complexation may be the factor which makes it possible to tell whether a compound is aromatic or aliphatic even though the functional group may not be recognized. For example, the subjects were uniformly successful in distinguishing the aliphatic 1,4-dichlorobutane from the aromatic β-bromostyrene, although the halogen atom imparted no recognizable odor of its own. Softness may also be the key to the fact that the upper oxidation states of nitrogen and sulfur are much less odorous than amines, sulfides, mercaptans etc. Nitro-methane and methyl sulfoxide, for example, are essentially odorless even though they have reasonably high vapor pressures. Other highly odorous compounds having soft electron pairs are the phosphines, organic selenides and arsines.

In general the data show that subjects conversant in the language of organic chemistry are moderately successful in recognizing the functional type of an unknown odorant. This ability does not result from familiarity with the individual compounds, because most of the odorants were unfamiliar to the subjects and only a small percentage were identified exactly. A second test kit consisting of 36 new odorants produced a similar pattern of replies in a set of 24 previously-untested subjects. In a third round of testing using our best subjects, it was found to be possible to mask recognizable functional groups with steric hindrance (5). These findings, based on the odors of small molecules, tend to run counter to purely morphological or spectroscopic theories of odor. We do not wish to propose yet another theory of odor, but suggest that future attempts along these lines should integrate the possible contribution of functional groups with other molecular attributes which may play a role in odor (6).

References

1. Amoore, J.E. 1970 Molecular Basis of Odor, C.C. Thomas, Springfield, Illinois; Wright, R.H. 1971. In G. Ohloff and A.F. Thomas (eds.), Gustation and Olfaction, Acad. Press, N.Y., pp. 134-144; Davies, J.T. 1965 J. Theor. Biol. 8: 1; Beets, M.G.J. 1971 In L.M. Beidler (ed.), Handbook of Sensory Physiology IV. Chemical Senses 1. Olfaction, Springer-Verlag, N.Y., pp. 257-321; Mozell, M.M. 1970. J. Gen. Physiol. 56: 46; Mazziotti, A. 1974 Nature 250: 645.
2. Ruzicka, 1920 Chemiker-Zeitung 44: 129; Dravnieks, A. 1966 Adv. Chem. Ser. 56: 29; Klopping, H.L. 1971 J. Agr. Fd. Chem. 19: 999.
3. Schafer, R., and K.R. Brower 1972 Science 177: 388.
4. Pearson, R.G. 1963 J. Am. Chem. Soc. 85: 3533.
5. Brower, K.R., and R. Schafer 1974 J. Chem. Ed., In Press.
6. We thank our subjects and the following individuals who contributed to the design and execution of this experiment: T.V. Sanchez, E.C. Palmer, G.T. Peake, O. Maller, M.R. Kare, R.H. Cormack, C.J. Buys, J.E. Amoore, B.R. Lucchesi, and S. Schafer. This experiment conformed to the guidelines set forth in The Institutional Guide to DHEW Policy on Protection of Human Subjects, U.S.D.H.E.W., U.S.P.H., N.I.H., DHEW Pub. No. (NIH) 72-102. Supported in part by NSF Institutional Grant G-3533 to the New Mexico Institute of Mining and Technology.

A Selective Centrifugal Control of Olfactory
Input Depending on Hunger Specific Arousal

J. Pager

Lab. d'Electrophysiologie, Universite Cl. Bernard,
69100 Lyon-Villeurbanne (France)

The possible mechanisms by which olfactory cues take part in the regulations of behaviour may be represented by the general schema of behavior elaborated by Le Magnen (1971) where the admission of any significant external cue at a convenient input of the nervous system might be submitted to a selective control from the corresponding internal state, thus allowing well adapted behavioral sequences to be achieved.

As a result of anatomical and electrophysiological investigations, mitral cells, second neurons of the ascending olfactory pathway appear as cross roads for upstream olfactory and downstream *centrifugal* information, and therefore could attractively materialize a strategic point of the olfactory regulation of a behavior like food intake. Their multiunit activity was recorded in hungry and satiated rats through chronically implanted bipolar electrodes. The probability of positive responses to an odorant stimulus, corresponding to an increase of electrical activity was considered to represent the mitral layer activation level in a given experimental situation: hunger or satiety/food odor or non alimentary odor. During a series of 10 identical stimulations, the recorded activation displayed a more or less developed peak for the first stimulations, and decreased then to a steady level, the whole time pattern characterizing the experimental situation. It was observed in hungry or insulin hypoglycemic rats stimulated with food odor that the activation level was significantly higher than in the same satiated animals, and than with amyl acetate in either alimentary state (Pager et al. 1972). Thus the basic theoretical hypothesis of a selective modulation exerted on olfactory input in correlation with the specific internal alimentary state was verified.

Histologically controlled unilateral surgical lesions established that hunger activating effects are conveyed by nervous *centrifugal* pathways modulating mitral activity. Transverse sections of the olfactory peduncle (O.P.) suppressed the selectively activated responses to food odor in hungry rats, as well as the initial activation peak, while the

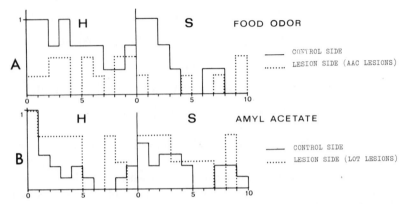

Figure 1. Activation level evolution of mitral cells (ordinate) as a function of the rank order of stimuli (abscissa) in hunger (H) and satiety (S).

non modulated amyl acetate responses displayed an increased activation level. After sections of the anterior limb of the anterior commissure (AAC), the mitral layer no longer displayed hunger activation with food odor, and the initial activation peak had also disappeared in all experimental situations (Fig. 1A; Pager, 1974 a). Sections restricted to the lateral olfactory tract (LOT) reproduced only the increase of the non modulated activation level observed with amyl acetate after OP lesions (Fig. 1B). All the lesions affected the ipsilateral activity without changing the contralateral one.

Further investigations ought to determine how the *centrifugal* modulating influences could be integrated into the pattern of mitral activity resulting from their proper resting activity, olfactory coding and various intrabulbar effects. As intact commissural *centrifugal afferences* to the olfactory bulb are necessary for a food odor to exert its activating feed-back, it may be asked on what criteria the odor constitutes an activation releaser specifically fitted to the alimentary arousal state. Eucalyptol odor, a non alimentary one, initially gives the same activation level in hunger and satiety. Its precocious and maintained adjunction to the diet supplied to the rats endowed it with transitorily modulated responses with a significant decrease of the initial satiation activation level (Pager 1974b; Fig. 2A).

Inversely, the first intake to the eucalyptol diet followed by an apomorphine disease further elicited a significant decrease of this palatability, in a parallel with the disappearance of the modulation of responses to the eucalyptol diet odor: the satiation activation level

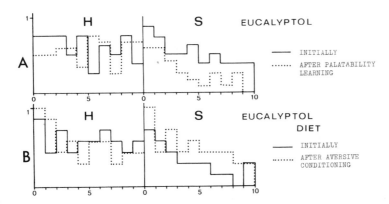

Figure 2. Activation level evolution of mitral cells (ordinate) as a function of the rank order of stimuli (abscissa) in hunger (H) and satiety (S).

was increased up to that reached in hungry rats (Fig. 2B).

Thus the alimentary character of an odor seems to be expressed by the shift of mitral activation level in satiety. In an alimentary point of view, odors could be tentatively classified according to the following electrophysiological criteria: — activation level significantly higher in hunger than in satiation: alimentary (palatable) odor; — no significant difference between the activation level in hunger and satiety: aversive (unpalatable) odors likely to acquire alimentary properties after a suitable treatment, provided that the unmodulated level is high enough for a satiation decrease to be elicited.

The whole data are inserted in an alimentary behavior schema where a message received at the olfactory input, if it can assume the alimentary character by reference to the animal's previous experience, is able to release a *centrifugal* activation of the mitral cells provided that the internal input had collected appropriate information about the hunger arousal state: the massive specific admission of the alimentary cues is likely not irrelevant of a food intake behavior achievement, as the output of such a system.

References

Le Magnen, J. (1971). Olfaction and nutrition in "Hb. of Sensory Physiology" (Ed. L.M. Beidler) vol. IV (I), pp 465-482, Springer Verlag, Berlin.
Pager, J., Giachetti, I., Holley, A. and Le Magnen, J. (1972). Physiol. Behav. 9, 573-579.
Pager, J (1947a). Physiol. Behav. 13 in Press.
Pager, J., (1974b). Physiol. Behav. 12, 189-195.

Human Temporal Lobe and Olfaction

Rebecca Rausch and E.A. Serafetinides

University of California

Brain Research Institute, Los Angeles, USA

Although substantial evidence exists to implicate the human temporal lobe in olfactory information processes, the specific nature of that role is not known. The temporal lobe structures have been shown to have both afferent and efferent connections with the olfactory bulb, the major synaptic junction for primary fibers from the olfactory receptors. [1, 2] In addition two studies report in-depth EEG evoked responses to olfactory stimuli in the human amygdala. [3, 4].

As early as 1875, Jackson reported that olfactory auras accompanied epileptic seizures originating in the temporal lobe. Many years later, electrical stimulation of the uncus and amygdala in conscious patients was found to produce odor sensations. [5] In spite of these suggestions of human temporal lobe involvement in olfaction, there has been only one systematic study of olfactory function in patients with temporal lobe lesions. Investigating the effects of bilateral amygdaloid lesions, Hughes et al. [3] report transient changes in olfactory judgements, including odor quality mistakes, increments in indifferent hedonic judgements to odors and increases in the number of odors considered weak in strength, as opposed to moderate or strong. Olfactory thresholds, however, were not measured in these patients.

In the present study two measurements of olfactory function, odor detection and odor recognition, were evaluated in patients with surgically verified excisions of the temporal lobe structures.

Methods

Fourteen patients who have undergone unilateral temporal lobectomy for control of drug-resistant epilepsy [6] were tested at least a year postsurgically. The performances of these patients were compared to 10 normal controls who had no history of CNS damage. Detection and recognition thresholds of pyridine, amyl acetate, and phenyl ethyl alcohol were measured by a forced-choice-three stimuli sniff technique. [7] Odors were diluted with Millipore ® -filtered, distilled water for each of 13 concentration levels. For odor presentations, 18 ml of each solution was present in individual 125 ml

Erlenmeyer flasks. Molar concentrations of pyridine were verified by spectro-photometric measurements at 260 nm.

The testing technique required the subject to sniff the vapor above each of three solutions. Two of the solutions contained the test substance. The subject was asked to select the test solution. As the concentration of the odor substance in successive trials was raised, a transition point occurred where the response changed from random selection to the correct choice. Re-tests were made at and around this point. The *detection* threshold was defined as the lowest concentration of test substance to which a subject gave 2 successive correct responses while giving 2 consecutive incorrect responses at the next lower concentration. (The probability that a threshold would be determined by chance by this technique is 1 in 2,500. [8]).

The *recognition* threshold was determined as follows. Before any of the tests were given, the subject was shown a short list of adjectives describing 4 odors. On each of the test trials, the subject was asked not only which solution was different from the other two, but also to identify its principal characteristics from the list. The recognition threshold was the level at which the subject could correctly identify the odor characteristic.

Results and Discussion

Although routine neurological examinations have not detected olfactory defects in the temporal lobectomy patients, increments in olfactory sensitivity were found by the present specialized technique, as illustrated in Figure 1. Temporal lobectomy patients had higher thresholds than controls for all odors tested, including pyridine ($p<.001$), amyl acetate ($p<.05$) and phenyl ethyl alcohol ($p<.001$).

Interestingly, detection thresholds for the three odors were not significantly different from the recognition thresholds for the temporal lobe patients. Conversely, for controls, the detection thresholds were significantly less than the recognition thresholds for pyridine ($p<.02$) amyl acetate ($p<.01$) and phenyl ethyl alcohol ($p<.05$). The average reduction was approximately 2 log units of concentrations.

Current electrophysiological studies in our laboratory demonstrate that wide spread changes in unit activity occurs in temporal lobe structures in response to olfactory stimuli. [9] Further studies, therefore, are needed to establish the relationship between behavioural and electrophysiological findings which suggest temporal lobe involvement in olfactory function.

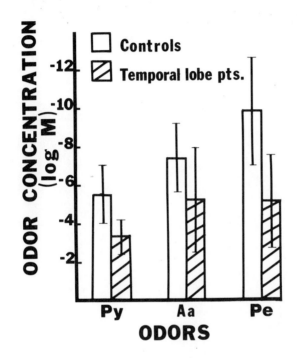

Figure 1. Effects of temporal lobe ablations on olfactory detection thresholds (Log Molar concentrations). Responses to pyridine (Py), amyl acetate (Aa) and phenyl ethyl alcohol (Pe) of 14 temporal lobe patients and 10 controls are presented as means ± S.D.

References

1. Allison, A.C. Biol. Rev. Cambridge Phil. Soc. 28, 195-244 (1953).
2. Powell, T.P.S. and W.M. Cowan. Nature 199, 1296-1297 (1963).
3. Hughes, J.R., D.E. Hendrix, O.J. Andy, C. Wang, D. Peeler, N. Wetzel. In: Neurophysiol. Studied in Man, (ed. G.G. Somjen) Excerpta Medica, Amsterdam, 1942.
4. Narabayashi, H. In: Neurobiol. of the Amygdala, (ed. B.E. Eleftheriou) Plenum Press, N.Y., 1972.
5. Penfield, W. Prog. Ass. Res. nerv.ment. Dis. 36, 210-226 (1958).
6. Crandall, P., R.D. Walter and R. Rand. J. Neurosurg. 20, 827-840 (1963).
7. Henkin, R.I., P.S. Schechter, R. Hoye and C.F.T. Mattern. J. Am. Med. Ass. 217, 434-440 (1971).

8. Henkin, R.I., J.R. Gill, Jr., and F.C. Bartter. J. Clin. Invest. 42, 727-735 (1963).
9. Halgren, E., R. Rausch, T.L. Babb, and P.H. Crandall. Soc. for Neuroscience Program, St. Louis (1974).

This research was supported by USPHS Grants NS 02808 and MH 06415-17.

CHEMORECEPTION IN INSECTS

Chairman: J. Boeckh

Chemical Communication in
Danaid Butterflies
By
Dietrich Schneider

Max-Planck-Institut, 8131 Seewiesen, Germany.

Males of the butterfly family Danaidae expand large abdominal scent organs (hair-pencils) when courting their females (1,6). In a number of species these structures have been found to contain volatile compounds (2,3,4,5). The main component, a pyrrolizidinone (I), serves as an arrestant pheromone to the female (6). Several species of Danaidae when reared in the laboratory were found to be deficient in (I) or lacking it completely (5,6,7). Recently, we observed *Danaus chrysippus* males in Kenya actively searching dry *Heliotropium.* (Boraginaceae) plants in an upwind flight (8). In the laboratory, the males assembled on these plants and sucked on re-moistened *Heliotropium.* Following earlier experiments and suggestions (2,3,4), we extracted a potential pheromone precursor (a pyrrolizidine ester alkaloid: II) from these plants (4,8). Hairpencils of laboratory-raised males of this butterfly were found to contain (I) in up to normal amounts after: a) access to the *Heliotropium* plants or b) feeding on an extract of this plant or c) feeding on the dissolved (II). – In electrophysiological experiments it was shown that (I), (II), and the plant elicit responses in antennal olfactory receptors of both sexes of the butterfly (7,8). Substance (II) thus appears to be the essential dietary factor for the biosynthesis of (I). Earlier field and new laboratory observations of other Danaids suggest that the precursor dependency of pyrrolizidine production is a widespread phenomenon in this butterfly family (5,8).

This work is the outcome of a collaboration with M. *Boppre* in my laboratory and the groups of J. *Meinwald* and T. *Eisner,* Cornell Univ., Ithaca, N.Y., U.S.A.

(I)   (II)

References

1. Brower, L.P. et al. 1965. Zoologica (N.Y.) 50, 1.
2. Edgar, J.A. et al. 1973. J. Austral. Ent. Soc. 12, 144.
3. Edgar, J.A. and C.C.J. Culvenor 1974. Nature 248, 614.
4. Meinwald, J. and Y.C. Meinwald 1966. J. Am. Chem. Soc. 88, 1305.
5. Meinwald, J. et al. 1974. Experientia 30, 721.
6. Pliske, T.E. and T. Eisner 1969. Science 164, 1170.
7. Schneider, D. and U. Seibt 1969. Science 164, 1173.
8. Schneider, D. et al. 1975. J. Comp. Physiol. (in preparation).

Orientation of the Male Silkmoth to the Sex Attractant Bombykol

by Ernst Kramer

Max-Planck-Institut, Abteilung Schneider

8131 Seewiesen, Germany

Anemotaxis plays an important role in orientation over long distances. However, in order to prevent lateral missing of the source, some additional mechanism is needed. For a long time most speculations on this mechanism concentrated on tropotaxis. But considering the small gradients occurring in extended odor fields and experimental data on the sensitivity of tropotaxis (Martin, 1964; Hangartner, 1967) it seems unlikely that tropotaxis can solve this task.

Wright (1958) discussed another mechanism based on the filamentary structure of natural odor plumes: The animal moves upwind in a zigzag pattern, the castings being controlled by the frequency of transitions between high and low odor intensities. Recently Kennedy (1974) showed that such a scanning mechanism based on optomotor anemotaxis exists in different moth species. However, it is still uncertain what kind of event releases changes of direction.

A locomotion compensator developed for the investigation of the orienting behavior of honeybees enabled the study of this question. Since this apparatus is designed for walking animals, *Bombyx mori,* which has lost the ability to fly, was chosen as the experimental animal.

Methods

The function of the locomotion compensator is to keep walking insects in a small experimental field and to precisely record the intended movements. A sphere 33 cm in diameter is mounted so that it can be rotated around two orthogonal horizontal axes by servomotors. The animal is put on its top. A position sensor, fixed above the sphere, evaluates the deviation of the insect from the upper pole. The deviation values are used to recenter the insect by driving the servomotors.

An almost laminar airstream is blown over the top of the sphere. It is composed of two halves which may be differently laden with odor. This is achieved by dividing the nozzle into two parallel ducts (see fig.

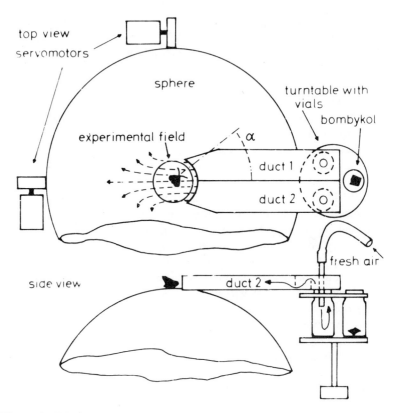

Figure 1. Schematic diagram of experimental apparatus.

1). The arrangement operates as follows: An insect walking upwind has the antennae in different halves of the airstream. If one half has a higher concentration, then a deviation from the exact upwind direction brings both antennae into the higher concentration. This is maintained as long as the insect walks askew of the wind. A return is immediately followed by a decrease of concentration on one antenna. A preference of the higher intensity therefore results in a transverse component in the path. This component increases with the distance the insect travels.

The tests consisted of series of 5 x 3 runs each run lasting one minute. The groups of 3 had the sequence: 1) higher concentration left 2) higher concentration right 3) no odor in both ducts. By this procedure all disturbing directional cues were eliminated. For each run the transverse component of the path was calculated (more precisely the y-component of the mean vector, the wind coming from +x). The mean of these values was tested to see if it was significantly different from zero. In the following, this is expressed as R which is defined as the mean value of the y-components of the mean vectors divided by the standard error. The wind velocity in all tests on the sphere was 20 cm/s.

In addition to this method male silkmoths were placed in a wind tunnel and filmed while approaching a bombykol source. The paths were reconstructed from the film, frame by frame, and correlated with the shape of the odor plume. This was to compare the results obtained with the sphere with paths in a more natural odor plume.

Results

A first series of 450 runs was made in a homogeneous odor field, i.e., both halves of the experimental field having the same concentration. Odor sources were pieces of filterpaper containing 0.01; 0.1 and 1.0μ g bombykol.

The paths show a conspicuous preference of angles between 30 and 50 degrees oblique to the direction of wind. There are no portions of the path longer than a few cm which point strictly upwind. An example is shown in fig. 2.

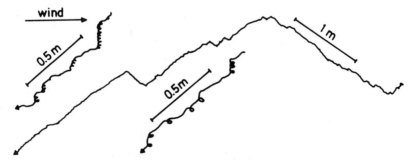

Figure 2. Paths of *Bombyx* in a homogeneous odor field, showing spontaneous changes of the angle to the wind and cycloids if this angle is large.

331

Further results of these experiments are:

1. Within single tests (15 min) there is neither a significant decrease in the walking speed nor in the directness D (D is defined as the straight line between the positions at start and after one minute/the actual distance travelled by the insect).

2. The walking speed is a function of the odor concentration. The average speeds were: 1.3 m/min with 0.01, 2.4 m/min with 0.1 and 3.1 m/min with 1.0μ g bombykol. Results 1 and 2 indicate that there is little or no adaptation in the whole system.

3. Many paths are characterized by loops of 360 degrees beginning with a *downwind* turn. The frequency of loops is correlated with the obliquity of the paths. Turns starting *upwind* come to a halt when the negative counterpart of the angle formerly maintained is reached.

The response to differences in concentration was studied in a second series. The ducts were supplied with different amounts of bombykol. The ratios applied and the significance of their discrimination are shown in the table below.

Bombykol-sources		Ratio	R	No. runs
0.1	0.0	1 : 0	11.4	40
0.01	0.0	1 : 0	11.7	30
1.5	0.5	3 : 1	10.0	40
0.15	0.05	3 : 1	3.9	30
0.015	0.005	3 : 1	3.6	80
1.0	0.5	2 : 1	6.1	60
0.01	0.005	2 : 1	1.9	80
1.25	0.75	5 : 3	7.1	50

Compared with runs made in a homogeneous field the number of loops is markedly higher. Upwind turns are immediately reversed after one antenna enters the lower concentration. The result is an oblique path with loops pointing downwind and flat bows with concave sides showing upwind. For the rest, the paths look very similar to those in the homogeneous odor field, the mean angle, the mean walking speed and D being about the same at the respective concentrations.

Since a concentration difference of 5 : 3 is still discriminated with the high R-value of 7, the minimal difference detectable is probably much smaller. The last few silkmoths available for this season were saved for experiments with unilaterally deantennated animals. Earlier tests in the wind tunnel showed that those also succeed in finding their goal. Circling did not occur as often as would be expected if tropotaxis were a predominant reaction. On the sphere, however, circling was increased, D was 0.13 (intact animals : 0.21). The walking speed was reduced to 0.8 m/min (1.9). In spite of that, the mean values of the y-components of all the 45 runs having the higher intensity on the intact side and those 45 having it on the operated side do not differ significantly (0.174 and 0.182). This result strongly contradicts a participation of tropotaxis. This is further supported by a direct observation of the moth's walk. If, e.g., a left-side operated moth, walking upwind to the right odorous half, turns to the left, this turn will be reversed when the intact right antenna reaches the left less-scenting half. An intact moth would already be deflected when the left antenna reaches the lower intensity.

All tests with operated animals were made with 0.1μ g bombykol against wind only.

Cybernetic Aspects.

Paths obtained by Wendler and Scharstein (1974) while studying the response of the corn weevil *(Calandra granaria)* to gravity show cycloids like *Bombyx*. They interpret their results by a theory first proposed by v. Holst and Mittelstaedt (1950). According to this theory the control of direction is accomplished by an *additive* effect of variables at least one of them being a function of angle.

The essence of this theory is sketched in fig. 3. The full sinusoidal curve represents the output of the gravity receptors. The system

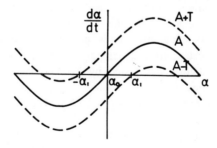

Figure 3. Angular velocity as a function of α. α is the actual angle of the animal to the stimulus. A is the output of that part of the system which evaluates this direction. A must be a sinusoidal function of α. T is an additive constant. It shifts the point of balance (α_0, $\alpha_1, -\alpha_1$) thus producing a menotactic angle.

comes to rest at $a = 0$ if positive values elicit turns to the left and negative to the right. A constant added to the curve — representing a turning tendency (T) — shift the point of balance to a_1, or if the constant is negative, to — a_1 (dashed lines).

This model successfully explains the behavior of *Bombyx* in a homogeneous field, if it is modified by the following assumptions: a) The output of the anemoreceptive system is sinusoidal (A), b) the sign of the turning tendency is changed from time to time by a stochastic process and c) there is some noise in the system. Such a system performs as follows:

High A_{max} and low T	small a.
Low A_{max} and high T	large a.
T sometimes A	large a, few loops.
T always A	very large a, subsequencing loops.
No wind	circling without any translation.

This is exactly what can be seen in the paths of Bomby. By a further modification this model becomes a complete guidance system: The perception of a drop in odor intensity influences the stochastic process which is responsible for the sign of the turning tendency such that the probability of a change of sign is greatly increased.

The features of this model then come very close to Wright's surmise. Out of several other models I tested by computer simulation this simple model turned out to be the most successful.

Taking into account Priesner's data on the properties of the bombykol receptors, the system could work without a central memory. He could distinguish three receptor types. One type is very sensitive and has a long fall time. The others are less sensitive but fast (short fall time). Evidently it is possible to draw all information needed by the system from a comparison of their actual excitation.

Finally, the theory was tested by a computer simulation using the following equations

$$a_t = a_{t-\Delta t} + (T \cdot s - f(u) \cdot f(C)\sin(a_{t-\Delta t}) + N) \cdot \Delta t$$
$$x_t = x_{t-\Delta t} + f(C) \cdot \cos(a_t)$$
$$y_t = y_{t-\Delta t} + (C) \cdot \sin(a_t)$$

a = angle of animal to the wind; T = turning tendency; u = wind velocity; x,y = actual position of the animal; C = concentration of odor; s determines the sign of T, it is a function of dC/dt; N = a random number, stands for all kinds of noise. The functions herein (f(C), f(u), s and the noise were, to start with, chosen so as to optimally fit the model with the data. A number of superimposed computed paths are depicted in fig. 4 and illustrate the potency of the model.

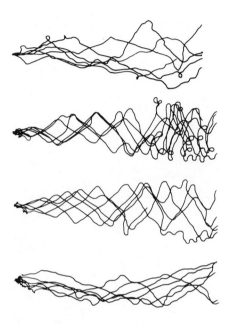

Figure 4. Six paths of *Bombyx* in the wind tunnel (1). 2, 3 and 4, paths obtained with the model in a computed odor field resembling that in the tunnel (1). 2, turning tendency and noise high. 3, like 2 but less noise. 4, medium turning tendency with high noise.

Summary
The orientation of *Bombyx mori* to the sex pheromone bombykol may be explained by an anemotaxis in which the sign of the angle to the wind is changed by decreasing odor concentrations.

References

1. Hangartner, W. (1967): Z. vergl. Physiol. 57, 103-136.
2. Holst, E.v. and H. Mittelstaedt (1950): Naturwiss. 37, 464-476.
3. Kennedy, J.S. and D. Marsh (1974): Science, 184, 999.
4. Martin, H. (1964): Z. vergl. Physiol., 48, 481-533.
5. Priesner, E.: unpublished
6. Wendler, G. and H. Scharstein (1974): Fortschir. d. Zool. 23, 33-44.
7. Wright, R.H. (1958): Can. Ent. 82, 81-85.

Localization and Specificity of Pheromone
Degrading Enzyme(s)
from Antennae of *Trichoplusia ni*

S.M. Ferkovich and M.S. Mayer

Insect Attractants, Behavior, and Basic Biology

Research Laboratory, Agric. Res. Serv., USDA

Gainesville, Florida 32604

Introduction

Attempts to relate electrophysiological data to the kinetics of pheromone-degrading enzymes in insects have been reported (1,2,3). The actual site of enzymatic degradation has not been established although some attempts have been made (4,5,6). Based on experiments *in vivo* (4-8) we decided that protein entities located in the chemosensory sensilla could best be studied *in vitro*. We reported the *in vitro* hydrolysis of the cabbage looper *(Trichoplusia ni (Hübner))* pheromone, (Z)-7-dodecen-1-ol acetate, to (Z)-7-dodecen-1-ol and acetate by antennal homogenates (9). Non-enzymatic binding of the alcohol, an olfactory stimulant but behavioral inhibitor, by antennal proteins was also observed. Subsequently, we found that the architecture of the antenna was favorable for the isolation of membranes and of the fluid bathing the dendritic nerve endings ("sensillum liquor") from chemosensory sensilla. This fluid contained enzymatic and alcohol binding activity similar to that of the antennal homogenates (10). Further *in vitro* studies with the antennal fluid suggested that the pheromone degrading enzyme(s) may be located within membranes of the sensillum (11). This report outlines results of research on this enzyme system and attempts to correlate *in vitro* with *in vivo* pheromone degradation.

Methods

In Vitro Techniques: Excised whole antennae from 2 to 3 day-old moths were placed in 0.5 ml of 0.5 M sucrose in 0.05 M Tris-HC1, pH 7.5, and then sonicated for 10 min in an ice bath to fracture the tips of the sensilla. The antennae were removed, and the solution was centrifuged at 45,000 g for 10 min to sediment the scales. The resultant supernatant was applied to a column containing Sephadex G-200 and enzyme assays were made with fractions of the eluant (11). The entire procedure was conducted at 4°C. All chemicals were at least 94+% pure with no alcohol

337

homolog detectable by gas-liquid chromatography (GLC)

In Vivo Techniques: Antennae of individual live insects were dipped into a sonicated aqueous solution of the pheromone or of the isomers and analogs. The antennae were excised into, and rinsed twice with 3-4 ml of diethyl ether. Total extraction time was about 3 min. The volume of the extract was reduced under nitrogen to 0.1 ml, and then analyzed by GLC (6).

Fig. 1. Elution of male antennal sonicate (0.2 mg protein) from a Sephadex G-200 column. Absorbance at 280μ m (●); percentage of pheromone degradation, 12 hr incubation (o). For details see (10).

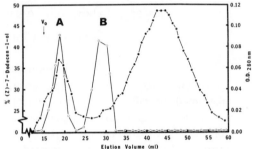

Results

In Vitro Hydrolysis: Gel filtration of the antennal sonicate resulted in two major UV absorption peaks (Fig. 1). The UV absorbing material eluting after 35 ml never degraded the pheromone or its isomers and contained no protein. Two peaks of enzyme activity were also observed: One coincided with the absorption peak eluting near the void volume (A); and another peak coincided with the front edge of the second absorption peak (B). This latter peak had an apparent molecular weight of 38,000 (11). Gel filtration of female antennal and male leg sonicates also resulted in two peaks of enzymatic activity with each source. Conversion of pheromone and isomers was essentially identical between the three sources and the two enzyme fractions, indication that the enzyme from all these sources had at least similar specificity and activity (Table 1).

We therefore speculated that the two peaks represented two forms of the same enzyme, the first peak (A) representing

Table 1. Percentage isomeric alcohol produced after 18 hr incubation with combined fractions containing enzyme peaks A and B (Fig. 1).

	Percentage conversion of acetate to isomeric alcohol by:					
	Male antennae		Female antennae		Male legs	
	Enzyme peak		Enzyme peak		Enzyme peak	
Chemical	A[1]/	B[3]/	A[1]/	B[3]/	A[2]/	B[3]/
Z-7-dodecen-1-ol acetate	51.5	56.9	54.6	45.8	5.3	27.5
Z-8-dodecen-1-ol acetate	99.2	92.9	97.6	89.5	25.3	55.8
Z-9-dodecen-1-ol acetate	39.0	74.7	59.6	40.7	1.6	8.7
E-7-dodecen-1-ol acetate	90.9	31.4	69.4	--	4.2	21.4
Dodecan-1-ol acetate	60.8	49.2	34.7	25.8	3.1	11.7

[1]/ 11 ug and [2]/ 3 ug protein, 100 ug pheromone/0.2 ml 0.05 M tris-HCl, pH 7.5 at 22°C.

[3]/ No protein detected by the method of Lowry et al. (10).

enzyme aggregates or membrane-bound enzyme; and the second peak (B) monomers of the enzyme. Treatment of the antennal preparation with Triton X followed by sonication gave only one peak of enzymatic activity (Fig. 2) with the same elution volume as the peak B in Fig. 1. The following evidence suggests that a portion of the enzymatic activity of peak A is membrane bound: Electron microscopy of material sedimented at 100,000 g from peak A for 2 hr revealed fragments of membranes and vesicles ranging in diameter from 50 to 150 μm; and (2) prior treatment of the antennal sonicate with phospholipase C revealed a reduction of enzymatic activity in peak A with a concomitant increase in the activity of peak B.

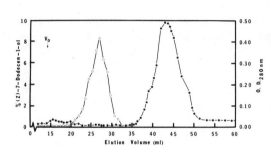

Fig. 2. Elution of Triton X-100 (0.6%) solubilized sonicate (1.9 ng protein) from a Sephadex G-200 column. Absorbance at 280 μm ($\bullet$); percentage of pheromone degradation, 1 hr incubation (o). For details see (10).

Other observations of the antennal sonicate revealed: (1) Inhibition of enzymatic activity with sulfhydryl group inhibitors (fluorescein mecuric acetate and iodoacetamide); (2) no reduction in activity following short-term incubations (10-15 min) with di-isopropyl phosphorofluoridate

Fig. 3. Conversion of pheromone to alcohol *in vivo* at various time intervals. Male antennae ($\bullet$, solid line left ordinate); female antennae (broken line, left ordinate); loss of pheromone by antennae of both sexes (heavy solid line, right ordinate).

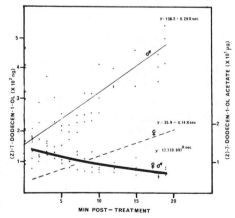

and eserine; and (3) no reduction of activity with DNAase and RNAase (11).

In Vivo Hydrolysis: The rate of degradation *in vivo* on the male antennae was two times faster than on female antennae (Fig. 3). No metabolites other than the isomeric alcohol were observed. The loss of pheromone substrate was exponential and was interpreted as being predominantly evaporation from scales as well as from some other areas. There was a large amount of degradation within 4 sec. The data from incubation times of 5-20 min of the pheromone, isomers and analogs were difficult to interpret in terms of enzyme specificity (Table 2); however, 4-sec incubations clearly demonstrated enzymatic specificity for the pheromone (Table 3). The amount of pheromone degradation on the legs was greater than on the antennae.

After an initial rapid rise the rate of reaction both *in vitro* and *in vivo* appeared similar, possibly because they represented the same enzyme(s) (Fig. 3).

Table 2. In vivo degradation of pheromone and 5 isomers and analogs on antennae of male T. ni after 4 time intervals.

Chemical	Amount of alcohol (ng/antenna*) after:			
	5 min	10 min	15 min	20 min
Z-5-dodecen-1-ol acetate	489.6	516.3	507.5	557.2
Z-7-dodecen-1-ol acetate	222.8	325.4	501.8	486.1
Z-8-dodecen-1-ol acetate	172.4	299.3	409.4	525.3
Z-9-dodecen-1-ol acetate	475.8	581.8	458.2	552.4
E-7-dodecen-1-ol acetate	152.1	274.6	305.9	208.8
Dodecan-1-ol acetate	152.3	393.6	364.5	296.6

*Three replications of 10 pair antennae.

Table 3. In vivo degradation of pheromone and 6 isomers and analogs on antennae and legs of male T. ni by 4 sec.

Chemical	Antennae*		Legs*	
	µg alcohol/ antenna	Ratio to pheromone	ng alcohol/ leg	Ratio to pheromone
Z-5-dodecen-1-ol acetate	69.4	0.6	610.1	0.7
Z-7-dodecen-1-ol acetate	118.1	1.0	865.8	1.0
Z-8-dodecen-1-ol acetate	36.4	0.3	183.5	0.2
Z-9-dodecen-1-ol acetate	>12.5	0.1	168.1	0.2
E-7-dodecen-1-ol acetate	37.8	0.3	138.4	0.2
Dodecan-1-ol acetate	60.8	0.5	107.9	0.1
Z-7-dodecen-1-ol butyrate	29.0	0.2	161.4	0.2

*Three replications of 10 pair antennae or 20 legs.

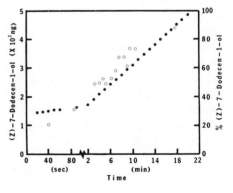

Fig. 4. Comparison of the rate of conversion of pheromone to alcohol *in vitro* (o, right ordinate); and *in vivo* (●, left ordinate), (calculated from $Y = 138.2 + 0.29 (X)$).

Discussion

The sonication technique employed in these experiments to obtain the enzyme(s) eliminated many potential sources of membranes and associated enzymes other than those of the primary olfactory receptor cells. It thus seems reasonable to ascribe some functionality of this enzyme(s) and alcohol binding protein to the olfactory process.

The preferential breakdown of the pheromone over its isomers *in vivo* (Table 3) is apparent only for short incubation intervals. As yet we have no explanation for why the pheromone was degraded less than the other isomers *in vitro* with long (1-12 hr) incubation intervals. These results were similar to those of Kasang (5) who found that bombykol was degraded less than the saturated analog on the antennae of *Bombyx mori* with incubation intervals longer than 4 sec.

These findings can be conceptually related to Kaissling's speculations on early and late inactivation of bombykol at "acceptors" in *Bombyx* (1). However, the conversion of the pheromone to the alcohol may have a greater importance than only inactivation of the stimulus. Single unit recordings (Mayer, unpublished data) indicate the existence of two primary receptor neurons in the sensilla trichodea of *T. ni*, one that responds to the pheromone, and a second that responds to the alcohol. (A third neuron has been observed for which no defined stimulus could be detected). The electrophysiological evidence, with the observed enzymatic pheromone binding and the non-enzymatic alcohol binding, suggests that we are monitoring at least a part of two specific olfactory receptor mechanisms at the membrane level.

Acknowledgements: We thank D.L. Silhacek for helpful discussions, R.R. Rutter and J.D. James for their technical assistance and enthusiasm, and D.M. Blackwell for typing this report.

References

1. Kaissling, K.-E., In Olfaction and Taste III. (C. Pfaffmann, Ed.). Rockefeller Univ. Press, New York, 1969, p. 42.
2. Kaissling, K.-E. In Handbook of Sensory Physiology and Chemistry Senses (L.M. Beidler, Ed.), Springer-Verlag, New York, 1971, p. 350.
3. Kaissling, K.-E. In Olfaction and Taste IV. (D. Schneider, Ed.). Wissenschaftliche GMBH, Stuttgart, 1972, p. 207.
4. Steinbrecht, R.A. and G. Kasang. In Olfaction and Taste IV. (D. Schneider, Ed.) Wissenschaftliche GMBH, Stuttgart, 1972, p. 193.
5. Kasang, G., Naturwiss. 60, 95 (1973).
6. Mayer, M.S. In press.
7. Kasang, G., In Gustation and Olfaction (G. Ohloff and A.F. Thomas, Eds.), Academic Press, New York, 1971, p. 245.
8. Kasang, G. and K.-E. Kaissling. In Olfaction and Taste IV (D. Schneider, Ed.), Wissenschaftliche GMBH, Stuttgart, 1972, p. 200.
9. Ferkovich, S.M., M.S. Mayer, and R.R. Rutter. Nature, Lond. 242, 53 (1972).
10. Ferkovich, S.M., M.S. Mayer, and R.R. Rutter. J. Insect. Physiol. 19, 2231 (1973).

11. Mayer, M.S., S.M. Ferkovich, and R.R. Rutter. Specificity and localization studies of a pheromone degradative enzyme isolated from an insect antenna. In preparation.
12. Lowry, O.H., N.J. Rosebrough, A.L. Farr, and R.J. Randall. J. Biol. Chem. 193, 265 (1951).

Chemoreception, Behaviour and Polyethism in Termites (Isoptera).

Alastair M. Stuart

Department of Zoology, University of Massachusetts,

Amherst, Massachusetts 01002 U.S.A.

While it has been known for many years (e.g. Dudley and Beaumont 1889-90) that termites use chemical cues in much of their behaviour, it is only within the last decade that serious analysis of behavioural and physiological phenomena associated with chemoreception has begun. In Switzerland, Lüscher and Müller (1960), and in the United States, Stuart (1961) independently found that trail-following in the primitive termite *Zootermopsis angusticollis* (Hagen) was a chemical phenomenon with a pheromone being produced in the sternal gland. The same gland in a more advanced termite, *Nasutitermes corniger* (Motschulsky) was shown to have the same function (Stuart, 1963). Subsequently further work (e.g. Noirot and Quennedey, 1974 : Stuart and Satir, 1968) confirmed the presence and elucidated the structure of the gland in other genera and families of termites. Chemical identification of the trail pheromone in various genera has been slow and in *Zootermopsis* it is only comparatively recently that the principal component of the trail pheromone has been identified as hexanoic acid (Hummel and Karlson, 1968). In *Reticulitermes flavipes* (Kollar) a different chemical (n-cis-3, cis-6, trans-8-dodecatriene-1-ol) has been claimed to be the pheromone in that species (Matsumura, Coppel and Tai, 1968).

Apart from further confirmation on other termites of the previous findings, work is proceeding along two main lines. The first is concerned with the behavioural analysis of phenomena associated with chemical communication and chemoreception. This includes work on the evolutionary and homeostatic significance of responses to chemical cues in the environment. The second is research into the structure and function of the glandular source of the pheromone. The following will attempt to give an account of some of the recent work and to place it in context with previous work, at the same time emphasizing how the behaviour elicited is a function of the total situation during the time the chemical is perceived.

Sternal Gland

The original work on the sternal gland showed unequivocally its role in producing a pheromone used in trail following. In its most obvious function, workers of a termite such as *Nasutitermes* lay trails between their discrete arboreal carton nest and their food source on the forest floor: the trail might, therefore, be thought of as a device for foraging. Some termites, however, in which it is known that the sternal gland produces a trail pheromone (e.g. *Zootermopsis*) do not forage at great distances from the epicentre of the colony nor do they construct arcades over their trails. It was found, (Stuart 1967) that in these termites the trail was used in conjunction with mechanical excitation to recruit individuals to various sites of general excitation; such as caused by breaks in the sealed colony workings and by the presence of intruders. In general terms, the trail pheromone acts as a directional vector and the number of termites following it is a function of the amount of excitation communicated to other individuals by the initially excited termites. The excitation is expressed by a characteristic zig-zag movement and by accelerated motion. If food is regarded as an excitatory substance, it can be seen that recruitment in foraging is only a special case of "alarm". Foraging recruitment has thus been shown experimentally (Stuart, 1967) to be indeed a "futteralarm" as Goetsch (1953) suggested.

Recently, however, a second function has been demonstrated for the sternal gland (Pasteels, 1972; Stuart, unpublished) that of being involved in the pairing of adults. There is a differential response when de-alates of each sex of *Zootermopsis nevadensis* (Hagen) are exposed to extracts or the squashed sternal gland of adult males and females. A modified rectangular Wilson nest was used as a choice chamber; the extract being placed at one end, while the control was placed at the other. In this situation female de-alates were attracted to an ether extract of the sternal gland of adult males and to some extent the ether extract from adult females. With the males there was an attraction to the females but no significant attraction to extracts from the sternal glands of males.

In *Zootermopsis*, therefore, the gland first functions as a trail pheromone producer in the nymphal instars and then functions in the adult as a producer of sex attractants in both male and female. Current studies on the ultrastructure (Potswald and Stuart, unpublished) show a sexual dimorphism in the distribution and form of the cells of the adult sternal gland, as well as a difference between the larval and adult forms. This dimorphism supports the experimental behavioural findings. At this time it is not known what substances are involved in this attraction, and it is possible that the trail pheromone is supplemented in the adults by sex specific chemicals. One other possibility that could explain the phenomenon is that there is differential reception in the male and female.

At present this possibility is being investigated using scanning electron microscopy and electro-antennagram (E.A.G.) techniques.

Cephalic Gland

Many glands in insects have an alarm function. In termites an example of this is found in the cephalic gland of *Nasutitermes exitiosus*. Moore (1969) considers that the secretion from the gland, in addition to its obvious defensive function, (Ernst, 1959), acts as an alarm substance. The secretions from the Australian nasutes have been identified as limonene and α-pinene, common terpenes. Moore's observations were qualitative and it is possible that the excitation could have been produced secondarily. Recent experiments (Stuart, unpublished) have shown that there is, indeed, and "alarm" effect of the cephalic gland secretion, at least in the Panamanian *Nasutitermes corniger*. The secretion seems to have a differential effect on the soldiers and there is a significant increase in the numbers of that caste leaving the cover of their artificial nest and moving into the open (Table I #2). The numbers in the open wane quickly without further stimulation and, all in all, the effect is similar to that occuring during a general mechanical disturbance (Table I #1) except that there is not an initial drop due to the initial thigmotaxis associated with general alarm (Stuart, 1967). In both instances it is only the soldiers who leave cover; the workers remain in the nest. The excitatory function, therefore, of the cephalic gland seems to be involved in short-term rapid recruitment of soldiers.

Colony Odour and Environmental Odours

Colony odour is a significant olfactory phenomenon in termite ecology. At present not much is known about its physiological basis, but there is no doubt that members of a certain colony are able to recognize their own nest-mates by their odour as well as being able to distinguish individuals of the same species but from a different colony. A suggestion has been made (Stuart, 1970) that the termite becomes habituated gradually to the medley of odours in its own colony derived from food, pheromones whose quality and quantity would be a factor of the colony composition, species specific odours and odours adsorbed from other sources in the environment such as fungi. Information on the phenomenon has been obtained entirely from noting behavioural reactions, such as actual attack and vigorous antennal examination of foreign individuals and from experimentation where attempts have been made to alter the odour by manipulation. More information is needed about reception and it is hoped that E.A.G. and single cell electrophysiological recording will be used for the analysis of the phenomenon.

Termites have evolved a chemical discrimination for certain substances produced by fungi. This can be seen in the dramatic preference shown by *Reticulitermes flavipes* for a substance, n-cis-3, cis-6, trans-8-dodecatriene-1-ol, found in *Lenzites trabea* infested wood (Matsumura, Coppel and Tai, 1968: Smythe, Coppel, Lipton and Strong, 1967), and in the preference shown by some termites for wood attacked by brown-rot fungus (Becker, 1965).

The mode of action of the chemicals in the behaviour of the animals is in some dispute, some authors (e.g. Smythe, Coppel, Lipton and Strong, 1967) believing that the animals follow an odour gradient (similar in some ways to trail-pheromone) to the source of fungus attacked wood. Experiments (Stuart, unpublished) have shown that the stimulus actually causes high excitation in any termite encountering it. The excitation is transmitted to other termites and a trail is laid back towards the epicentre of the colony just as in a specific alarm (Stuart, 1963) caused by an intruder or a breach in a gallery. The numbers recruited in this instance will be a factor of the degree of excitation produced by the chemical, and initially will be proportional to the stimulus intensity "I", the duration of the stimulus "T", the number of termites encountering the initial stimulus "N", and inversely proportional to the distance "D" from the colony epicentre. The relationship may be expressed by the following equation:

$$N_i = K \; \frac{I \times T \times N}{D}$$

where K is a constant related to the species of termite and the colony size and N_i is equal to numbers recruited initially. The formula will hold only if other factors affecting recruitment are held constant. These will include the physiological state of the insects (e.g. recently fed or starved), habituation to the odour and the chance of excitations of a higher intensity occurring elsewhere.

Polyethic Responses to Pheromones and Environmental Chemical Stimuli

A start has been made in analyzing what effect naturally occurring odours have on the behaviour of termites when presented in various ways and in various situations. This work has been carried out on *Nasutitermes corniger* (Motschulsky) from Panama (Stuart, unpublished) and will be reported fully elsewhere. The chemicals involved were the trail pheromone, the cephalic gland secretion (see above) and the unidentified chemicals in a piece of wood. The following experimental situations were involved:

1. The mechanical excitation stimulus produced by tapping the exit of the artificial nest.

2. The presentation of cephalic gland secretion at the exit.

3. The presentation of trail pheromone as a point source at the exit.

4. The presentation of trail pheromone as a 5.5 cm. trail drawn from the exit.

5. The presentation of trail pheromone as a 5.5 cm. trail drawn from the exit but with a piece of wood at the distal end of the trail.

The results are presented in Table I.

TABLE I

Numbers of termites leaving nest in response to various experimental stimuli

Situation	Max. Nos. of termites appearing within 30 minutes of start of experiment		Nos. remaining out at 60 minutes	
	Soldiers	Workers	Soldiers	Workers
#1	17	0	2	0
#2	19	0	1	0
#3	26	0	15	0
#4	56	1	10	0
#5	91	17	40	2

See text for explanation

The animals used in the study were part of a colony supplied with wood and confined in an upturned plastic petri dish on a clean glass plate. A small 3 mm. radius semi-circular exit hole was bored in the petri dish to allow the animals egress. Temperature remained at 27°C ± 1° and the relative humidity outside the nest ranged from 65% to 75%.

When the trial pheromone was presented without being drawn out (Situation #3) an initial outpouring of soldiers occurred as with the cephalic gland secretion when similarly presented, but there was less decay in the numbers remaining outside the nest. The soldiers took up positions in a tight semi-circle and they did not venture more than 2.0 cm. from the exit after the initial surge.

When the trial was drawn out (Situation #4) the number of termites drawn out increased greatly (see Table I #4). Except for one worker, all were soldiers. In situation #5 where a piece of wood was placed at the end of the trail the results were dramatic. There was a further increase in the numbers of soldiers drawn out and for the first time workers also appeared in significant numbers. (see Table I #5).

It is obvious that the reaction to the various pheromones and attractants is dependent on the context and that the chemicals have differential effects on the workers and soldiers. Such reactions have an important effect in maintaining the social homeostasis of the colony (Stuart, 1972) and in foraging behaviour. Now that the behavioural phenomena have been discovered work will continue on analysis at the physiological level to determine whether the basis of the polyethism lies in the differences in the receptors, the C.N.S., or whether secondary communication is involved. Electro-physiological techniques will play an important role in such analysis and should also provide information on the method of chemoreception in the Isoptera.

Acknowledgement
The original research reported here was supported by a grant from the U.S. National Science Foundation (Psychobiology Section). Part support of travel was provided by a Faculty Research Grant from the University of Massachusetts.

References
Becker, G. 1965. later. Organismen 1, (Berlin), 95-156.
Dudley P.H., and Beaumont, J. 1889-90. Trans. N.Y. Acad. Sci. 8, 85-114;9, 85-114; 9. 157-180.
Ernst, E. 1959. Revue Suisse Zool., 66, 289-295.
Goetsch, W. 1953. Vergleichende biologie der Insektenstaaten, Leipzig: Gest & Portig, pp. 482.
Hummel, H., and Karlson, P. 1968. Hoppe-Seyler's Z. physiol. Chem., 349, 725-727.
Lüscher, M., and Müller B. 1960. Naturwissenschaften, 27, 503.
Matsumura, F., Coppel, H.C. and Tai, A. 1968. Nature (London) 219, 963-964.
Moore, B.P. 1969. In Krishna, K, and Lechtleitner, F., eds. Biology of Termites, N.Y. Academic Press. p. 407.
Noirot, C. and Quennedey, A. 1974. Ann. Rev. Ent., 19, 61-80.
Pasteels, J. 1972. Experientia 28, 105-106.
Smythe, R.V., Coppel, H.C., Lipton, S.H. and Strong, F.M. 1967. J. Econ. Entom. 60: 228-233.
Stuart, A.M. 1961. Nature (London), 189, 419.
Stuart, A.M. 1963a. Physiol. Zool., 36, 69-84.
Stuart, A.M. 1963b. Physiol. Zool., 36, 85-96.
Stuart, A.M. 1967. Science 156, 1123-1125.
Stuart, A.M. 1970. Advances in Chemoreception I, 79-106.
Stuart, A.M. 1972. Am. Zoologist 12, 584-594.
Stuart, A.M. and Satir, P. 1968. J. Cell Biol., 36, 527-549.

OLFACTION IN BIRDS

Chairmen: J. Le Magnen and F. Bell
Moderator: B. Wenzel

New Data on the Influence of Olfactory Deprivation
on the Homing Behavior of Pigeons[1]

N.E. Baldaccini, S. Benvenuti, V. Fiaschi, & F. Papi

Institute of General Biology

University of Pisa, Pisa, Italy

Papi et al. (1971, 1972) have reported that homing pigeons deprived of olfactory perception are incapable of correct initial orientation and either lose their way completely or return to the aviary much later than the control birds. The pigeons were deprived of olfactory perception by bilateral section of the olfactory nerves, by occlusion of both nostrils, or by section of one nerve and occlusion of the contralateral nostril. The last method seems best since the experimental birds can be compared with controls who have also undergone unilateral nerve section and occlusion of one nostril, in their case ipsilateral to nerve section. This method has produced differences for initial orientation but not for homing, possibly because experimental pigeons lost their nostril plug. The present study repeated this experiment on larger groups.

Materials and Methods

Twenty eight 2-yr. old homing pigeons, housed at S. Piero a Grado near Pisa, were classified into four groups according to number of previous releases and maximum distance as follows: A (pigeons 1-9), 28 releases, 255 km; B (10-16), 56 releases, 255 km; C (17-21), 7 releases, 131 km; D (22-28) 40-43 releases, 145 km. Release direction had varied for each group,.

Under Equithesin anesthesia, the left or right olfactory nerve was cut in all pigeons on Mar. 21-26, 1974. On April 1-4, the birds were given 4 training releases up to 57 km from the aviary. On May 6, all were released 87.1 km ENE of the aviary with both nostrils patent to provide comparative data for the test release on May 9. On May 8, one nostril of each pigeon was occluded with cement "Gron 505". The nostril ipsilateral to nerve section was treated in 12, the contralateral in 16. The groups were matched for homing experience, side of nerve section, and accuracy of initial orientation and homing performance in the control release on May 6. The test release site was 130.8 km ESE of the aviary. Groups B, C, and D had been released there once in May, 1973 and Group D once in Aug.,

[1] Supported by the Consiglio Nazionale delle Ricerche. We thank Prof. S. Dijkgraaf for stimulating discussions prompting us to repeat this experiment.

1972 also. The pigeons were always released singly. In the test, 2 controls were released alternatively with 3 experimental birds.

Results

In the control release on May 6, the birds were well-oriented and homing performance was normal. As expected, no differences were observed between the birds operated on the right and left sides. Initial orientation and homing performance of the future control and experimental birds are shown in Fig. 1. By the V-test, vanishing points cluster significantly around the expected direction in the case of both the control (u = 4.4826, p<.0001) and experimental birds (u=4.7896, p<.0001). By the Mardia, Watson, and Wheeler test (Batschelet, 1972), the distributions of the vanishing points do not differ significantly (U= 3.6006, p<.10) nor do the homing performances of the two groups (Mann-Whitney U=82, p>.05).

In the test release on May 9, there was a definite difference both in initial orientation and homing performance between the control and

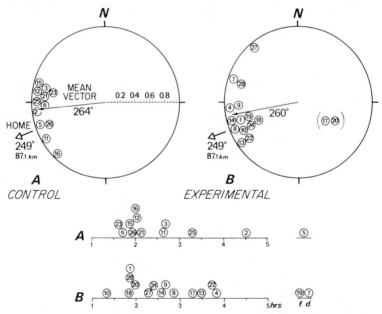

Fig. 1. Vanishing points (top) and homing time (bottom) in control release. Numbers in parentheses refer to birds that disappeared behind objects near the release point or that perched. (f.d. = following day).

experimental pigeons (Fig. 2). There is a significant clustering of vanishing points of the control birds around the home direction (u=3.0573, p<.001) which is not true for the experimental birds (u=0.1956, p>.10). Group bearings differed significantly (U=6.8120, p<.05). Seven experimental birds were lost but all controls came back. Birds 4 and 13 returned without nostril plugs and no. 28 arrived with control pigeons 25 and 26. The difference in homing performance is significant whether all of the birds are taken into account (U=24, p<.001, one-tailed) or pigeons 4, 13, 25, 26 and 28 are excluded (U=19, p<.01).

These results support the hypothesis of olfactory navigation (Baldaccini et al., 1974). Cutting one olfactory nerve does not alter homing performance but complete loss of olfactory perception prevents correct orientation and greatly reduces homing capacity.

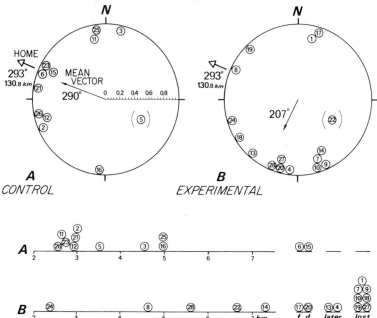

Fig. 2. Vanishing points (top) and homing time (bottom) in test release. See Fig. 1 for details.

References

Baldaccini, N.E. Benvenuti, S., Fiaschi, V., Ioale', P., & Papi, F. J. comp. Physiol. 1974, 94, 85-96.

Batschelet, E. In S.R. Galler et al. (Eds.) Animal orientation and navigation. NASA, Washington, D.C., 1972.

Papi, F., Fiore, L., Fiaschi, V. & Benvenuti, S. Monitore zool. ital. (N.S.) 1971, 5, 265-267.

Papi, F., Fiore, L., Fiaschi, V. & Benvenuti, S. Monitore zool. ital. (N.S.) 1972, 6, 85-95.

Olfactory Use in the Wedge-Tailed Shearwater
(Puffinus pacificus) on Manana Is. Hawaii

Robert J. Shallenberger

Dept. of Biology, University of California

Los Angeles, California

Introduction

Various sensory cues, including olfaction, have been implicated in behavioral observations of procellariiform birds, yet the literature abounds with far more speculation than actual field research. A notable exception is recent work with Leach's Petrel *(Oceanodroma leucorrhoa)* by Grubb (1,2). Results of similar work with birds of other orders suggests the need to expand our efforts in this area of investigation (3,4,5).

The two olfaction experiments reported here represent part of a comprehensive study of communication patterns in the Wedge-tailed Shearwater, conducted in Hawaii between 1969-1972 (6).

Deolfaction Release Experiments

On three separate nights, a total of 44 incubating birds were taken from their burrows on Manana and transported to a laboratory onshore Oahu (one mile from the nesting colony). Olfactory nerves of the first 14 birds were severed under Equithesin anesthesia. Twelve of these birds were used in the release experiment. Birds of the second group (n=10) were also anesthetized, and skull holes were drilled, but nerves were not severed. An additional 20 birds were brought to the laboratory as controls, with no surgery done.

All experimental, sham and control birds were released from the laboratory site between 2000 and 2200 hours on the night after capture. Their burrows were checked nightly for seven days, beginning the night of release.

Results

Three deolfacted birds (25%) successfully returned to their burrows during the first week, one of these on the night of release. By comparison, seven (70%) of the sham-operated birds and eighteen (90%) of the control birds located their burrows. The three deolfacted birds, all of which appeared healthy, were captured and dissected. Nerve transection appeared complete in all cases.

Discussion

These results suggest that olfactory ability is not critical for successful burrow homing in *Puffinus pacificus* when released in sight of the nesting colony at Manana. In contrast, a similar experiment with *Oceanodroma leucorrhoa* provided more convincing evidence of olfactory use (1,2). Using a larger sample, 91% of Grubb's control birds and 74% of his sham-operated birds successfully homed, while none of his deolfacted birds (n=23) returned. Similar results were obtained using birds with plugged nares.

The results of these experiments can be interpreted in several ways. Grubb assumed that "the physiological and psychological trauma of the operation cannot have differed greatly from that of the sham operation." (2). This assumption seems premature until more rigorous examination of behavior of deolfacted birds is completed. Certainly, it is possible that subtle behavioral and physiological side effects of nerve transection and/or plugging of the nares have remained undetected thus far, and could account for these results.

Also, in absence of significant predation, the shearwaters on Manana are not nearly so bound to a nocturnal cycle of return to the colony as the petrels of Kent Island (1). Clouds over the colonies in Hawaii are usually scattered, and Manana is rarely obscured by rain or fog for extended periods. The vegetative cover on Manana is generally less than 3-4 feet in height, and visual landmarks are plentiful. In contrast, returning petrels at Kent Is. must drop to their nesting grounds through a thick spruce-fir canopy. These significant differences in biology and nesting conditions may be responsible, at least in part, for apparent differences in sensory use.

Laboratory Olfactometer Experiments

In experiments with a variety of avian species, Wenzel (7) demonstrated measurable heart rate and/or respiration rate changes in response to laboratory odorants. Using similar laboratory techniques, I attempted to measure physiological responses to odors of more natural significance to *Puffinus pacificus*.

In the olfactometer apparatus (Figure 1), bottled air was piped through silica gel and activated charcoal for drying and deodorization, then split to pass through two cylinders and one pure air line. A valve permitted selection of one air flow, with simultaneous exhaustion of the other two. The selected air flow passed through a flowmeter and into the test chamber.

Paired birds were collected from Manana and tested in the laboratory. One bird of a pair was placed into the darkened test chamber, with

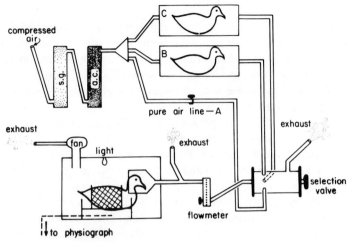

Figure 1: Olfactometer Design

TEST	MEAN MATE	MEAN NON MATE	M - NM SIG.	MEAN AIR	MEAN LIGHT
7	- 6.1	+ 11.5	P ≤ 0.05	+ 7.5	+ 12.2
8	- 3.4	+ 15.6	"	+12.2	+ 17.4
12	-12.5	+ 1.7	"	+ 9.5	+ 7.7
14	- 2.7	+ 26.6	"	+ 9.0	+ 9.5
9	+ 3.5	+ 10.1	P > 0.05	+ 8.3	+ 6.5
10	+ 4.4	+ 10.2	"	- 2.8	+ 12.1
11	+ 4.4	+ 0.5	"	- 3.4	+ 35.9
13	+ 8.0	+ 5.2	"	- 4.8	+ 11.6
16	+ 6.5	+ 6.7	"	+ 1.8	+ 9.1
	X = + 0.23	X = + 9.8	P < 0.05	X = + 4.1	X = + 13.5
			(t test)		

Table 1: Olfactometer Results (Mate vs. Non-Mate).

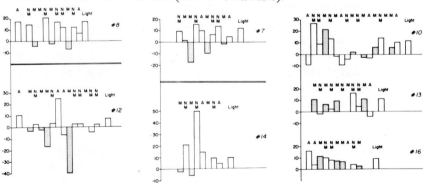

Figure 2: Mate vs. Non-Mate Test Results (Sample).

monitoring electrodes attached, and left to sit undisturbed for twenty minutes or more, until its heart rate and respiration rate had stabilized at a resting level. During this period, the mate of the test bird was placed into one of the airflow cylinders in the system. Choice of containers was varied randomly in successive tests. Another adult of the same sex as the "mate" was placed in the second cylinder.

After the initial rest period, a 15 SCFH pure air flow was established through the system for five minutes, after which the test bird received its first dosage of odorized air. To avoid the problem of pressure buildup in the lines, system air was exhausted during switching. Thus, the test bird was exposed to a consistent drop in air flow with each switch. During each test session, the bird received a series (8-12) of 20 second stimulus periods, separated by 2-5 minutes of pure air flow. Three different air flows were used as stimuli: pure air "A", odor cylinder "B" and odor cylinder "C". Order of "B" and "C" was randomized, while "A" was used intermittently for comparison. At the end of each test session, a light was turned on in the chamber as a control for level of responsiveness. Also, in a separate set of tests, nest material from the burrow of the test bird was placed in one cylinder and material from an unfamiliar burrow was placed in the other.

Results

For analysis, heart rate and respiration rate were compared during the 15 seconds prior to presentation of the odor stimulus and the last 15 seconds of the stimulus period. As was the case in similar work with *Puffinus puffinus opisthomelas* (7), respiration rate changed more consistently than heart rate in response to odorous air flows. When mean values of response (% shift) are examined, respiration rates of four birds decreased in response to mate odors, yet increased significantly during exposure to non-mate odors (Table 1, Figure 2). For the remaining five birds, respiration rate increased for both mate and non-mate odors, with no significant difference between the two.

Of four birds tested for response to familiar and unfamiliar nest material odors, three showed a consistent increase in respiratory rate in response to both odorants. The final bird displayed a respiratory rate reduction (mean = 9.7%) in response to odor of his own material and an increase (mean = 22.7%) in rate when receiving air odorized with unfamiliar material.

Discussion

While a larger sample size would have been desirable, these results appear to break down into two categories: those birds which did demonstrate a significantly different response to familiar and unfamiliar odors, and those which did not. It should also be noted that increased respiratory rate in response to unfamiliar odors is in keeping with Wenzel's results with laboratory odorants (7).

Future experimentation should be designed to reduce the stress caused by this unnatural testing situation, which, in fact, may have been responsible for the lack of differential response in several birds. Demonstrated use of visual and auditory cues in this species (6) suggests the need to devote more effort to those species in which olfactory use has been implicated by other research, such as Leach's Petrel. Attempt should also be made to isolate the odor sources (i.e. stomach oil, uropygial oil) and utilize these chemicals in olfactometer research.

Acknowledgements

I gratefully acknowledge the advice and help of Dr. Bernice Wenzel, Dr. Kenneth Dormer and Mr. William Mautner in this phase of my shearwater research.

References

1. Grubb, T., Unpub. Ph.D. Dissertation, Univ. of Wisconsin, Madison (1971).
2. Grubb, T., Animal Behavior. 22(1):192-202 (1974).
3. Stager, K., Los Angeles Co. Mus. Contrib. Science 81:63 (1964).
4. Papi, F., L. Fiore, V. Fiaschi, S. Benvenuti, Monit. Zool. Ital. 6:85-95 (1972).
5. Benvenuti, S., V. Fiaschi, L. Fiore, F. Papi, J. Comp. Physiol. 83:81-92 (1973).
6. Shallenberger, R., Unpub. Ph.D. Dissertation Univ. of Calif., Los Angeles (1973).
7. Wenzel, B. & M. Sieck. Physiol. & Behavior. 9 (3):287-293 (1972).

The Ipsilateral Olfactory Projection Field in the Pigeon[1]

Garl K. Rieke[2] and Bernice M. Wenzel

Department of Physiology and Brain Research Institute
University of California, Los Angeles
California 90024, U.S.A.

The olfactory projection field in the class Aves has not previously been examined experimentally by electrophysiological or neuroanatomical techniques. Normal brains of several species have been studied but the resulting descriptions of the olfactory structures are often inadequate (Huber & Crosby, 1929; Craigie, 1940). Avian orders vary widely in olfactory ratio, i.e., the diameter of the olfactory bulbs relative to that of the hemisphere (Bang & Cobb, 1968), with the pigeon *(Columba livia)* occupying an intermediate position. Thus, the pigeon olfactory system may be representative of many orders but may also differ widely from those at more extreme positions.

Odorous stimuli are known to initiate effector activity in the pigeon, and recent ethological and laboratory experiments have suggested that olfaction may play a role in the survival of at least some avian species (Wenzel, 1973). Interpretations of these results are limited, however, by the lack of a clear understanding of the olfactory projection field in any avian order. This report describes the ipsilateral olfactory projection field in the pigeon's telencephalon.

Methods

A total of 57 adult pigeons of both sexes in the weight range of 350-500 g were used, in electrophysiological and 31 in neuroanatomical experiments.

For electrophysiology, the pigeons were anesthetized with Equithesin (2 ml/Kg i.m.) and placed in a stereotaxic instrument. The skull was carefully removed over the dorsal telencephalon and the stimulating electrode was passed through the sinus overlying the bulbs without otherwise damaging the sinus. All exposed neural tissue was covered with

[1]Supported by NIH grant no. NS 10353. We thank Pamela Cummings, Irene Jones, Harold McCaffery, and James Michaud for technical assistance.
[2]Now at: Dept. of Anatomy, Hahnemann Medical College, 230 N. Broad St., Philadelphia, Penna. 19102.

mineral oil to prevent dessication. A small screw embedded in the skull over the cerebellum served as the indifferent electrode for recording evoked responses. The ipsilateral forebrain was then systematically explored with recording electrodes.

All electrodes were constructed from stainless steel size 00 insect pins held in seamless hypodermic tubing (0.028" OD; 0.00625" wall) and were insulated with Insl-X except at the tip. Bipolar stimulating electrodes consisted of two separate insulated electrodes held together. Resistances ranged from 100-150 KΩ. Test stimuli were monopolar square-wave pulses (1 msec. 2-10 V, 0.5-50 Hz) from a Grass S88 stimulator. To test for post-tetanic changes, 1-msec, 10-V pulses were presented at 10 Hz for 10-15 sec. Frequency following was measured over a range of 0.5-50 Hz in stimulus frequency using 1-msec, 10-V pulses. These two tests provided an estimate of the relative number of synapses between the olfactory bulb and the recording sites (Patton, 1965). All stimuli were passed through a stimulus isolation unit. Responses were amplified by Grass P15 amplifiers with a band pass of 1 Hz to 3 KHz and wired for single-ended recording, displayed on an oscilloscope, and photographed. Onset latencies were measured on traces recorded at a sweep speed of 5 msec/division for fast responses and at 10 or 20 msec/div. for slower ones. In all records an upward deflection was positive in polarity.

The locations of all electrode tips were marked by passing 100μ A dc current for 10-15 sec through the desired electrode. The pigeon was then perfused through the heart with a mixture of 10% formalin and 1% potassium ferrocyanide. The tissues were processed histologically (10-μ paraffin sections stained with carbolfuchsin and luxol fast blue) and examined microscopically to confirm electrode sites using the nomenclature of Karten and Hodos (1967).

The Fink-Heimer I method (1967) was used for detection of degenerating axon cylinders and axon terminals. In all birds except the sham-operated controls, one olfactory bulb was partially ablated by a sharp pointed scalpel inserted through the exposed sinus. The anterior two-thirds of the bulb was removed, leaving intact the posterior portion which contains the anterior olfactory nucleus. In the controls, only the sinus was broken. After bleeding had stopped, the exposed areas were covered with Gelfoam and the incision was closed. Predetermined survival periods ranged from 6 hrs. to 8 days. At the appropriate time, each bird was anesthetized, exsanguinated through the left ventricle with physiological saline, and perfused with 10% formalin. The brains were removed from the cranial vault and stored in 10% formalin for two weeks when they were placed in a mixture of 10% formalin and 35% sucrose for

362

at least 3 days. The tissues were frozen and cut at 25μ in either the coronal or sagittal plane. Every fifth section was processed by the Fink-Heimer I technique, while adjacent sections were stained with a nuclear stain (carbolfuchsin). The patterns of degeneration were recorded on camera lucida drawings of the sections.

Results and Discussion

Single pulse stimulation of the olfactory bulb elicited two classes of responses from the ipsilateral forebrain, designated Type I and Type II. Type I responses were recorded from the ipsilateral cortex prepiriformis (CPP), the hyperstriatum ventrale (HV), and the lobus parolfactorius (LPO) (Fig. 1). These responses were primarily positive in polarity and had onset latencies ranging from 4-8 msec. They could not be driven faithfully beyond 20 Hz and they demonstrated short duration post-tetanic depression of 30 sec.

The Type II responses (Fig. 2) were recorded from the hyperstriatum dorsale (HD), the hyperstriatum accessorium (HA), the nucleus septalis (S), the paleostriatum augmentatum (PA) and primitivum (PP), and the neostriatum (N). These responses, unlike Type I, showed a marked reduction in amplitude over the stimulus frequency range of 4-8 Hz and

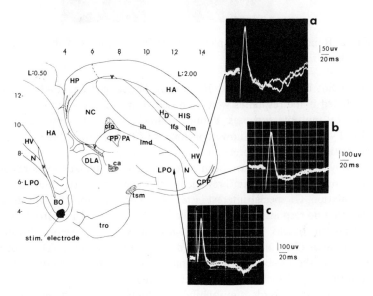

Fig. 1. Representative examples of Type I responses and the areas where they were recorded.

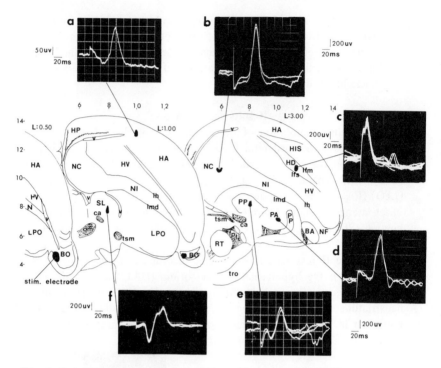

Fig. 2. Representative examples of Type II responses and the areas where they were recorded.

could not be driven beyond 16-18 Hz. The reduced amplitude would be predicted if the responses were recorded from centers separated from the bulb by multiple synapses. Poorest following, to 5 Hz only, occurred in HA. Type II responses showed prolonged (40-90 sec) post-tetanic changes, consisting mainly of depression.

Only positive sites have been reported here. Systematic exploration of the ipsilateral forebrain revealed no other response loci to bulb stimulation. Of special interest in terms of earlier anatomical reports is the fact that no responses could be evoked in the ipsilateral archistriatum or in the nucleus basalis. Whether Type 1 responses could be recorded from the tuberculum olfactorium (TO) was difficult to determine. TO is not well formed in the pigeon and blends imperceptibly with the overlying LPO, a structure from which Type I responses were obtained.

In all lesioned birds, argyrophilic granules were observed upon the soma and initial processes of neurons in the ipsilateral CPP, HV, LPO, and the nucleus accumbens septi (Ac). Fine degenerating fibers were traced as three loosely organized pathways – to CPP, to HV, and to LPO and Ac. Degenerating fibers associated with the third pathway were traced into the anterior commissure and were observed to enter the contralateral forebrain. This projection will be considered in a later paper.

Fine dense degeneration was observed throughout the glomerular lamina of the ipsilateral olfactory bulb in all preparations because the olfactory nerve was transected when the lesion was produced. Many of the mitral cells in the uninjured portion of the lesioned olfactory bulb had dense argyrophilic granules on their somas probably representing the degenerating collaterals of mitral cells destroyed by the lesion. Fine degenerating fibers were traced through the granule cell core of the olfactory bulb into more caudal ipsilateral structures as described below.

As early as 6 hrs. following damage to the bulb, degenerating fibers could be traced from the bulb into the adjacent CPP, with a light scattering to the HV and LPO. There was no evidence at this stage or at later stages of any well formed, compact olfactory tracts. The severity of degeneration had increased by 18 hrs., being prominent in the CPP and more extensive in the HV and LPO. Degeneration patterns were distinct 24 hrs, after the lesion, with three loosely organized pathways clearly obvious. The density of degenerating fibers and the numbers of obvious terminals reached a maximum at 4 to 6 days (Fig. 3). The intensity of degeneration decreased markedly by the eighth postoperative day.

There was only slight degeneration in the paleostriatum, consisting of fibers at the boundary between the LPO-Ac complex and the PA with no terminal degeneration in PA. A light scattering of degenerating fibers and an occasional cell with argyrophilic granules was seen inconsistently in the septal region. In those birds where such degeneration was present, the degenerating fibers were concentrated most heavily in the vicinity of the anterior commissure, which suggests that the degeneration represented fibers to the commissure and that these fibers of passage spread loosely around the commissure as they converge on it (Zeier & Karten, 1973).

In other areas of the ipsilateral forebrain, there was a clear demarcation between those centers with and without degeneration. For example, the neostriatum (NF, NI, NC) contained no signs of degeneration while the LPO, which is separated from N by the lamina medullaris dorsale, contained many degenerating fibers and terminals. The HV was likewise distinctly bounded, being separated sharply from N ventrally and from HD and HA dorsally. In addition to the lack of degeneration in N, HD, and HA, there was no evidence of degeneration in the nucleus basalis and only a few silver grains appeared in the archistriatum.

The neuroanatomical and electrophysiological data suggest the

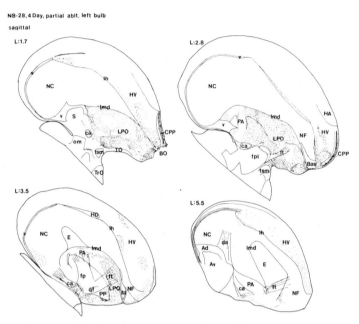

Fig. 3. Sagittal sections through the left forebrain 96 hrs. postoperative. BO = olfactory bulb.

following model for the ipsilateral component of the pigeon's olfactory projection field. Those areas from which Type I responses were obtained following bulb stimulation (the CPP, the HV, and LPO-Ac) appear to constitute the primary olfactory projection field. This hypothesis is strongly supported by the observation of degenerating fibers to, and terminal degeneration within, these areas following lesions to the ipsilateral olfactory bulb. Those areas from which Type II responses were elicited following ipsilateral bulb stimulation (S, N, PA-PP, HA, HD) constitute secondary or higher-order areas of olfactory input. The absence of degeneration in these structures supports this conclusion, showing that they are removed from the olfactory bulb (mitral cells) by at least one extrabulbar synapse, which is probably in the CPP, the HV, or the LPO-Ac.

There may be fundamental differences between the olfactory projection fields of various avian orders. For example, the absence of a well-formed olfactory tract in the pigeon is in contrast to the kiwi *(Apteryx australis)* (Craigie, 1930), a bird with very large olfactory bulbs. The projection to the CPP in the pigeon agrees with the report of Huber and Crosby (1929) for the sparrow (presumably *Passer domesticus*), but the absence of degeneration in the ipsilateral nucleus basalis and archistriatum of the pigeon conflicts with their description.

The wide distribution of olfactory fibers in the pigeon is striking. Unlike most reptiles and mammals, birds have no accessory olfactory bulb and the projection field described here originates entirely in the main bulb. Such a broad distribution field suggests that the system may be involved in more than direct olfactory activities. In mammals, its role in behavior patterns usually associated with the limbic system has been increasingly well documented and comparable effects have been described for pigeons (Wenzel, in press).

It now seems that olfaction plays a role in species-sustaining activities of birds such as feeding (Stager, 1964; Wenzel, 1971), homing (Papi et al., 1973), and burrow location and recognition of young (Grubb, 1971). Exactly how the olfactory system contributes to these behaviors is difficult to specify until the nature of the inputs to each primary olfactory center is known. The question is further complicated by our limited understanding of more central connections of these primary centers. Olfactory-related inputs may reach the hypothalamus via the medial forebrain bundle after a synapse in the LPO (Karten & Dubbeldam, 1973). As in the rat (Heimer & Wilson, in press), ipsilateral olfactory input reaches the basal ganglia, in this case through unidentified relays to PA and PP. Olfaction, as well as other special senses (Karten et al., 1973; Perisic et al., 1971), reaches the Wulst as evidenced by Type II responses in two of its components, the HD and HA. Fuller discussion of the role of primary olfactory centers in complex behaviours will accompany our presentation of the projections to the contralateral telencephalon.

References

Bang, B.G. & Cobb, S. Auk, 1968, 85, 55-61.
Craigie, E.H. J. Comp. Neurol., 1930, 49, 223-357.
Craigie, E.H. J. Comp. Neurol., 1940, 73, 179-234.
Fink, R.P. & Heimer, L. Brain Res., 1967, 4, 369-374.
Grubb, T.C. Jr. Olfactory navigation by Leach's petrell and other procellariiform birds. Doctoral Dissertation, Univ. of Wisconsin, Madison, 1971.
Heimer, L. & Wilson, R.D. In M. Santini (Ed.), Golgi Centennial Symposium, Raven Press, N.Y., in press.
Huber, G.C. & Crosby, E.C. J. Comp. Neurol., 1929, 48, 1-225.
Karten, H.J. & Dubbeldam, J.L. J. Comp. Neurol., 1973, 148, 61-89.
Karten, H.J. & Hodos, W. A Stereotaxic Atlas of the Brain of the Pigeon *(Columba liva)*. John Hopkins University Press, Baltimore, 1967.
Karten, H.J., Hodos, W., Nauta, W.J.H., & Revzin, A.M. J. Comp. Neurol., 1973, 150, 253-278.
Papi, F., Fiore, L., Fiaschi, V., & Benvenuti, S. J. Comp. Physiol., 1973, 83, 93-102.
Patton, H.D. In Ruch, T.C. & Patton, H.D. (Eds.), Physiology and Biophysics, 19th edn., Saunders, Philadelphia, 1965.

Perisic, M., Hihailovic, J., & Cuenod, M. Intern. J. Neuroscience, 1971, 2, 7-14.
Stager, K.E. L.A. County Museum Contributions in Science, No. 81, 1964.
Wenzel, B.M. Ann. N.Y. Acad. Sci., 1971, 188, 183-193.
Wenzel, B.M. Chemoreception. In Farner, D.S. & King, J.R. (Eds.), Avian Biology, vol. 3, Academic Press, N.Y., 1973.
Wenzel, B.M. In DiCara, L.V. (Ed.), Limbic and Autonomic Nervous System Research, Plenum, N.Y., in press.
Zeier, H.J. & Karten, H.J. J. Comp. Neurol., 1973, 150, 201-216.

Recovery of Olfactory Function in Pigeons after Bilateral Transection of the Olfactory Nerves

Don Tucker and Pasquale P.C. Graziadei

Department of Biological Science

and

James C. Smith

Department of Psychology

Florida State University

A decade ago birds were being included in the comparison of mammals, amphibians and reptiles with electrophysiological recording techniques applied to the peripheral olfactory system. The results were similar for species of birds that covered a wide range of development of the olfactory system (Tucker, 1965).

The influence of Audubon was still strong and it was felt incumbent to prove that birds can "smell". Application of a modified conditioned suppression technique in this laboratory showed that pigeons can make odor quality (Henton, 1969) and intensity (Shumake et. al., 1969) discriminations. An effort was made to discern a role for olfaction in natural behaviour. The late Daniel Lehrman tested ring doves in 1971, which had had the olfactory nerves sectioned by Smith, to no avail for deficiencies in courtship, mating, nest building, egg laying and young rearing. However, postmortem examination by Graziadei revealed that the nerves either had not been cut (unbelievable) or had regenerated. The result has been a continuing investigation of the phenomenon in pigeons in this laboratory (Oley et al., 1975).

We now report that pigeons have recovered from bilateral section of the olfactory nerves as judged by the application of electrophysiological, behavioral and ultrastructural methods of study.

Methods

The pigeon was selected because of the previous experience in this laboratory with electrical recording and behavioral experimentation in this species.

Surgery. Olfactory nerves were visualized by scraping away pneumatized bone between the eyes and removing small portions of the bony orbital walls. A nerve was dissected free, lifted with a hook and cut with scissors. The cut ends were approximated by returning them to the sheath.

Electrophysiology. For recording, a nerve was exposed as described above, connecting air spaces were closed off with bone wax and a small twig was dissected free under Ringer solution. The nerve twig was placed on Pt-Ir wire electrodes under mineral oil and the asynchronous neural spike activity was processed with an AC-to-DC converter to record a running average of the baseline and response magnitudes. Stimuli of controlled concentrations of amyl acetate were alternated with a clean air background by means of an olfactometer.

Behavior. Pigeons were trained in a breathing chamber to peck a key at a high rate by intermittently rewarding them with grain. While pecking, odor, the conditioned stimulus (CS), was presented for 20 seconds and terminated by a brief electric shock, the unconditioned stimulus (US), delivered to the pubic bones. During the 20 sec CS the pigeon suppressed key pecking, apparently in anticipation of the US delivered at the termination of the CS. The magnitude of the response suppression was measured by comparing CS responses with responses in a comparable time prior to CS onset. To control for nonolfactory cues, possibly produced by switching of the air lines to the breathing chamber, zero concentration control trials were presented during which the odorized air stream was adjusted for zero concentration. Birds were tested for olfactory discrimination before and after bilateral sectioning of the olfactory nerves.

Morphology. Some of the behavioral birds were used for anatomical observation. In another series, unilateral nerve sections were performed and sacrifices were made at varying time intervals. Gross examination with the dissection microscope was made during the procedure for securing pieces of olfactory mucosae, nerves and bulbs, after intravascular perfusion with buffered aldehydes. Tissue specimens were prepared and embedded in Araldite. Thick (1μm) sections for light microscopy (LM) and thin (800 A) sections for transmission electron microscopy (TEM) were cut and stained appropriately.

Methods are described in much greater detail by Oley et al. (1975).

Results

An ineluctable fact is that all the olfactory nerves sectioned more than a month previously grew back. Healed nerves in the early stages often bore neuromas, were smaller than contralateral control nerves and were enmeshed in extensive scar tissue. The gross anatomy approached normality within a year except for the occasional appearance of decussation in bilaterally sectioned birds, which is an abnormality. The arrangement of small blood vessels within the nerve, seen in the almost transparent tissue of the electrical recording preparations, was less regular in the healed nerves.

The electrophysiological results did not differ between controls and nerves cut 6 months or more before. In particular, autonomic reflex responses to odorant stimulation (Tucker, 1971) were reflected in the records.

The time for recovery of behavioral discrimination varied and there was a suggestion of correlation with the extent of neural and bulbar regeneration. Four pigeons were trained to criterion with a CS of amyl acetate at 3.2% vapor saturation, the nerves were resected and the birds were tested until they regained criterion performance in 16, 52, 76 and 82 days. The sacrifice was performed too late to ascertain whether the nerves had been completely sectioned in the 16-day subject.

Another group of 4 naive birds had their nerves cut first, conditioning was begun after about 150 days to allow regeneration and the CS was presented for the first time 211-231 days later. Two acquired the olfactory discrimination during a single session and the other pair within 4 sessions. About a year after the initial lesion, the nerves were sectioned again and the birds were tested with the CS for 10 sessions. Their suppression was nil, indicating no perception of odor, and ruling out trigeminal involvement.

A group of 3 (the 4th never met criterion) was put on a differential odor discrimination problem. Amyl acetate was the CS and presentation of butyl acetate was not followed by shock. The initial acquisition times were 24, 63 and 69 days; after sectioning and a recovery period the reacquisition times were 100, 75 and 108 days, respectively.

As was suggested above, reconstituted olfactory nerves and bulbs were often smaller than their contralateral controls. Massive degeneration of receptor cells in the mucosa, their axons in the nerve and their axon terminals in the bulbar glomeruli appeared soon after nerve section. Then, extensive mitotic activity appeared from stem cells in the basal cell region of the mucosa and the olfactory glomeruli disintegrated. Later the new cells differentiated into new olfactory receptors, their neurites grew back to the bulb and glomeruli reformed.

The initial rejoining of cut nerve ends may have been accomplished by proliferation of Schwann cells and connective tissue elements. A greater density of Schwann cell nuclei was revealed by LM in healed nerve. The new axons appeared fewer and their average diameter was smaller by TEM. Receptor cell density in the olfactory mucosa also appeared to be less than normal. The reorganized glomeruli were smaller by LM. However, the fine

371

structural details of the axonal synapses on mitral cell dendrites did not appear to differ from normal.

Discussion

The biological basis for the degeneration and regeneration phenomena (see earlier review by Takagi, 1971) briefly described here has been worked out chiefly in the frog (Graziadei, 1973; Graziadei & DeHan, 1973; Graziadei & Metcalf, 1971). It can be described by a set of 3 interlocking statements. I) Severance of olfactory nerve fibers causes retrograde degeneration of the parent receptor cell bodies. II) Sectioned olfactory nerves can heal and this entails appearance of a completely new population of receptor axons. III) There is a normal turnover of olfactory receptors, which is stimulated massively by simultaneous axotomy of many neurons. Electron microscopic autoradiography with the use of tritiated thymidine was the methodological basis.

The electrical recording approach in the pigeon was a quick way of showing that the new receptors were indeed responsive to natural stimulation.

The recovery of behavioral function showed that at least a required amount of order was imposed on the reorganization of the system of numerous elements in the olfactory bulb. Functional testing, only, can prove this point. Odor quality discrimination, as opposed to mere detection, was regained. The introduction of an auditory discrimination problem in the experiment in which the nerves had been sectioned the second time resulted in a promptly learned auditory discrimination; therefore, a general learning deficit of the type described by Wenzel et al. (1969) for pigeons with the olfactory nerves sectioned or the olfactory bulbs ablated was not revealed.

Acknowledgments

This research was supported in part by NIH Grants NS 01884, NS 05258, NS 08943, NS357-74, MH 11218 and a contract with the Division of Biology and Medicine, USAEC, AT-40-2903.

References

Graziadei, P.P.C., 1973. Tissue Cell 5, 113.
Graziadei, P.P.C., and R.S. DeHan, 1973. J. Cell Biol. 59, 525.
Graziadei, P.P.C., and J.F. Metcalf, 1971. Z. Zellforsch. 116, 305.
Henton, W.W., 1969. J. Exp. Anal. Behav. 12, 175.
Oley, N., R.S. DeHan, D. Tucker, J.C. Smith and P.P.C. Graziadei, J. Comp. Physiol. Psychol. (in press).
Shumake, S.A., J.C. Smith and D. Tucker, 1969. J. Comp. Physiol. Psychol. 67, 64.
Takagi, S.F. in Handbook of Sensory Physiology, Vol. 4, Pt. 1, L.M. Beidler, Ed. (Springer-Verlag, New York, 1971), pp. 75-94.
Tucker, D., ibid, pp. 151-181.
Tucker, D., 1965. Nature 207, 34.
Wenzel, B.M., P.F. Albritton, A. Salzman and T.E. Oberjat, in Olfaction and Taste lll, C. Pfaffmann, Ed. (Rockefeller Univ. Press, New York, 1969), pp. 278-287.

Olfaction in Birds

Chairmen's Summary

J. Le Magnen and F. Bell

College de France, Paris

Royal Veterinary College, London

The moderator, Bernice Wenzel, opened the session with a brief discussion of the possible significance of the wide variation in the relative size of olfactory bulbs among avian species. As true vertebrate wings began to develop during the upper Jurassic and lower Cretaceous periods, it is easy to imagine that eyes and inner ears became extremely important for guidance and communication. Visual and auditory signals are so obviously significant for existing birds that the nose has been largely overlooked as a distance receptor. The assumption of a positive correlation between evolutionary age and olfactory bulb size is not well supported if olfactory ratios for extant representatives of avian orders (Bang & Cobb, 1968) are superimposed on a diagram showing their times of emergence and development (Austin, 1961). Such a comparison may be fallible but is the only approach unless good paleontological information about brain structure becomes available. If bulb sizes are distributed instead according to such items as habitat, preferred food, location of nest, etc. (Bang, 1971), some striking agreements appear which imply that the olfactory system has adapted to certain aspects of avian behavior. Stager's work on vultures supports this idea (1964). He showed that one species, *Cathartes aura,* is characterized by large olfactory bulbs in both the Pleistocene and current periods, in contrast to other vultures, and that this form depends most heavily on odor cues for orienting toward a food source. Such evidence suggests differing degrees of olfactory reliance in closely related species as a result of early variation possibly associated with different behavioral tendencies.

During the session, evidence was presented concerning the significance of olfaction for homing pigeons and for wedge-tailed shearwaters, the extensive distribution of secondary and higher-order olfactory neurons in the pigeon's telencephalon, and the ability of the pigeon's olfactory nerve to regenerate to preoperative functional capacity. Several topics were clarified, as follows, in the limited discussion period.

In reference to Papi's work:

Stuart (USA) – Does Papi know the nature of the experimental pigeons' disorientation? There is a question of how to relate his hypothesis to those of other workers in the field of migration.

Wenzel — Papi's hypothesis refers to the area surrounding the aviary. His releases have been made at distances up to 150 km away. Once the pigeons have disappeared from the release point, he has no information about their behavior except that they do or do not return to the loft. Keeton at Cornell University is engaged in repeating some of Papi's experiments and his results on the effects of nerve section should be available next year. Such replication is very important.

In reference to Shallenberger's work:

Laing (Australia) — In using respiration and heart rate as an index of odor perception, are there some odors that elicit no response?

Shallenberger — Yes. In Wenzel's earlier experiments, only laboratory chemicals were used as stimuli with most birds and little attempt was made to present odors associated with natural situations. In my own work, only natural odorants served as stimuli with no identification of the critical components, although I have ideas about what they are in some cases.

In reference to Tucker's report:

Halpern (USA) — Do I understand that, during regeneration of the olfactory nerve, new receptor cells are differentiated from the basal cells and send their axons toward the olfactory bulb?

Tucker — That is correct.

Miller (USA) — To what extent is there somatotopic representation of the olfactory mucosa in the olfactory bulb and is it different after regeneration? This relationship is relevant to Mozell's chromatographic hypothesis of odor coding.

Tucker — We have no definite answer. There seems to be rough agreement in the normal animal but we have not studied it after regeneration. Crossing the cut nerves will give interesting results on this point.

Zippel (Germany) — We have studied regeneration of the olfactory nerve in the goldfish where a somatotopic relationship does occur and we find that it is reduced after regeneration but we have worked only with relatively old fish. Our preliminary results agree that the receptor cells disappear completely and then regenerate from basal cells. If the olfactory bulbs have been removed, the axons grow toward the forebrain but we do not know to which part nor do we know anything about their functional status.

Oakley (USA) — Can the pigeons discriminate between two odors after regeneration and have any birds been trained preoperatively and tested after regeneration?

Tucker — Yes, to both questions. Of 4 birds trained postoperatively,

3 met criterion for discrimination between amyl and butyl acetate, showing that functional recovery permits more than mere odor detection. In the case of 4 pigeons that were trained before operation, 1 reached postoperative criterion in 15 days, which was suspiciously early, while the others reached it in 60-80 days. Because postoperative testing was repeated until criterion performance was achieved, it is still possible that improvement was due to new learning rather than to restoration of a habit acquired previously.

Pfaffmann (USA) — Do you think the memory trace was present after regeneration?

Tucker — We cannot say at present.

References

Austin, O.L., Jr. 1961. Birds of the World, New York, Golden Press.
Bang, B.C. 1971. Acta Anatomica, Suppl. 58 = 1 ad Vol. 79, 1-76.
Bang, B.G. and Cobb, S. 1968. The Auk, 85, 55-61.
Stager, K.E. 1964. *(Cathartes Aura)*. L.A. County Museum, Contributions in Science. No. 81.

OLFACTION AND PHEROMONES IN ANIMAL BEHAVIOUR

Chairmen: R. Bradley and J. Coghlan

Olfactory-Endocrine Interactions and Regulatory Behaviours

J. Le Magnen

Laboratoire de Neurophysiologie Sensorielle

et Comportementale. College de France. Paris

In the phylogenetic evolution of animal species through natural selection, defensive, ingestive and sexual behaviours and their respective neuroendocrine controls have been the most "selective" physiological functions. In every species, the survival of young, from birth to maturity, is critically dependent on the physiological mechanisms underlying feeding and drinking and active defense against predators and various physical stresses. When safe at maturity, the reproduction of each individual and the genetic transmission of its selected characteristics are only permitted by an efficient sexual behaviour and its appropriate interaction with endocrine events. This high survival value and selection pressure of the three behaviours favoured the differentiation of highly adapted sensory systems and of the most efficient neural organization of these behavioural mechanisms. So strongly selected, the particular sensory systems and sensory cues driving such behaviours may be postulated to be themselves of the highest survival value for the living animal. In almost all animal families, the predominant use of chemosensory systems in the control of feeding, in defensive reactions as in sexual and other social behaviours, is thus suggested to be the result of a critical biological adaptation.

The vital advantage provided by a chemosensory control of behaviour lies in four different characteristics of chemical stimuli and sensory systems.

1. by a chemical analysis of its environment the animal is allowed to exhibit or to develop anticipatory reactions towards chemicals and their sources, in relation to their biochemical properties. At the receptor level, the analysis may be already biochemical in nature with the discrimination of groups of substances identical in some of the biochemical functions (carbohydrates or toxic compounds, for instance).

2. when the chemical is an excreted body odor, it conveys as a sensory stimulus, information not only on the identity of the source but also on its various physiological conditions (sexual receptivity, for example).

3. as volatile or dissolved compounds, chemical stimuli possess specific and advantageous properties in their production and propagation. As such, they form an indispensable communication system.

4. the neuronal organization of the central projections of olfactory and gustatory afferents provides basis for chemosensory, endocrine, reciprocal

interactions which are a common and essential feature of ingestive, defensive and sexual behaviours.

The first three points have been discussed elsewhere (13). In this introductory overview, particular attention will be focused on the one hand on recent advances in the knowledge of the neural organization of chemosensory and particularly of the olfactory input and, on the other hand, on new findings of olfactory endocrine reciprocal interactions.

The neuronal organization of central olfactory projections

For a long time, it was thought that the olfactory input was essentially an extrathalamic system and that this characteristic was unique among sensory modalities. The two assertions have been now ruled out. Confirming Allison's pioneer work (1), the presence of an olfactory thalamic relay and of a neocortical projection is now well documented (21-5). In other sensory modalities, extrathalamic pathways have been recognized or suspected. Via the accessory optic tract and direct pathways to the telencephalon, visual afferents are involved in behavioural and endocrine responses synchronized, for example, by an external photic cycle (9-19). The inferior colliculus projection system seems primarily concerned in the participation of acoustic signals in basic behavioural responses such as spatial localization and echolocation (3). In the gustatory system, a direct and extrathalamic pontotelencephalic projection is suspected and suggested to be primarily involved in feeding and drinking systems (22-33). From this data emerges the notion of a generalized dual system of sensory projections in the C.N.S. subserving two different levels of behavioural functions. Extrathalamic projections, prevailing in the olfactory system, would be primarily involved in fixed behavioural responses to emergency situations in the internal or external environment. Thalamic projections and thalamo-neocortical interactions would be specifically the basis for anticipatory and accurately adaptive responses in various behaviours.

In the last decades, neuroanatomists using new histological techniques have provided new decisive insight into the neuronal organization of connections between the olfactory bulb and diencephalon. The lateral olfactory tract and its terminal projections in the olfactory tubercle, amygdala and the prepiriform cortex and the olfactory anterior nuclei and anterior commissure provide ipsi and contralateral pathways to various sites in the hypothalamus that are now well recognized. The lateral and medial olfactory tracts contain centrifugal fibers afferent to the bulb suggesting a neuroanatomical basis for a feedback loop from the hypothalamus to the bulb (4-5-28-29-7-27).

382

Olfacto-endocrine interactions and the behavioural arousal

Considerable evidence exists that the hypothalamus is at least a crossing road of the external and internal input of information to the C.N.S. leading to behavioural arousal or motivational states. Via the humoral route, peripheral factors acting on hormonosensitive sites in the hypothalamus bring about the specific arousal (hunger or thirst for instance) in a complex relation with the external sensory input. Typically and in relation to the established arousal, reciprocal positive or negative feedbacks are permitted by the hypothalamic confluence of external and internal inputs. Endocrine or humoral factors centrifugally modulate the sensory input. External sensory afferences stimulate or repress endocrine responses. New data on these two way sensory endocrine interactions have been recently provided.

The centrifugal processing of the incoming olfactory information

The motivation dependent centrifugal processing of the olfactory incoming information, early suggested in the sexual system, has been recently demonstrated by electro-physiological and lesion studies in the field of feeding behaviour.

Twenty five years ago, it was shown that the odor sensitivity fluctuated in woman subjects in relation to the endocrine events of the menstrual cycle. A maximal sensitivity was observed at the time of ovulation corresponding to the highest level of estrogen secretion. The magnitude of the cyclic variation of threshold was the highest for the musky odor or pentadecalactone (exaltolide) and of some steroids (androstane). Hypogonadism was shown in some patients to be associated with a general hypoosmia which recovered under estrogen therapy. In addition it was found that the human sensitivity to some odors was more sex dependent in mature female subjects than in mature male or immature male or female subjects (10-11-12). This early data has been recently reexamined, fully confirmed and extended (10-8-6-30-20). The author failed to extend the finding of a fluctuation of the olfactory threshold as a function of the estrous cycle to an animal species. This demonstration was recently and beautifully performed in rats by Pietras & Moulton (26).

The role of centrifugal pathways afferent to the olfactory bulb in such hormonal modulations of the odor sensitivity was suggested early. Its electrophysiological demonstration was recently provided in analogous phenomena related to feeding and defensive behaviours. A detailed account of these experiments will be given elsewhere in this volume. Briefly it was shown in rats that the integrated multiunit discharge recorded in the mitral cell layer, in response to and only to a "palatable" food odor, was enhanced when the rat was made hungry either by food deprivation or by insulin administration (4-24-25). This selective facilitation of the olfactory input which seems to be a "disinhibition" at the mitral cell level, was suppressed after the section of the anterior commissure and the lateral olfactory tract containing centrifugal fibers afferent to the bulb. In a preliminary study, the same phenomenon was found in the response to aversive odors involved in defensive behaviour (2). In rats stimulated by the fox urine odor (behaviourally highly aversive to the rat) the number of inhibited units by the same odor was considerably reduced in the olfactory bulb isolated from higher centers by a circumscribed section.

These data provide clear cut evidence that in behavioural arousal such as hunger and fear initiated by neuro-endocrine interaction at specific hypothalamic sites, there is a feedback processing of the transmission of specific olfactory messages at the bulbar level. These facts must lead to a reconsideration of the classical concepts of peripheral and central coding of a stimulus specific message and of its behavioural significance (preference and aversion).

Effects of odor stimulation on neuro-endocrine events underlying feeding behaviour

The converse effect of olfactory input upon neuroendocrine events in the sexual system demonstrated by the now classical works of Whitten, Bruce and others, was recently studied in the field of feeding behaviour.

In unanesthetized rats, it was shown that the oral free intake of 1 ml of 50% glucose solution is followed within 2 or 3 minutes by a sharp rise of the plasma insulin concentration and by a concomitant fall of the blood glucose level. This reflexly induced insulin release is followed, after 4 or 5 minutes, by the slow rise of the blood glucose level and sustained insulin release subsequent to the intestinal absorption of glucose.

The early phase was still present and the last phase absent with an isosweet oral saccharine intake. The preabsorptive insulin release disappeared after a bilateral subdiaphragmatic vagotomy. Interestingly enough the early peak of insulin release in response to the oral glucose intake is considerably exaggerated after the electrolytic V.M.H. lesion (15).

The implication of these findings in the interpretation of the interaction between oropharyngeal stimulation and neuroendocrine events in the control of the feeding pattern is of importance. Some light on the nature of such interactions and of their presumable role in the behavioural effect of a food odor was obtained in another experiment (16).

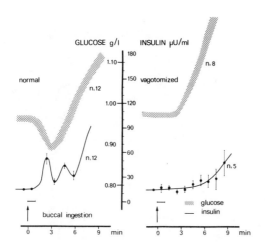

Fig. 1

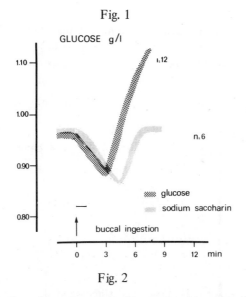

Fig. 2

It was known for a long time that after repeated insulin administration to the rat, the injection procedure alone, acting as a conditioned stimulus, induced (according to the dose of insulin used during the conditioning) either a rise or a fall of the blood glucose level. In the experiment, unanesthetized and freely moving rats, bearing a permanently implanted cardiac catheter, received, during a conditioning period, odorous stimulation by citral which was repeatedly paired with a glucose load, used as the unconditioned stimulus. Citral was previously shown ineffective

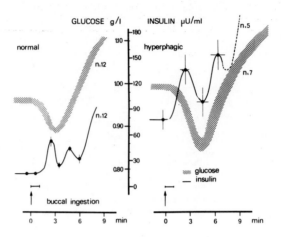

Fig. 3

when added to the familiar food to modify the ad libitum feeding pattern and also ineffective on the blood glucose level. After six daily combinations of the odor (C.S.) and glucose load (U.S.), the odor stimulation alone induced a rise of the plasma insulin concentration and a fall in the blood glucose level.

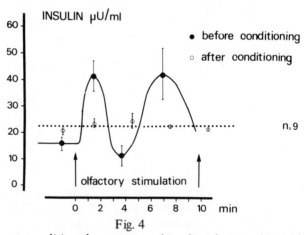

Fig. 4

This odorant, conditioned as an external insulin releaser, when added in minute quantity to the familiar food, induced a striking modification of the ad libitum food intake. The eating rate within each meal was augmented and the cumulative daily intake transiently elevated.

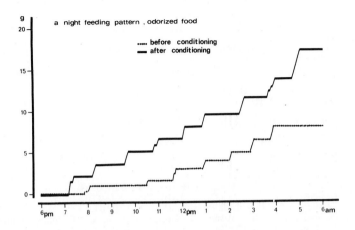

Fig. 5

These rats, before extinction, exhibit some characteristics of the feeding pattern of hypothalamic hyperphagic rats.

Thus, a chemosensory neuroendocrine positive feedback is suggested to be involved in feeding behaviour as it is in the sexual control system. An olfactory stimulus emanating from the male may enhance, in a receptive female, the gonadal secretory level which, in turn, increases the sexual arousal and thereby the behavioural efficacy of the stimulus. The same positive feedback seems to occur in the normal control of feeding behaviour. In a hungry animal, the food odor, by acting on insulin release, may reflexly exaggerate the humoral condition underlying the hunger arousal which, in turn, increases the efficient palatibility of the food and even, as shown above, the neural transmission of the external signal. Evidence has been accumulated that the electrolytic lesion of an ascending pathway in the ventromedial hypothalamus is responsible for hyperphagia and obesity by a disinhibition of a central vagus mediated control of insulin secretion by pancreatic islets. It is presumed that the V.M.H. lesion brings about an accelerated positive feedback. As a result of the considerable increase in the prandial insulin release in response to oropharyngeal cues, the short term control system of feeding would be over stimulated. The fact already shown (14), that after the V.M.H. lesion the elevated insulin response is induced by the same load given intravenously, is in agreement with this interpretation.

Thus, from such new evidence, it is confirmed that chemosensory-endocrine reciprocal interactions are a common feature of chemosensory directed behaviours permitted by the characteristic neuronal organization of central control systems.

References

1. Allison, A.C. – J. Compar. Neurol., 1953, 98, 309- 355.
2. Cattarelli, M. – Role du bulbe olfactif dans l'integration de certains messages odorants chez le rat : etude comportementale et electrophysiologique. – Thesis, 1973.
3. Erulkar, S.D. – Physiol. Rev., 1972, 52, (1), 237-360.
4. Giachetti, I., MacLeod, P. & Le Magnen, J. – C.R. Soc. Biol., 1970, 164 (4), 841-846.
5. Heimer, L. – Brain, Behav. & Evolut., 1972, 6, 484-523.
6. Koelega, H.S. & Koster, E.P. – Conf. on Odors, Evaluation, Utilization and Control, N.Y. Acad. Sci., October 1973..
7. Komisaruk, B.R. & Beyer, C. – Brain Res., 1972, 36, 153-170.
8. Koster, E.P. – Rhinol. Int., III, 1965, 1, 57-64.
9. Kostovic, I. – Brain Res., 1971, 31, 202-206.
10. Le Magnen, J. – C.R. Acad. Sci. (Paris), 1948, 226, 694-695.
11. Le Magnen, J. – C.R. Acad. Sci. (Paris), 1950, 230, 1103-1105.
12. Le Magnen, J. – Arch. Sci. Physiol., 1952, 6, 125-167.
13. Le Magnen, J. – in : Advances in Chemoreception, vol. 1., Johnston, Moulton & Turk, ed., Appleton Century Crofts, N.Y. 1970, p. 393-404.
14. Louis-Sylvestre, J. – 4th Intern. Confer. on Regulation of Food & Water Intake, Cambridge, 1971.
15. Louis-Sylvestre, J. – Unpublished.
16. Louis-Sylvestre, J. & Reynier-Rebuffel, A.M. – Unpublished.
17. Lundberg, P.O. – Acta Physiol., Scand., 1960, 49, suppl. vol. 171.
18. Lundberg, P.O. – J. Compar. Neurol., 1962, 119, 311-316.
19. Marg, E. – Ann. N.Y. Acad. Sci., 1964, 117, 35-52.
20. Meixner, C.H. – M.A. Thesis, Providence, R.I. Brown University, 1955.
21. Motokizawa, F. – Brain Res., 1974, 67, 34-337.
22. Norgren, R. & Leonard, C.M. – Science, 1971, 173, 1136-1139.
23. Norgren, R. & Pfaffmann, C. – Soc. for Neuroscience, 2nd annual meeting, Houston, 1972.
24. Pager, J., Giachetti, I., Holley, A. & Le Magnen, J. – Physiol. Behav., 1972, 9, 573-580.
25. Pager, J. – Physiol. Behav. 1974, 12, (2) 189-196.
26. Pietras, R.J. & Moulton, D.G. – Physiol. & Behav., 1974, 12, 475-492.
27. Powell, T.P., Cowan, W.M. & Raisman, G. – J. Anat. 1965, 99, 791-813.
28. Price, J.L. & Powell, T.P. – J. Anat., 1970, 107, 215-237.
29. Price, J.L. & Powell, T.P. – J. Anat. 1970, 107, 239-256.
30. Vierling, J.S. & Rock, J. – J. Appl. Physiol., 1967, 22, 311-315.

Responses to Pheromones by Mammals

W.K. Whitten

The Jackson Laboratory, Bar Harbor, Maine 04609

Introduction

Pheromones have been classified according to the type of response they induce. Those that cause a relatively slow endocrine type of response are referred to as primer pheromones, while those that induce rapid behavioural reactions are called releaser pheromes. Examples of both types of pheromones have been found in mammals, but the classification may be too simple for such complex animals. In mammals it is notoriously difficult to identify a single stimulus with a response because there are several alternate sensory pathways and the same pathway may convey alternate clues. In addition these animals learn rapidly and a single transmitter may affect both endocrine and motor responses as Pfaff (1973) has observed with LRH. We are beginning to understand some of the reactions to pheromones but as yet we do not know what role imprinting plays nor do we know how important learning is in these responses.

Releaser or Behavioral Pheromones

The first report of experimental evidence indicative of a mammalian behavioral pheromone was made by Kelley in 1937 with sheep. He showed that vaginal secretions from an estrous ewe would induce a ram to attempt to copulate with an inappropriate partner, a pregnant ewe. This evidence may be somewhat suspect because Lindsay (1965) observed that a small series of anosmic rams attempted to mate, probably following a trial and error course of behavior. Nevertheless Kelley's observations were made in ideal conditions. With the careful use of sheepdogs he arranged encounters between ram and ewe in the field, remote from observers and contamination.

Recent evidence from well controlled experiments in several laboratories (Murphy and Schneider, 1970, Lisk, Zeiss and Ciaccio 1972) have shown that mating behavior in naive and experienced male hamsters is totally dependent on olfactory clues. The pheromone is secreted by the female genitalia — the vagina, the clitoral glands or the lateral pouches — and will induce males to attempt to copulate with inappropriate partners — anosmic or anaesthetized males and castrates of either sex. Males which have been unilaterally bulbectomized respond to the pheromone when the nostril on the opposite side is open but not when it is closed (Devor and

389

Murphy, 1973). There have been no reports on the chemical nature of the pheromone but no doubt its identity will soon be known.

The above experiments clearly show that mating behavior of male hamsters is dependent on a pheromone from an estrus female and that this dependence is innate. Similar results have been obtained with mice (Rowe and Edwards, 1972). However, many years of experimentation with rats, cats, dogs, rabbits, and sheep have not yielded such clear-cut findings. There may be some survival value in this highly specific response by male hamsters because he is smaller than the female and when she is not receptive she is the chauvinist.

In some very elegant experiments, Dr. Richard Michael and his associates (1972), have identified the pheromone from the vaginal secretions of primates. As one criterion for response to the pheromone they used the willingness of trained rhesus males to work by bar pressing, for access to castrate females anointed with the pheromone − normally they would ignore such castrates. We will hear more of this interesting work and I hope be informed if the pheromone is the product of the microbial flora of the vagina and if the response to this pheromone is innate or learned.

Not all chemical communications during courtship is from the female to the male. Signoret et al, (1961) has shown that a dialogue occurs at least in pigs where the odor of the boar causes immobilization of the sow and thus permits mounting at estrus. The odor comes from the salivary glands and contains androstenol and androstenone. An aerosol mixture of these compounds is commercially available to detect estrus in sows. The proportion of sows immobilized is not one hundred percent but it can be increased if the odor is accompanied by the playing of a recording of a boar's vocalization. These observations illustrate to some extent the complexity of mammalian responses. First the pheromone is a mixture of compounds and second the response is greater when two sensory pathways, olfactory and auditory, are simultaneously stimulated.

Many other responses to pheromones have been documented for laboratory rodents. A pheromone in male mouse urine initiates aggression between males (Mackintosh and Grant, 1966) and may be the same compound which attracts or influences females. Female mice produce a substance in their urine which inhibits this aggression in males (Mugford, 1973). Other pheromones may communicate fear or stress (Valenta and Rigby, 1968). Young appear to find the nest and the nipple by olfactory clues (Moltz, Leidahl and Rowland, 1974) and retrieval of young is absent in anosmic females, (Gandleman et al. 1971). It is also probable that amniotic fluids play a role in the initiation of maternal behavior and this could be mediated in part by the sense of taste.

Much has been written and we will hear more at this meeting about territorial marking. I think that we must be aware of two common pitfalls in this field. First, some of the compounds used by animals for marking are inodorous to us and the best example is the secretion of the chin gland of the rabbit (Mykytowycz, 1962). Second, we must be careful not to ascribe a pheromonal function to every odorous substance produced by an animal. At ISOT III I asked the question, What message can civetone have for the civet? It is reported to be produced by both sexes and by related species. Perhaps the answer is that civetone is used by perfumers as a fixative for floral scents. It probably does this by reducing the vapor pressure and by augmenting the peripheral response. It could serve the same function for the civet and extend and accentuate more subtle odors such as those related to gonad function. If this is correct it could be considered a co-pheromone.

Another possibility is that civetone is used for marking home territory so that when an individual wanders from base the concentration of this substance is reduced. Thus a critical signal, reducing aggression for example, may not be conveyed by civetone but by its absence.

Primer Pheromones

In retrospect, the first clue that a primer pheromone functioned in mammals was the observation by Andervont (1944) that female mice housed in groups had a lower incidence of mammary tumor and fewer periods of estrus than did their sisters who were caged singly. In 1958 I confirmed the second of these observations and found that when the population was 30 per cage many of the animals became anestrus. I concluded that the causal stimulus was probably olfactory because experiments had shown that it was not visual, auditory, or tactile, and bulbectomy induced a similar anestrus. Since then, Champlin (1971) has shown that soiled bedding from females inhibits the cycle of certain mouse strains. There is also considerable variation between strains in their response to bulbectomy.

The first evidence to suggest that primer pheromones from the male may influence female cycles was the observation by Coleman (1950) that maiden ewes bred earlier if they had been running with vasectomized rams beforehand. This evidence is incomplete but similar findings have been made with many species. However few animals have been examined critically for such responses and such experiments are difficult to execute.

The best documented evidence for a pheromone from males that influences female reproductive performance has been obtained with the laboratory mouse. (For review see Whitten and Champlin, 1973). Females that have had their cycles inhibited by grouping show a synchronization of estrus when placed with males. Estrus cycles are shorter and more regular and puberty is advanced. When a recently mated female of a suitable strain is placed with an alien male, her pregnancy is blocked and she continues to ovulate and show estrus as if she had not been pregnant. This is now referred to as the Bruce effect after the discoverer, Bruce, (1959).

In all of the responses to the male primer pheromone there appears to be a release of gonadotrophin. If this conclusion is correct then a comparable dose of exogenous gonadotrophin should block pregnancy. To test this hypothesis Dr. Hoppe and I (1972) administered a normal ovulating dose of 5 IU of pregnant mare serum gonadotrophin to mice on various days after mating. On the first and second day pregnancy was completely blocked. When hormone was administered on the third day the number of implantations was reduced and no effect was observed following treatment on the fourth day. The median effective dose for a strain which is susceptible to normal pregnancy block was 0.6 IU and for a resistant strain it was 1.8 IU. Like pregnancy block the hormone treatment induced behavioral estrus 72 hours later, increased the rate of egg transport through the oviduct and reduced the decidual response to oil injected into the uterine lumen. Similar treatment blocked pregnancy in rats showing that even in a non susceptible species added gonadotrophin blocks pregnancy.

Further evidence indicating that pregnancy block is associated with a surge of gonadotrophin was obtained by my colleague Dr. Wesley Heamer who while a student with Geschwind at Davis developed radioimmunoassay for LH, FSH, and prolactin in mouse plasma (Beamer et al. 1972). In his experiment, females from the Parkes strain were isolated from males 24 hours after mating. Then approximately half were exposed to alien males housed behind a wire screen. Orbital plexus blood samples were obtained from these the the control animals at various intervals and individual plasma samples assayed for prolactin and pooled samples for LH and FSH. The data show that exposure to the olfactory stimuli from the alien males resulted in 1) a small but significant elevation in LH, 2) an immediate significant decline in FSH lasting approximately 48 hours, and 3) an immediate and significant decline in prolactin through the period of exposure. These data partly support the early prolactin hypothesis of Bruce and Parkes (1961) and the more recent finding of Chapman, Desjardins and Whitten (1970) showing LH release. It also indicates that FSH is involved in this complex of hormonal events. Beamer also observed an earlier release of LH and a greater surge of FSH in normal cycling females exposed to male odor.

A recent study by Bronson and Desjardins (1974) shows that when suitable prepubertal female mice were exposed to stimuli from males there was a 4-5 fold increase in plasma LH 1-3 hours after exposure and a 15-20 fold increase in plasma estradiol at 3-6 hours. The level of circulating FSH and progesterone did not change, but there was another peak of estradiol on the second day. "Normal adult-like preovulatory changes in serum LH and FSH and in plasma progesterone began late on the afternoon of the third day." The initial sequential changes in LH and estradiol during the first 12 hours were not sufficient to ensure the completion of a pubertal cycle. Thirty-six-48 hours of male exposure were necessary to accomplish this.

These findings indicate that Beamer may have missed an early major peak in LH and that he should have taken samples at shorter intervals after exposure.

From the preceding work it is clear that gonadotrophins are released in response to the male pheromone. Therefore it is safe to assume that a secretion of gonadotrophin releasing hormone (GnRH) from the hypothalamus would precede any rise in circulating LH. Thus depletion of the median eminence of GnRH may constitute the primary endocrine response to the pheromone. Eskay et al., (1974) have shown that it is possible to assay the GnRH content of the hypothalamus by using the recently prepared antiserum in a radioimmunoassay. My associate Dr. Melba Wilson is pursuing this project with F_1 hybrid mice which show good suppression and a spectacular synchronization by male pheromones.

Depletion of the hypothalamus of GnRH should occur in the first few hours of exposure and promises to be a suitable endpoint for an assay for the male pheromone. Previous attempts to assay this substance have required prolonged exposure during which it is difficult to maintain the vapor pressure and the chemical stability and at the same time prevent sensory accommodation.

We hope to use this rapid bioassay in conjunction with some chemical analysis of male mouse urine being performed by Dr. Zlatkis and Dr. Bertsch of the University of Houston and Dr. Leibich of Tubingen, Germany. In this study the methods developed for the analysis of volatile metabolites of normal and diabetic urine (Zlatkis et al., 1973) have been applied to mouse urine. The headspace of vessels containing the warmed specimen is swept with helium, which then is passed through a condenser at 4ºC, and the volatiles trapped on a porous polymer Tenax GC. The trap is inserted into a modified injector port of a Perkins Elmer Model 900 gas chromatograph and the sample eluted at 300ºC into a precolumn and finally onto the separating column. The flame ionization profiles are determined and samples simultaneously withdrawn for mass spectroscopy, the detection of sulphur compounds and for bioassay. So far over 40 compounds have been identified and two major sulphur containing compounds partially characterized. However the biological assays must await the validation of our proposed GnRH depletion assay.

Conclusions

From the foregoing it is clear that we know very little about mammalian pheromones and the responses they induce, but modern chemical and behavioral methods promise early identification of several further pheromones. From this information synthetic labelled pheromones should soon follow when we can predict rapid progress in the understanding of the responses and the pathways involved.

Acknowledgements

The preparation of this chapter was supported by NIH Research Grant HD-04083 from the National Institute of Child Health and Human Development.

References

Andervont, H.B. (1944) J. Nat. Cancer Inst. 4:579-581.
Beamer, W.G., and Murr. W.G. and Geschwind, I. I. (1972) Endocrinology, 90:823-826.
Bronson, F.H. and Desjardin, C. (1974) J. Endocrinol. 94:1658-1668.
Bruce, H.M. (1959) Nature 184:105.
Bruce, H.M. and Parkes, A.S. (1961) J. Endocrinol. 22:vi-vii.
Champlin, A.K. (1971) J. Reprod. Fert. 27:233-241.
Chapman, V.M., Desjardins, C. and Whitten, W.K. (1970), J. Reprod. Fert. 21:333-337.
Coleman, J.M. (1950) Agric. Gaz. N.S.W. 61:440.
Devor, M. and M.R. Murphy. (1973) Behav. Biol. 9:31-42.
Eskay, R.L., Oliver, C. Grollman, A. and Porter, J.C. (1974) Endocrine Soc. Abstr. 83.
Gandelman, R., Zarrow, M.X. and Denenberg, V.H. (1971) Science 171:210-211.
Hoppe, P.C. and Whitten, W.K. (1972) Biol. Reprod. 7:254-259.
Kelley, R.B. (1937) Common. Aust. CSIRO Bull. 112.
Lindsay, D.R. (1965) Animal Behav. 13:75-78.
Lisk, R.D., Zeiss, J. and Ciaccio, L.A. (1972) J. Exp. Zool. 181:69-78.
Mackintosh, J.H. and Grant, E.C. (1966) Zeitschrift fur Turpsychologie 23:584-587.
Michael, R.P., Zumpe, P., Keverne, E.B. and Bonsall, R.W. (1972) Prog. Hor. Res. 28:665-706.
Moltz, H., Ludahl, L. and Rowland, D. (1974) Physiol. and Behav. 12:409-412.
Mugford, R.A. (1973) J. Comp. Physiol. Psychol. 84:289-295.
Murphy, M.R. and Schneider, G.E. (1970) Science 167:302-303.
Mykytowycz, R. (1962) Nature (Lond) 193:799.
Pfaff, D.W. (1973) Science 182:1148-1149.
Powers, J.B. and Winans, S.S. (1973) Physiol. Behav. 10:361-368.
Rowe, F.A. and Edwards, D.A. (1972) Physiol. and Behav. 8:37-41.
Signoret, J.P. and du Mesnil du Buisson, F. (1961) In: Proc. 4th Intern. Congr. Animal Reprod. 2:171-175.

Valenta, J.G. and Rigby, M.K. (1968) Science 161:599-601.
Whitten, W.K. (1959) J. Endocrinol 18:102-107.
Whitten, W.K. and Champlin, A.K. (1973) In: R.O. Greep (ed), The Female Reproductive System. Sec. on Endocrinology. Handbook of Physiology. Amer. Physiology Society.
Zlatkis, A., Bertsch, W., Lichtenstein, H.A. Tisbee, and Shumbo, F. (1973) Anal. Chem. 45:763-767.

Olfactory Stimuli and Their Role in the Regulation of Oestrous Cycle Duration and Sexual Receptivity in the Rat
by Claude Aron

Introduction

At the present time the mechanisms whereby olfactory stimuli control sexual receptivity are still poorly understood. According to Orbach and Kling (1966) and to Edwards and Warner (1972) peripheral olfactory deprivation has no influence on sexual receptivity. Whereas Kling (1964) reported a slight decrease in mating behaviour following olfactory tract transection, Moss (1971) and Edwards and Warner (1972) claimed olfactory bulb deprivation enhanced the lordosis reflex in receptive females. Finally, some form of oestrous synchronization in 5-day cyclic rats has been described by Hughes (1964).

My aim in this brief survey will be to clarify this situation with reference to observations made in my laboratory over the last ten years on two strains of Wistar rats (W I and W II), which have proved eminently suitable for this purpose.

Firstly, both strains have been shown to display precocious receptivity when caged with a male on the night following dioestrus 3 of 5-day cycles, although early mating frequency was lower in W II, than in W I females (Asch et al., 1964; Aron et al., 1966; Aron et al., 1968; Roos et al., 1971). In W I 4-day cyclic females oestradiol priming was necessary to enhance the weak early receptivity spontaneously occuring in untreated females during the night following dioestrus 2 (Aron et al., 1961; Aron et al., 1966). Thus the olfactory control of mating could be studied in conditions of both early and oestrous receptivity in the rat.

Secondly, olfactory stimuli were proved to be involved in the control of oestrous duration, a shortening of the cycle from 5 to 4 days resulting from grouping of W I strain females (Aron et al., 1972) or exposing those of the W II strain to the odor of urine (Aron and Chateau et al., 1973). Thence it was of interest to determine to what extent changes in oestrous duration caused by pheromones would also affect receptivity in the rat.

Methodology

Three-to-four month old virgin females were used in all our experiments. Kept under a normal rhythm of natural lighting, each showed 2 to 3 regular 4-or 5-day cycles prior to use.

The following categories of animals were used: a) Intact 4-or 5-day cyclic W I females, caged in groups of 40-50 from weaning to 3 months, then either isolated or caged in groups of 9; b) Females, caged in groups of 9 as above, after bulbectomy at 2-2½ months: c) 5-day cyclic intact W II females, constantly caged in groups of 3, exposed or not to the action of pheromones by twice daily sprinkling of cage litter with male or female urine; d) 4-or 5-day cyclic W II females, bulbectomized at 6 weeks or 2-3 months.

Early receptivity was controlled either on the afternoon of dioestrous 2, for 4-day cycles, or the night following dioestrous 2 or 3, for 4-or 5-day cycles, respectively. Oestrous receptivity was controlled during the night following prooestrous (day 3 or 4 of 4-or 5-day cycles, respectively). Only males of proven fertility were used. The vaginal smear was examined following the night of cohabitation for the presence of sperm. The number of females mating in each category (mating frequency) was computed.

In some cases blood samples were obtained by cardiac puncture, and 17 β oestradiol levels estimated by radioimmunoassay.

Results

A. Olfactory bulb deprivation and sexual receptivity in the rat.

1°) *Precocious receptivity*

A statistically significant decrease in early mating frequency was apparent in W I females, bulbectomized at 2-3 months, when compared to unoperated animals (4/50 and 35/87 respectively χ^2 = 16.1; p<0.001). But oestradiol priming, on the afternoon of dioestrous 2 led to a restoration of early receptivity (17/30 and 42/70, respectively, in operated and non-operated females; χ^2 = 0.09 NS) (Aron et al., 1970). The question now was whether the decrease in early receptivity following bulbectomy could result from hormonal unbalance. 17 β oestradiol was then estimated at 5-6 p.m. on dioestrous 3 in the peripheral plasma of 10 bulbectomized and 55 unoperated 5-day cyclic females, 17 of the latter, while none of the former, mated precociously during the following night (Roser, 1974). No significant difference was observed between bulbectomized and unoperated females (41.08 ± 4.55 and 45.00 ± 2.98 pg/ml, respectively). Neither was there any difference in peripheral blood oestradiol concentration between receptive and non-receptive unoperated females (49.04 ± 6.30 and 43.19 ± 3.28 pg/ml, respectively).

2°) *Oestrous receptivity*

No significant difference in oestrous mating frequency could be observed in: a) intact, and b) bulbectomized 5-day cyclic W I females [a)

36/42 and b) 31/38; a) 26/35 and b) 23/27, respectively] (Aron et al., unpubl. obs.). The same results were obtained in unoperated and operated W II females [5-day cycles: a) 22/34 and b) 17/24, respeicitvely, NS; 4-day cycles: a) 13/16 and b) 8/10 respectively, NS] : A decrease in oestrous mating frequency occurred in both 4-and 5-day cyclic females inflicted with small stereotaxic lesions of the hypothalamic ventromedial nucleus (SVMN) following olfactory bulb deprivation (Chateau and Aron, unpubl. obs.). Only 5/21 4-day cyclic SVMN lesioned females mated against 7/10 sham operated and 10/15 control females (χ^2 = 8.76; p<0.01; 1 df). Similarly only 6/15 5-day cyclic SVMN lesioned females mated against 22/34 intact control and 17/24 bulbectomized control animals. Statistical analysis confirmed this data (χ^2 = 4.22, p< 0.05, 1 df).

B. Olfactory stimuli and sexual receptivity in the rat
$1°$) *Precocious receptivity*

5-day cyclic W II females, maintaining 5-day cyclicity depsite exposure to the odor of male or female urine (a) exhibited no increase in precocious receptivity when compared with untreated females (b) [a) 3/38, b) 7/40; a) 1/14, b) 3/21; a) 1/20, b) 6/40] (Aron and Chateau 1972; unpubl. obs.). 4-day cyclic W I females, oestrogen primed, when grouped showed a decrease in early receptivity on the afternoon of dioestrous 2 in comparison to isolated animals (63/235 against 78/176) (χ^2 = 13.65, p<0.001, 1 df). This grouping effect was completely suppressed by olfactory bulb deprivation (33/69 in bulbectomized, grouped females; 45/102 in isolated females; χ^2 = 0.22, NS). Moreover, olfactory bulb deprivation increased mating frequency in grouped females compared to unoperated animals (20/49 against 11/74; χ^2 = 10.67, p<0.01, 1 df) (Aron et al., 1969).

$2°$) *Oestrous receptivity*

There was no significant difference in oestrous receptivity between grouped and isolated W I 4-day cyclic females (121/145 and respectively). Neither was there any significant difference in oestrous mating frequency between 4-day cyclic W II females, exposed to the odor of female urine, and their untreated counterparts (11/16 against 12/16) (Aron et al., unpubl. obs.). Likewise, mating frequency remained unchanged in W II females maintaining their 5-day cyclicity despite exposure to the odor of a) male and b) female urine compared to 5-day cyclic controls [a) 30/33; 28/33, respectively, NS; b) 12/13; 14/18, respectively, NS] (Aron and Chateau, 1972; unpubl. obs.). However, quite different observations were made concerning oestrous receptivity in 5-day cyclic females, with cycles reduced to 4 days following exposure to the odor of urine. Although no significant decrease in oestrous receptivity took place in a first experimental series, a subsequent series revealed a significant decrease in comparison to 4-day cyclic control females (5/29 against 42/65; X^2 = 10.80; p<0.01) (Aron and Chateau, unpubl. obs.).

Discussion

At first sight, our findings, regarding the effects of olfactory bulb deprivation on oestrous mating in the rat, are not consistent with those of Moss (1971) and Edwards and Warner (1972), who demonstrated an enhanced lordosis reflex in olfactory bulb deprived females, whereas our own experiments showed no change in oestrous mating frequency in 4-or 5-day cyclic bulbectomized females. That the nervous structures involved in the control of mating receptivity may function in the absence of olfactory bulbs is supported by the fact that a decrease in oestrous mating frequency occured following a lesion of the hypothalamic ventromedial nucleus in unoperated as well as in bulbectomized animals. Thus, we are led to make a clear distinction between sexual receptivity i.e. the willingness of the female to accept the male, which is outside the control of the olfactory bulbs in female oestrous rats, and mating performance, which is estimated by the lordosis quotient, and is submissive to olfactory bulb inhibitive control. However, with regard to precocious mating, stimulatory or inhibitory effects may result from olfactory bulb activity. In further support of this statement is the decrease in early mating frequency in 5-day cyclic bulbectomized rats, indicating that olfactory structures exert a facilitory effect on early receptivity. The mechanism involved has been described. Oestradiol priming fully restored precocious receptivity in bulbectomized females. There being no difference in peripheral blood 17 β oestradiol concentration between unoperated and bulbectomized females the olfactory bulbs may be supposed to regulate the threshold at which behavioural nervous structures respond to oestrogens.

Of course, our experiments do not answer the question of whether olfactory stimuli, as such, are implicated in the control of early receptivity in the rat. However, the interaction shown above, between olfactory structures and hormonal factors, provides a satisfactory interpretation of the mechanisms controlling sexual receptivity in the rat. In oestrous females hormonal requirements for full receptivity are fulfilled, and there is no necessity for olfactory bulb stimulation. On the contrary early mating, which takes place at a stage of the cycle before blood oestradiol and progesterone have reached their optimal level, needs the action of these stimuli. Observations on W I females confirm this finding. However, no increase in early receptivity was observed in 5-day cyclic W II strain female rats who maintained their 5-day cyclicity on exposure to the odor of urine. This draws our attention to the role of neural structures in precocious receptivity control in the rat. Probably the olfactory stimuli failed to sufficiently lower the threshold of reactivity to oestrogens of these structures to allow early receptivity in these females exposed to pheromones.

Now, to consider the decrease in precocious sexual receptivity caused by grouping in oestradiol primed 4-day cyclics − an effect suppressed by bulbectomy − and the diminution in oestrous receptivity noted in experimental 4-day cycles following exposure to pheromones, it would appear certain that olfactory stimuli, as such, may inhibit mating behaviour in the rat. This was rather surprising, as the occurrence of normal oestrous receptivity had been shown to be independent of any olfactory influence. Again, quite unexpected were the changes in sexual receptivity observed in grouped females, where oestradiol priming resulted in an oestrous like condition, early in 4-day cycles.

Perhaps the discrepancy between these two series of results is only apparent. There is a strong suggestion that the decrease in oestrous receptivity, noted in 5-day cycles shortened to 4 days in W II females exposed to the odor of urine, did not result from a direct action of pheromones on the mechanisms controlling sexual behaviour, since similar treatment was without any effect on oestrous mating frequency in natural 4-day cyclic females. Moreover, changes have been detected in the hormonal balance of W II females whose cycles were reduced to 4 days following exposure to pheromones (Roser and Chateau, at press; unpubl. obs.). Therefore it is quite possible that, in 4-day cycles originating from 5-day cycles, the rhythm of nervous activity involved in sexual receptivity control is not adjusted to the new rhythm of steroid hormone production. Further investigation of this subject is necessary. However this conception accounts for observations made in W I females. In grouped females, as against isolated females, there is a relationship between the decrease in the number of early receptive 4-day cyclic females and the increase in the number of 4-day cycles (unpubl. obs.). This would suggest that the increased number of early non-receptive females corresponds to those whose cycles were shortened by pheromones. Certainly we do not completely understand the mechanisms by which olfactory stimuli are involved in the control of sexual behavior in the rat. However, our observations clearly indicate that oestrous receptivity, expressed in terms of mating frequency, is independent of olfactory control. They also demonstrate that olfactory bulbs may serve as a facilitory center for the occurrence of early receptivity in the rat. Finally they throw some light on the still open question of whether olfactory bulbs, as such, are involved in mating behaviour control in the rat. Certainly there is such a control as our experiments show but the action of the olfactory stimuli on sexual receptivity is probably only secondary.

References

Aron, Cl. and Chateau, D. 1971. Horm. Behav., 2, 315-323.
Aron, C.l., Asch, G. and Asch, L. 1961. C.R. Acad. Sci., 253, 1864-1866.
Aron, Cl., Asch, G. and Roos, J. 1966. Int. Rev. Cytol., 30, 139-172.
Aron, Cl., Roos, J. and Asch, G. 1968. Rev. Comp. Anim., 2, 52-71.
Aron, Cl., Roos, J. and Asch, G. 1970. Neuroendocrinology, 6, 109-117.
Aron, Cl., Roos, J. and Roos, M. 1971. J. Interdiscip. Cycle Res., 2, 239-246.
Aron, Cl., Roos, J., Asch, G. and Roos, M. 1969. C.R. Soc. Biol., 163, 2691-2694.
Aron, Cl., Roos, J., Chateau, D. and Roos, M. 1972. Ann. Endocrinol., 33, 23-34.
Asch, G., Roos, J. and Aron, Cl. 1964. C.R. Soc. Biol., 158, 838-841.
Chateau, D., Roos, J. and Aron, Cl. 1973. C.R. Acad. Sci., 276, 2823-2826.
Edwards, D.A. and Warner, P. 1972. Horm. Behav., 3, 321-332.
Hughes, R.L. 1964. CSIRO Wildl Res., 9, 115-122.
Kling, A. 1964. Am. J. Physiol., 206, 1395-1400.
Moss, R.L. 1971. J. Comp. Physiol. Psychol. 74, 374-382.
Orbach, J. and Kling, A. 1966. Brain. Res., 3, 141-149.
Roos, J., Chateau, D. and Aron, Cl. 1971. C.R. Soc. Biol., 165, 2181-2183.
Roser, S. 1974. Thèse de Doctorat ès-Sciences, 202 p., Université Louis Pasteur, Strasbourg.
Roser, S. and Chateau, D. C. R. Soc. Biol., in press.

Effects on Gonadal Steroids on Odor Detection Performance in the Rat

R.J. Pietras* and D.G. Moulton

Monell Chemical Senses Center and Department of
Physiology, University of Pennsylvania and
Veterans Administration Hospital, Philadelphia,
Philadelphia, U.S.A.

The performance of women in detecting at least nine odors fluctuates significantly during the menstrual cycle (Le Magnen, 1952; Schneider et al., 1958; Koster, 1965; Vierling & Rock, 1967; Guerrier et al., 1969; Masmaianian and Voskanian, 1972; Henkin, 1974). Apart from the study by Schneider et al (1958), the peak performance in all cases occurs around the time of ovulation. Although Le Magnen (1952) found no comparable variations in odor detection performance of female rats he did find differences in this respect between hormone-treated, castrate and normal rats. More recently we reported marked fluctuations in odor detection performance correlated with the estrus cycle in rats (Pietras & Moulton, 1974). Females, but not males, showed cyclicity in performance non-specific to the odor. The responses were maximal around ovulation and corresponded with fluctuations in the plasma concentrations of ovarian steroids. Induction of pseudopregnancy or ovariectomy of cyclic females eliminated these performance variations, but female response levels comparable to those of normal males were only obtained following treatment of ovariectomized rats with testosterone.

Although the gonadal steroids clearly influence olfactory function, the mode of hormonal action in controlling performance is unclear. In fact, target mechanisms abound between the point of arrival of odorous molecules in the nasal airways and the final efferent output. Of these, the first in sequence are the mechanisms controlling odor access to receptor sites. The patency (or degree of constriction) of the nasal airway appears to be a factor of primary importance (Tucker & Beidler, 1956) since flow rate is a first order variable in the olfactory stimulation process (Tucker, 1963). Patency, in turn, is under control of the autonomic nervous system. The sympathetic tone usually keeps the blood vessels of an organ constricted to about half their maximum diameter (Jackson 1970). Thus,

*Present address: University of California School of Medicine, Department of Physiology, Los Angeles, California, 90024.

stimulation of the sympathetic supply to the nasal mucosa results in a marked increase in the patency of the nasal airway — an increase which is associated with the enhancement in the response of the primary olfactory neurones to odors (Tucker & Beidler, 1956).

The possibility that autonomic control may, in turn come under hormonal control is supported by several lines of evidence of which the most immediately relevant are the reports that, in women, the degree of engorgement of the nasal mucosa fluctuates during the course of the menstrual cycle — the maximum congestion occurring at menstruation (Schneider & Wolf, 1960). Hypogonadal women also show changes in nasal mucous membrane function while receiving androgen and estrogen (Mortimer et al., 1936; Schneider et al., 1958).

Thus, the cyclicity in performance of female rats on an odor detection task might be determined by variations in odorant access to receptors, reflecting fluctuations in sympathetic activity. An experiment involving elimination of the sympathetic supply to the nasal region was designed to test this hypothesis.

Methods

Female Long-Evans rats (about 5 months old) were grouped 2 to a cage in a controlled 15/10 hr, light/dark cycle. Food was available *ad lib*. Water intake was restricted to 30 min daily following testing. Vaginal smears were taken daily at the same time, and only females with a regular 4-day cycle were used.

Eight rats were divided into two groups — one with three rats served as sham operated controls while the remaining five formed the experimental group. Each member of this last group was anesthetized with equithesin and the cervical sympathetic ganglion exposed and sectioned bilaterally. It lies in the same sheath as the carotid arteries (Farris & Griffith, 1963; Zeman & Innes, 1963). Postoperative examination showed bilateral ptosis of the eyelids and miosis of the pupils (cf. Cannon & Rosenbluth, 1949).

Members of the control group underwent the same operation except for the sectioning of the cervical sympathetic supply. Rats were allowed three days to recover. Odor detection testing, as described previously (Pietras & Moulton, 1974), was completed within 14 days of removal to ensure that no significant collateral sprouting could have occurred from other sympathetic neurones in this region. The rats were presented with cyclopentanone (10^{-3} of saturation) and each was given ten trials per day for eight consecutive days (in the afternoon).

In what follows we use the terms "odor detection performance" and "odor responses" (and not "olfactory sensitivity"). These terms involve no assumptions about the level at which any observed changes are initiated.

Results and Discussion

It is clear from Figure 1 that — following sympathectomy — the cyclic fluctuations in performance during the course of the estrous cycle persist. The scores of the experimental rats do not differ significantly from those of the sham-operated controls. There is, therefore, no evidence that the basic cyclicity in scores is imparted by hormonal influences on sympathetic outflow to the nasal mucosa.

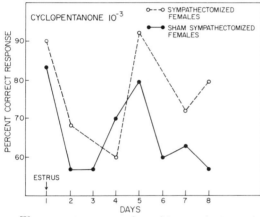

Fig. 1. Influence of bilateral cervical sympathectomy on performance of female rats detecting cyclopentanone during the course of the estrus cycle.

We cannot assume that this conclusion necessarily applies to other species. The rat is said to lack cavernous tissue in the nasal mucosa (Taylor, 1961) so there is less possibility that vasomotor changes could markedly alter nasal airway resistance. Nor is it clear to what extent a rat, as opposed to some other species, can compensate for increased airway resistance by increased intensity of sniffing.

On the other hand, the overall performance of the sympathectomized rats was superior to that of the controls. It is therefore possible that sympathectomy has some effect on the level of performance, at least during certain stages of the estrous cycle. To examine this possibility, data from another group of normal rats (Experiment 1, Pietras & Moulton, 1974), was compared with the data obtained in this experiment (see Figure 2).

The comparison shows that the performance scores of the sympathectomized rats were significantly higher than this normal group on the days corresponding to vaginal estrus ($p < .05$) and metestrus ($p < .05$). This could be taken as offering some limited support for the view that sympathetic control of nasal mucus membrane function influences performance in an odor detection task. However, there is one difficulty

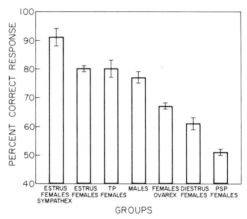

Fig. 2. Comparison of odor detection responses (Mean ± S.E. of mean) of rats in various hormonal states. Cervical sympathectomized (sympathex) females at estrus, normal females at estrus, ovariectomized females treated with 0.4 mg testosterone propionate (TP)/100gm, normal males, ovariectomized (ovarex) females, females at diestrus and pseudopregnant females were all tested on their response to cyclopentanone (10^{-3}). Data are from Pietras & Moulton (1974) and present study.

with this interpretation. Stimulation of the cervical sympathetic nerve in the rabbit increases the diameter of the nasal airways (Tucker & Beidler, 1956). Sectioning of the nerve eliminates this capacity and presumably results in a sustained reduction in odorant access to receptors. Consequently, if the rat is similar to the rabbit in this respect, we would anticipate a lowering and not an enhancement of performance on an odor detection task. This raises the possibility that sympathetic activation may be more significant for olfaction in decreasing nasal mucus secretion than in inducing vasoconstriction of nasal vessels. Since the effects are marginal, however, it is premature to speculate on the basis of this apparent anomaly, and further studies are, in any case needed to establish the magnitude of the effect.

Since we conclude that autonomic influences on the nasal mucosa (acting through the cervical sympathetic supply) cannot account for cyclicity in performance what mechanisms remain? We can see no other hormonal control of odorant access to receptors (see Pietras & Moulton, 1974). Steroid hormones, however, may act directly on central olfactory structures, non-olfactory sites such as the reticular activating system of the brain stem or some combination of these targets. Thus Curry (1971) has shown that circulating levels of estrogens can increase the excitability of the olfactory system independently of action at the receptor level. Figure 3 compares Curry's data with pooled detection responses of normal rats to the odors of cyclopentanone, eugenol and α-ionone obtained during the estrus cycle (data from Pietras & Moulton, 1974, Fig. 6). This reveals that mean thresholds for potentials in the prepyriform cortex (PCEP) evoked by stimulation of the lateral olfactory tract, peak latencies of the second negative wave N_2 of the PCEP and odor detection responses all vary in a similar sine wave fashion during the estrus cycle.

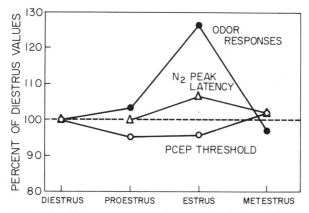

Fig. 3. Comparison of mean values for threshold of prepyriform cortex evoked potentials (PCEP), peak latency of the second negative wave, N_2 of the PCEP and odor detection performance of normal rats during the estrus cycle. All data points are presented as percentage of diestrus values.

The studies of Oshima and Gorbman (1960) also show that administration of estradiol in the frog elicits increments in the amplitude of olfactory bulb potentials evoked by electrical stimulation of the olfactory nerve. Conversely, response amplitudes were depressed by progesterone given alone or in combination with estradiol. In addition, Pfaff and Pfaffmann (1969) found an increased frequency of single unit firing in the olfactory bulb in response to odors follow injection of testosterone; changes in the rate of nasal air flow could not account for the hormone effect. Collectively, these results suggest that both behavioral and central electrophysiological responses to odor exhibit a comparable dependence on hormonal status.

We have previously shown that the cyclicity in odor detection performance in normal females varies with circulating levels of ovarian steroids (Pietras & Moulton, 1974). Elimination of these fluctuations occurred following ovariectomy or the induction of pseudopregnancy. The performance level of the pseudopregnant rats, however, was significantly less ($p<0.001$) than that of the ovariectomized females (see Fig. 2). When the circulating titer of estrogen is low and progesterone is elevated (as occurs during pseudopregnancy (Gilmore & McDonald, 1969; Hashimoto et al., 1968, performance is low. Conversely, when estrogen levels are elevated and progesterone is low (as occurs before ovulation), performance levels are high. When the levels of both gonadal steroids are low following ovariectomy, the females perform at a constant level which is almost exactly intermediate to the extremes above, but significantly less than that of normal males. The performance of ovariectomized females treated with testosterone, however, is not significantly different from that of normal males. Similarly the studies of Kawakami and Sawyer (1959; 1967) on the EEG arousal threshold show that progesterone and pregnancy raise the threshold while estrogen and androgen decrease it. The correlation of these changes with those of odor detection performance under corresponding hormonal conditions is striking.

In general, the increasing number of reports that associate hormonal changes with fluctuations in the levels of various sensory and behavioral responses (Kenshalo, 1966; Bell & Zucker, 1971; Banarjee, 1971; Pietras & Wenzel, 1974) suggest that some element of non-specific behavioral arousal, induced by the central action of sex hormones, may underlie the expression of certain functional changes observed during the sexual cycle. We cannot, therefore, conclude that the observed fluctuations in odor detection performance are specific to the olfactory system.

This research was supported in part by Grant No. 73-2425, U.S. Air Force Office of Scientific Research. We acknowledge the technical assistance and advice of Drs. D.A. Marshall and G. Celebi.

References

Banarjee, V. Neuroendocrinol. 7: 278-290, 1971.
Bell, D.D. and Zucker, I. Physiol. Behav. 7: 27-34, 1971.
Curry, J. In: Influence of Hormones on the Nervous System, edited by D.H. Ford. Basel: Krager, 1971, pp. 255-268.
Farris, E.J. and J.O. Griffith, Jr., editors. The Rat in Laboratory Investigation. New York; Hafner Publishing Company, 1963.
Gilmore, D.P. and P.G. McDonald. Endocrinology 85: 946-948, 1969.
Guerrier, Y., R. Azemar, Ch. Morineau, J. Mirouze, and H. Martin. Olfactologia 1: 7-71, 1969.

Hashimoto, I., D.M. Hendricks and L.L. Anderson. Endocrinology 82: 333-341, 1968
Henkin, R.I. In: Biorhythms and Human Reproduction, edited by M. Ferin, F. Halberg, R.M. Richard, and L.V. Wicle. New York: J. Wiley and Sons, 1974, pp. 277-285.
Jackson, R.T. Ann. Otol. Rhinol. Laryngol. 79: 461-467, 1970.
Kawakami, M. and C.H. Sawyer. Endocrinology 65: 652-668, 1959.
Kawakami, M. and C.H. Sawyer. Endocrinology 80: 857-871, 1967.
Kenshalo, D.R. J. Appl. Physiol. 21: 1031-1039, 1966.
Koster, E.P. International rhinology, Rhinologie International 3: 57-64, 1965.
Le Magnen, J. Archiv. Sci. Physiol. 6: 295-332, 1952b.
Masmainian, M.A. and P.S. Voskanian. XX Int. Cong. Psych. Abstract 473, 1972.
Mortimer, H., R.P. Wright, C. Bachmann and J.B. Collip. Proc. Soc. Exptl. Biol. Med. 34: 535-58, 1936.
Oshima, K. and A. Gorbman. Acta Endocrinol. 62: 537-545, 1969.
Pfaff, D.W. and C. Pfaffman, Brian Res. 15: 136-156, 1969.
Pietras, R.J. and D.G. Moulton, Physiol. Behav. 12: 475-491, 1974.
Pietras, R.J. and B.M. Wenzel. Horm. Behav. (in press).
Schneider, R.A., J.P. Costiloe, R.P. Howard and S. Wolf. J. clin. Endocrinol. and Metab. 18: 379-390, 1958.
Schneider, R.A. and S. Wolf. J. Appl. Physiol. 15: 914-920, 1960.
Taylor, M.J. Laryngol. 75: 972-977, 1961.
Tucker, D.J. gen. Physiol. 46: 453-489, 1963.
Tucker, D. and L.M. Beidler, Am. J. Physiol. 187: 637 (Abst.), 1956.
Vierling, J.S. and J. Rick, J. Appl. Physiol. 22: 331-315, 1967.
Zeman, W. and R.M. Innes. Craigie's Neuroanatomy of the Rat. New York: Academic Press, 1963.

Olfaction and Aggressive Behaviour in Female Mice

Ropartz Ph. and Haug M.

(Laboratoire de Psychophysiologie,

Strasbourg, France)

In 1968, we have proved that it was possible to decrease the intensity of fighting behaviour in male mice by masking the natural scent of the opponents with perfume; it was then possible to obtain a complete disappearance of aggressive behavior by bilateral removal of the olfactory bulbs. Other authors (Mackintosh and Grant, 1966; Ropartz, 1968; Mugford and Nowell, 1970) showed that the loss of olfactory informations suppressed or, at least, reduced the intraspecific aggressive behaviour of male mice. It seemed therefore likely that olfactory cues or pheromones (Karlson and Butenandt, 1959) were involved in the aggressive behaviour of mice. Several studies suggest that at least two pheromones are involved in communication during aggressive encounters in mice. The first one is an aggression-promoting pheromone which stimulates intermale aggression and may be released by males (Haug, 1970; Lee and Brake, 1971); the second one is an aggression-inhibiting pheromone which inhihits aggressive interactions between males and females and which is released by females (Mugford and Nowell, 1971; Dixion and Mackintosh, 1971).

A few studies have stressed the lack of aggressive response to female mice when introduced in a group of males. But, by chance, we have found a new pattern of aggressive behaviour in female mice. When a female is introduced in a small group of female mice (3 or 4 subjects), the latter attacks the first one (Haug, 1972).

We have been working with Swiss female mice; at the weaning, these female mice are grouped by three in transparent cages (32 x 12 x 18 cm). Eight weeks later the tests begin. During five successive days, each group receives in its cage an intact female mouse of the same age and the same strain; the duration of the test varied from 30 min to 1 h. During the test, the cage is placed in a soundproof box with an isotropic light. When the strange female is wildly bitten during the first 30 min the latency is recorded. Then the number of attacks from the grouped females is recorded during a second period of 30 min. If the strange female is not attacked, the test is stopped after 30 min.

We have noticed that the aggressive response of the grouped female mice is faster and stronger when the strange female is a lactating one. We

411

have been able to show that this agonistic behaviour of the grouped female mice varies as a function of the stage of lactation of the intruder; it is when the mother is at its second week of suckling that it provokes the strongest aggressive response from the grouped female mice.

In another study, we have tried to investigate the determinism of this effect; the grouped female mice have been rendered anosmic by surgical ablation of their olfactory bulbs. Such anosmic female mice do not attack at all the lactating intruder. It was possible that the control of this behavioural pattern involves chemical cues. As the method used, the surgical ablation, provokes an aversive effect on aggressive behaviour (Ropartz, 1968; Rowe and Edwards, 1971), we have preferred to use another one.

We have compared (Haug, 1972) the aggressive behaviour of grouped female mice when an intact female mouse, a spayed one or a spayed female coated with urine collected from intact females is introduced in their cage. The results show that the scent of the urine of the strange female releases the attacks from other grouped female mice. Moreover, this effect disapears after ovarian ablation of the female mouse from which the urine is collected (Haug, 1972).

In another experiment (Haug, 1973), we have found that a virgin female which has been placed several days in the cage of a lactating female is more attacked by a group of female mice than a virgin female which has been placed in the cage of another virgin female.

There is at least one pheromone responsible for this agonistic pattern between female mice and this pheromone is localized in the urine of the female.

Now we are trying to identify chemically the pheromone with the help of the Centre Neurochimie de Strasbourg.

References

Dixion A.K. and Mackintosh, J.H. 1971. Anim. Behav., 19, 138-140.
Haug, M. 1970. C.R. Acad. Sc., Paris, 271, 1567-1570.
Haug, M. 1972. C.R. Acad. Sc. Paris, 275, 995-998
Haug, M. 1972. C.R. Acad. Sc. Paris, 275, 2729-2732.
Haug, M. 1973. C.R. Acad. Sc. Paris, 276, 3457-3460.
Karlson, P. and Butenandt, A. 1959. Ann. Rev. Entom., 4, 39-58.
Lee, C.T. and Brake, S.C. 1971. Psychon. Sc., 24, 209-211.
Mackintosh, J.H. and Grant, E.C. 1966. Z. Tierpsychol., 23, 584-587.
Mugford, R.A. and Nowell, N.W. 1970. Nature (London), 266, 967-968.
Mugford, R.A. and Nowell, N.W. 1971. Anim. Behav., 19, 153-155.
Ropartz, Ph. 1968. Anim. Behav., 16, 97-100.
Rowe, F.A. and Edwards, D.A. 1971. Physiol. Behav., 7, 889-892.

Olfaction and Pheromones in Animal Behaviour

Chairman's Summary

R. Bradley

Dept. Oral Biology, Univ. of Michigan,

Ann Arbor, Michigan

The behavioral effects of odors is a subject of increasing importance, not only in an academic sense but practically for pest control and animal breeding. This session reviewed two topics: (1) the effect of odors on hormone levels, and (2) the production of odors as cues to initiate or block behavior. The session opened with a review by J. LeMagnen of his work demonstrating changes in olfactory sensitivity during the menstrual cycle in women. He postulated that this effect was due to olfactory — hypothalamic loops — the olfactory input can affect the hypothalamus and the hypothalamus in turn can affect olfactory sensitivity by a neural feedback mechanism to the olfactory bulb. He also described recent experiments in which odors were shown to affect insulin release and blood glucose levels. R. Pietras (read by D.G. Moulton) described experiments which demonstrated that circulating hormones in female rats could lead to changes in olfactory sensitivity. W. Whitten reviewed the existing literature on mammalian pheremones and, in particular, the pregnancy block produced by the odor of male urine (the Bruce effect).

The discussion of this session centered on clarification of methodology and requests for further information. None of the participants could suggest a possible biological significance for the reported altered olfactory sensitivity with changing hormonal levels. It was pointed out that not only the olfactory organ is involved in the behavioral effects, but also the vomeronasal organ. Whitten was questioned on the site of production of the odor in male urine. He answered that bladder urine was just as effective as micturated urine and that the site of production was probably in an area close to Bowman's capsule.

OLFACTION AND PHEROMONES IN ANIMAL BEHAVIOUR

Chairman: J. Amoore

Primate Sexual Pheromones

Richard P. Michael, R.W. Bonsall & P. Warner

Department of Psychiatry, Emory University

School of Medicine, Atlanta, Georgia 30322, U.S.A.

Introduction

The role of olfactory mechanisms in the control of reproductive processes in lower mammals is well-established and beyond dispute. Substances with the properties of pheromones are involved, although their chemical identification in many cases lags behind the demonstration of their physiological effects. Among primates, especially the higher, anthropoid primates, we are on less firm ground: information is sparse, and several significant questions are without clear answers. This state of affairs is, perhaps, to be anticipated when one considers the greater complexity and richness of primate behaviour. The great development of neocortex and of neocortical mechanisms implies a lessened dependence on phylogenetically primitive functions, and a greater influence of individual life experiences that depend on learning and memory. Thus, primate behaviour is characterized by marked individual variations, it is much less stereotyped and repetitive than that of lower mammals, and separated from the behaviour of invertebrates by several orders of magnitude. Since the notion of a pheromone was developed in the context of insect physiology, there need be no surprise that, as originally defined, it is no longer entirely apposite.

Communication systems have developed in the course of evolution by which primates convey information about their behavioural state to conspecifics. These take the form of certain gestures and behavioural patterns, called Displays, and much work has been done by ethologists, both in identifying the motor patterns themselves, and in clarifying their information content (Altmann, 1962; van Hooff, 1962; Andrew, 1963; Michael & Zumpe, 1970; Zumpe & Michael, 1970a,b). Displays, which are visual signals, are very important in primate communication but cannot be considered further here. Along with vision, there persists another form of distance communication that depends on olfactory pathways. Among primates, the prosimians provide many examples of communication between the sexes by olfaction. They possess specialized, apocrine scent glands used for territorial marking, for self-marking and for marking each other. However, the possible role of olfaction in sexual communication in microsmatic, higher primates has only recently been investigated experimentally.

Pheromones and Female Sexual Attractiveness

Field observations have been made on the frequent scenting of the female's genital region by males in several species of macaques, and similar behaviour has been observed in the laboratory (Michael & Zumpe, 1971). Michael & Saayman (1968) found that when small doses of oestrogen were administered intravaginally they were more effective in stimulating the sexual interest of males than similar doses given subcutaneously. This pointed to an oestrogen-dependent vaginal mechanism capable of stimulating male sexual activity. Use was then made of an operant-conditioning situation in which male rhesus monkeys were required to press a lever 250 times to obtain access to a female partner. Males would not work consistently for access to an ovariectomized, untreated female, but regular, high rates of pressing occurred for access to a female injected with oestrogen. If males were made anosmic by inserting plugs into the nasal olfactory area, they failed to lever-press when the female partner was injected with oestrogen, but commenced doing so when the nasal plugs were removed. It appeared, therefore, that males were unable to detect the change in the endocrine condition of the female when anosmic but did so when their sense of smell had been restored (Michael & Keverne, 1968). The operant performance of the males depended, therefore, on the integrity of their olfactory pathways, and the data suggested that the attractiveness of the females was due to an olfactorily-acting pheromone. Ether extracts of vaginal secretions collected from oestrogen-treated females were shown to stimulate the sexual behaviour of males when applied to the area of the sexual skin of long-ovariectomized females (Keverne & Michael, 1971). In order to determine the chemical nature of the substances responsible for the powerful stimulation of male sexual behaviour, it was necessary to use extraction and fractionation procedures in conjunction with behavioural assays (Michael, Keverne & Bonsall, 1971). Ether-soluble acidic components were examined by ion-exchange chromatography (Curtis et al., 1971), and the fraction corresponding to that of short-chain fatty acids was further examined by analytical gas chromatography and mass spectrometry. By these means it was established that a mixture of the following volatile fatty acids was responsible for the sex-attractant properties of the vaginal secretions of oestrogen-treated rhesus females: acetic, propanoic, methylpropanoic, butanoic, methylbutanoic and methylpentanoic acids. A mixture of authentic acids was then made up to match the concentrations present in a pool of 24 washings collected from

418

oestrogen-treated females. The mixture also possessed powerful sex-attractant properties, and this provided further evidence that the attractant properties of vaginal secretions depended in large part on their fatty acid content (Michael et al., 1972).

To ascertain if the production of aliphatic acids in vaginal secretions was an isolated phenomenon in the rhesus monkey, we turned our attention to other primate species and conducted similar gas chromatographic studies. Figure 1 shows that the same aliphatic acids are present in much the same proportions in the vaginal secretions of both New and Old World monkeys, in other species of macaques and also in the anubis baboon: it will be observed that the patterns of acid production seem to differ somewhat for different species. However, their occurrence in primates is quite general.

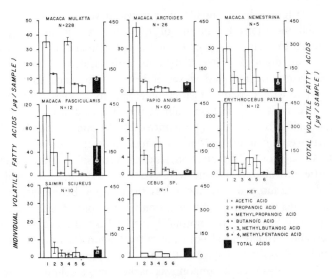

Fig. 1. Volatile fatty acids in the vaginal secretions of different primate species. Total acid contents are drawn to the same scale.

Volatile Acids in the Vaginal Secretions of Women

In a preliminary study on the cooperative wives of colleagues, we demonstrated the presence of short-chain fatty acids in vaginal secretions. To obtain quantitative data on a larger human sample we collected secretions from women attending an infertility clinic, but very few contained acids. Because the negative results from a clinic population contrasted with our preliminary findings, we felt it necessary to investigate a population of healthy women with normal menstrual cycles. University women were recruited by means of a leaflet informing them that a study on the physiology of women was being undertaken involving measurements of the biochemical composition of body fluids, and this would require participants to wear vaginal tampons. Women willing to cooperate were then contacted individually by a female psychologist, and were provided with a box containing special tampons and bottles. Identities were protected by code numbers. A month after their distribution, boxes were collected under conditions preserving confidentiality, menstruation dates were re-checked, and questions asked about the use of oral contraceptives, feminine hygiene products, and the incidence of sexual intercourse. Subjects were not aware of the type of assays being conducted, and there was no direct contact between the biochemists and subjects.

The following acceptable method was developed for collecting human vaginal secretions by means of tampons. A commercial tampon was reduced to 1 cm in length, washed in hot methanol in a Soxhlet extractor for 2 hr, dried at 110°C and hermetically sealed in a polyethylene bag. This procedure removed waxes and other extractable matter that would interfere with subsequent gas chromatography and also protected tampons from contamination. Each subject was provided with a convenient box containing 16 tampons (1 a control blank) and 15 numbered, snap-cap bottles each containing 20 ml methanol. Subjects wore each tampon in the usual way for 6-8 hr and, on removal, dropped it immediately into the bottle provided. Tampons were worn by day or night, whichever was preferred, and a day or night was then missed before inserting the next one. Using very small tampons for only 6-8 hr out of each 48 hr minimized any disturbance to the normal bacterial flora. Immediate immersion in methanol after removal both prevented bacterial action outside the body and initiated the first stage of the extraction process.

For analysis, each tampon was packed into a column and washed with methanol in chromatographic fashion. The eluate was combined with the methanol from the sample bottle, mixed with 0.1 ml N/10 sodium hydroxide and evaporated to dryness. The residue was taken up in 1 ml water, washed with 4 ml ether (to remove basic and neutral components), and the aqueous layer was acidified and extracted with 4 ml ether containing n-pentanol as a concentration marker. The extract was concentrated to 50 μl and analyzed on 10% FFAP columns in a Perkin Elmer gas chromatograph programmed from 50 - 220°C. The area of the pentanol peak and calibration constants, determined with authentic compounds, were used to calculate recoveries and amounts of acids present. Overall coefficients of variation ranged from 6% (butanoic acid) to 18% (acetic acid). Control tampons contained low levels (about 12.0 μg) of acetic acid only; this blank was not subtracted.

Results

Only when assays of all 682 samples had been completed were data grouped in relation to the stage of the menstrual cycle of the donors. Of the original 50 women, data from 3 were excluded because of irregular bleeding and unreliable menstruation data. The remaining 47 subjects (mean age, 20.4 ± 2.3 years, ± S.D.) (mean cycle length, 29.2 ± 3.8 days, ± S.D.) gave data for 86 menstrual cycles (635 samples). Figure 2 shows the

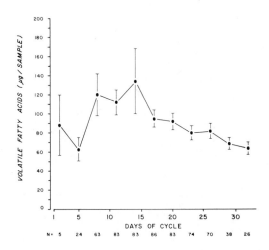

Fig. 2. Total volatile fatty acids (means ± S.E.) in the vaginal secretions of 47 women (86 cycles) during 3-day periods of the menstrual cycle. N = number of samples. The mid-cycle increase was statistically significant (P < 0.02).

421

total volatile aliphatic acid (acetic, propanoic, methylpropanoic, butanoic, methylbutanoic, methylpentanoic) content of the vaginal secretions of all 47 women arranged according to successive 3-day periods of the menstrual cycle (Day 1 being the 1st day of menstruation). Secretions showed high levels of volatile acids in the late follicular phase, and there was a progressive decline during the luteal phase of the menstrual cycle. The mid-cycle increase in acid content was significant (P <0.02) as tested by a two-way analysis of variance for non-orthogonal data: preliminary tests having shown a lack of subject-time interaction. Variance between subjects was highly significant (P < 0.001). There were 32 subjects not taking oral contraceptives (61 cycles, 449 samples) and 15 subjects using oral contraceptives (25 cycles, 186 samples). Figure 3 shows the total volatile aliphatic acid content of vaginal secretions of women not taking (non-pill cycles) and taking (pill cycles) oral contraceptives, arranged according to successive 3-day periods of the menstrual cycle. Women with normal cycles showed high levels of volatile acids in the follicular phase, and a progressive decline during the luteal phase of the cycle. The rhythmic change in content was completely absent in women using oral

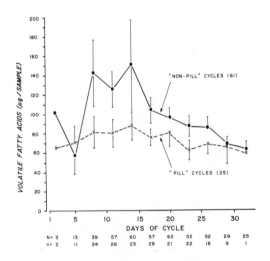

Fig. 3. Volatile fatty acid content (means ± S.E.) of the vaginal secretions of women not taking (non-pill cycles) and taking (pill cycles) oral contraceptives during successive 3-day periods of the menstural cycle. N = number of non-pill samples. n = number of pill samples. (*Science* – in press).

422

contraceptives. The mean acid content of all 'non-pill' samples was 105.6 ± 7.8 (± S.E.) μg per sample (N = 449), that of all 'pill' samples was 76.3 ± 4.3 (± S.E.) μg per sample (N = 186). The difference between these means was highly significant (P < 0.001) using a logarithmic transformation to eliminate skew and Student's t test (one-tailed).

All women produced acetic acid in some of their samples, and 34% produced acetic acid only. This acid has a relatively high olfactory threshold in the human (Amoore, 1967) when compared with methylpropanoic and methylbutanoic acids, and generally contributes little to the odour of the secretion. Thirty-six per cent of women produced samples containing acetic acid and small amounts of other volatile acids in addition: they were arbitrarily designated "non-producers". The remaining 30% of women produced samples containing acetic acid and amounts greater than 10 μg of other volatile acids in addition: they were designated "producers". The increase in acid levels during the follicular phase was very pronounced in women "producers", while the remaining women showed no evidence of rhythmicity.

These findings demonstrate that volatile aliphatic acids (copulins) are normal, physiological constituents of the vaginal secretions of healthy young women with regular menstrual cycles. Amounts of these acids increased near mid-cycle as they do in infra-human primates (Michael et al., 1972). However, in women taking oral contraceptives the increase was abolished and levels of acids were significantly lower. The precise role of olfactory mechanisms in human sexual behaviour needs to be clarified. We have scant information on the physiological importance, if any, of these volatile constituents in the human, although it is of interest that the same substances possess sex-excitant properties in other primates. Furthermore, human vaginal secretions have been demonstrated to possess this activity in cross-species experiments with rhesus monkeys (Michael, 1972). The sample population we studied fell into two groups. About one-third of the women produced high levels of acids and were responsible for the rhythmic changes in the acid content of vaginal secretions during the menstrual cycle of the group as a whole. It appeared, therefore, that some women, like the infra-human primates studied, have a mechanism for regulating the level of odoriferous substances in their vaginal secretions such that they increase during the most fertile period of the menstrual cycle. Although this was a small sample of 50 women, 57% of women "producers" and 21% of women "non-producers" reported having participated in sexual intercourse during the periods they were providing samples. This difference was significant (P = 0.017), using the Fisher exact probability test. The statistical association between sexual activity and the level of copulins in human vaginal secretions has no causal significance, but certainly suggests the need for further careful studies on a larger group of women.

We have data demonstrating that the production of copulins both in rhesus monkey and in woman depends upon the bacterial flora of the vagina, and that gonadal steroids exert their action indirectly by determining the availability of substrate (Michael et al., 1972); a rather novel symbiotic relationship. Constant vaginal douching eventually results in the destruction of the normal bacterial flora upon which the production of these volatile acids depends. The current popularity this practice enjoys is based on widely felt apprehensions about genital odours which may in fact, be entirely without justification and an artifact of our society.

We thank Dr. M. Kutner, Department of Biometry, Emory University, for statistical help, and give our thanks to the women who generously participated in the study. Work was supported by NIMH grant no. MH19506, and the Grant Foundation provided equipment.

References

Altmann, S.A. 1962. Ann. N.Y. Acad. Sci. 102, 338.

Amoore, J.E. 1967. Nature (London) 214, 1095.

Andrew, R.J. 1963. Behaviour 20, 1.

Curtis, R.F., Ballantine, J.A., Keverne, E.B., Bonsall, R.W. and Michael, R.P. 1971. Nature (London) 232, 396.

Keverne, E.B., and Michael, R.P. 1971. J. Endocrinol. 51, 313.

Michael, R.P. 1972. In "The Use of Non-Human Primates in Research on Human Reproduction", WHO Symposium (E. Diczfalusy and C.C. Standley, eds.), Acta endocrinologica Suppl. 166, 322.

Michael, R.P., and Keverne, E.B. 1968. Nature (London) 218, 746.

Michael, R.P., and Saayman, G. 1968. J. Endocrinol. 41, 231.

Michael, R.P., and Zumpe, D. 1970. Behaviour 36, 168.

Michael, R.P., and Zumpe, D. 1971. In "Comparative Reproduction of Non-human Primates" (E.S.E. Hafez, ed.), p. 205. Thomas, Springfield, Illinois.

Michael, R.P., Bonsall, R.W., Warner, P., Science – in press.

Michael, R.P., Keverne, E.B., and Bonsall, R.W. 1971. Science 172, 964.

Michael, R.P., Zumpe, D., Keverne, E.B., and Bonsall, R.W. 1972. Recent Progr. Horm. Res. 28, 665.

van Hooff, J.A.R. 1962. Symp. Zool. Soc. London 8, 97.

Zumpe, D., and Michael, R.P. 1970a. Anim. Behav. 18, 293.

Zumpe, D., and Michael, R.P. 1970b. Anim. Behav. 18, 11.

Activation of Territorial Behaviour in the Rabbit,
Oryctolagus cuniculus, by Stimulation with its own
Chin Gland Secretion

R. Mykytowycz

Division of Wildlife Research

CSIRO, Canberra, A.C.T.

Territoriality, because of the overt aggression associated with it, is the most conspicuous of the many ways in which animals utilise space.

Odours derived from faeces, urine and special externally secreting skin glands play an important role in the demarcation of territories (Mykytowycz, 1974).

In discussions of scent marking and of territoriality in the life of animals, their function in repelling con-specifics is usually emphasised. There are in fact many reports based on experimental and circumstantial evidence to support this view. On the other hand the confidence-giving effect of familiar space is much less frequently mentioned and although there is circumstantial evidence to support this idea, there has been no publication of experimental systematically collected data to demonstrate it.

In a series of earlier publications it has been shown that both sexes of the European wild rabbit, *Oryctolagus cuniculus,* display very strong territoriality. An individual always attempts to exclude intruders from the space it owns and inhibits the behaviour of subordinate members of its own social group. Aggression at different levels of intensity, not infrequently resulting in the killing of the intruder, accompanies the establishment of territorial claims. The European rabbit, to mark its territory, uses secretions from the anal and the chin glands (Mykytowycz, 1968). Here the results are reported on an experiment designed to show a confidence-giving effect of the chin gland secretion.

Materials and Methods

Animals

Two groups of fully grown, adult wild rabbits were used. The first consisted of 25 males and 25 females which were born and reared in the laboratory. The second comprised 106 males and 100 females caught in the field and used in experiments within 24 hours.

Testing Procedures

The tests, each lasting 10 minutes, were conducted in an indoor pen, 5m² in area. The pen was of a standard shape and size identical to those in which all animals were normally housed.

Before each test the floor of the pen was covered with a clean sheet of paper on which eight filter pads saturated with the chin gland secretion were scattered randomly.

The rabbits were introduced in pairs in different combinations. Each pair of individuals was tested twice; once in the presence of the chin secretion of one animal and again in the presence of the secretion from the other.

When tested in the presence of its own secretion the rabbit is referred to as being in the "home situation" or as the "home" or "donor" animal, and when in the opposite situation as the "away" animal or as being in the "away situation".

As chin secretion could not be obtained from laboratory reared females they could not be tested in the presence of their own secretions. Instead they were tested in the presence of secretions obtained from their permanent mates of the opposite sex.

An observer noted the incidence and duration of selected forms of behaviour which reflect the degree of confidence of the animals, and recorded these on a multi-channel event recorder.

Preparation of Stimuli

To prepare the stimuli the chin glands of individual rabbits were appropriately massaged to cause drops of secretion to appear on the surface of the skin through the secretory pores. These drops were collected into glass capillaries. Shortly before the tests the secretions were diluted (1:1000) with distilled water and two drops were placed on each of the eight filter papers giving approximately 0.5 mg of secretion per test.

Statistical Treatment of Data

The tests were designed to allow comparisons to be made between the performances of the same individual in the "home" and "away" situations. The differences between the logarithms of the "home" and "away" values for the various forms of behaviour were found to yield the nearest to normal distributions and hence were used in the analysis. Since the differences between logarithms corresponds to a ratio, the values shown in the first 12 items listed in Table 1 represent percentage rather than actual differences. The levels of significance shown adjacent to the values in the Table refer to differences between "away" and "home" situations and those shown in the inter-column spaces relate to differences between male and female series.

Results

Behaviour of Rabbits in the Test Situation

The differences between the behaviour of the "home" and "away" animals were sufficiently distinct to allow an observer uninformed about the experimental manipulation to pick the animal which was in the presence of its own chin secretion from the one which was not.

In the presence of its own secretion a rabbit usually initiated contact with the other individual. It seemed that right from the beginning the "home" animal was aware of the presence of its own marking. In 153 tests involving males, contact between the two rabbits was made by the "donor" animal on 85 per cent of occasions. The "donor" also initiated aggression and usually won in cases when a physical contest occurred.

In the preliminary stages of a physical contest, circling behaviour often occurred and the tail was curled back to expose the anal and inguinal gland areas. This probably assists the dissemination of the animal's own odour (Mykytowycz, 1974). When physical skirmishing was absent the movements of an individual in the presence of its own secretion were uninhibited, ranging over the whole of the pen. Thus, while the "home" animal appeared confident, the "away" rabbit was inhibited even in confrontation with an individual which it had dominated earlier in its own "home" situation. Although the "away" rabbit explored the pen it moved very cautiously and avoided the "home" animal, or when approached by it attempted to escape from the pen or adopted a submissive posture. The values shown in Table 1 reflect the importance of the different components of behaviour mentioned above.

Approaches

From Table 1 it is clear that for all animals the incidence and duration of approaches were consistently highly when they were in the "home" rather than in the "away" situation. The differences were statistically highly significant (P < 0.001). The differences between "home" and "away" situations were particularly pronounced in the case of the field males (incidence 412%; duration 493%) and least in field females (78% for both measures). This difference between field males and field females was highly significant (P<0.001).

Following and Chasing

Following and chasing are two distinct forms of behaviour, the first

Table 1. – Differences in the Incidence (inc.) and Duration (dur.) of Some Forms of Behaviour of the Wild Rabbits in the Presence and Absence of their own Chin Gland Secretion (Explanation of values, see Statistical Treatment. ♀ (♂)= females in presence of their mates' secretion.

		CAPTIVE RABBITS					FIELD-CAUGHT			
		% of tests	♂:♂ n=100		♀(♂):♀ n=100	% of tests	♂:♂ n=53		♀:♀ n=51	% of tests
Approaches	inc.	99	159***	ns	161***	100	412***	***	78***	100
	dur.		219***	ns	202***		493***	***	78***	
Following & chasing	inc.	67	175***	ns	191***	73	479***	**	151***	84
	dur.		208***	ns	216***		4316***	***	696***	
Aggression	inc.	54	111***	*	233***	56	544***	***	93**	78
	dur.		126***	*	246***		594***	***	95**	
Examination	inc.	91	148***	*	240***	98	406***	**	146***	100
	dur.		245***	*	466***		914***	***	260***	
Movement	dur.	100	127***	ns	165***	100	112***	ns	77***	100
Examination stimuli	inc.	89	-4^{ns}	ns	14^{ns}	83	3^{ns}	ns	4^{ns}	75
	dur.		-4^{ns}	ns	-15^{ns}		10^{ns}	ns	-4^{ns}	
Stamping	inc.	74	-12^{ns}	ns	1^{ns}	47	-56***	ns	-44*	63
Dominance	Won	77***		ns		83***		ns		70***
	Lost	15				9				20***

428

being a component of investigative and the second of agonistic behaviour. For the purposes of the present paper, however, they are regarded as one because both reflect the confidence of the animal and in any case are often difficult to separate.

After initial contact during which mutual identification took place, the subordinate individual commonly moved away and was followed by the confidently-behaving animal. Following often developed into chasing as a preliminary to aggression.

The incidence and duration of following and chasing were significantly higher for rabbits in the "home" situation (P<0.001) particularly in the field males (479% and 4316% respectively). The differences between males and females were highly significant for the field animals (P<0.001) but not significant for laboratory bred rabbits.

Aggression

Actual aggression involving biting, scratching and nipping occurred in the majority of tests and was more common in those involving females. Incidence and duration of aggression were again significantly higher for rabbits in the "home" as compared to the "away" situations. Differences between the sexes were significant in relation to duration and incidence (P 0.05 for laboratory and P<0.001 for field animals. Differences between rabbits in "home" and "away" situations were greatest in the field males (544% and 594%) and least in field females (93% and 95%).

Examination of other Animals

All groups of rabbits sniffed their test partners more frequently and for a longer time when in the "home" as compared with the "away" situation (P<0.001) and this was particularly evident in the field-caught males and laboratory does. There were significant differences between males and females in this respect.

Movement

All rabbits spent some time moving around the test pen but in the "home" situation they did this for a much longer period (P<0.001).

Examination of Stimuli

It is remarkable that while all other forms of behavior occurred most frequently and for the longer time in the "home" situation, there were no differences between the rabbits in each situation in respect of sniffing at the stimulus-pads.

Stamping

Stamping with the hind feet is involved in the communication of various messages, and there is no doubt that it is used by rabbits in some behavioral situations as an expression of fear and as a warning signal (Mykytowycz and Hesterman, in press).

While all the other forms of behavior, which obviously reflected the confidence of an individual, were displayed most frequently and for the longest time in the presence of a rabbit's own chin gland secretion, the frequency of stamping did not follow the same pattern. Thus the field rabbits stamped significantly less frequently in the presence of their own chin secretion (males $P < 0.001$; females $P < 0.05$). In the laboratory-reared animals the males showed the same tendency, but the differences did not reach statistically significant levels. The differences between males and females were also not significant.

Acquisition of Dominance by Rabbits in the Presence of their own or Familiar Chin Gland Secretion

In tests in which actual fighting occurred assessment of dominance could be made on the outcome of the physical contests. In the other cases, components of confident behavior (initiation of contact, following, chasing, uninhibited exploration) were used as an indication of victory.

The results shown in Table 1 indicate that in all groups, the rabbits which were accompanied by their own or familiar chin gland secretions emerged as dominants in a high proportion of tests ($P < 0.001$). In a small number of cases no clear-cut dominance was displayed by either animal, and hence there is a discrepancy between the total number of tests and the sum of victories and losses.

Discussion

In an earlier experiment on aggression in rabbits, it was shown that when an individual is introduced into a pen which has been permanently occupied by another one, irrespective of the sex of the animals involved, a fight will develop from which the permanent occupier will emerge victorious. Reversal of the positions of the two individuals reverses the outcome of the contest (Mykytowycz and Hesterman, in press).

The type of rabbits and the experimental procedures used during this earlier experiment and in the present one were identical. The incidence of acquisition of dominance by the "home" individual in the two experiments was also very similar — 62 to 78 per cent in the earlier and 69 to 83 per cent in the present study. Thus there seems to be no doubt that the introduction of a rabbit's own chin secretion or that of a member of its social group turns the given space into a "home".

The advantage to the "home" animal arising from the presence of its own secretion would be even more pronounced under natural conditions. The experimental situation forced a physical contest which often favoured the stronger individual, rather than the "home" rabbit.

The results reported here demonstrate clearly that during the first few minutes in a new situation an individual assesses its territorial position and that its own or its partner's chin secretion is an important factor in this process. The correct evaluation by the rabbit of its relationship to the space in which it lives or finds itself, regulates its relationship towards conspecifics. This minimises aggression and inhibits or stimulates reproduction.

Elsewhere it has been stressed that the behaviour of all animals is such that they saturate their environment with their own smell. This affects the mood of an individual, that is, its preparedness to engage in activities which would be inhibited in strange surroundings (Mykytowycz, 1973).

The comforting effect of an animal's own odour has been stressed before and sometimes utilised for man's benefit but never experimentally demonstrated (Johnson, 1973). It has been possible to do this in the case of the European wild rabbit. Brief information on the comfort-giving effect of the odour of the anal gland secretion has been given elsewhere (Mykytowycz, 1973) and a further report comparing the effects of the anal, chin, and inguinal secretions and urine will be published soon. It is interesting to see that the different sources of odour are not equally effective in influencing the territorial behaviour.

In the course of the present experiment the field-caught rabbits responded more strongly than the laboratory ones to the presence of their own chin secretion. This observation as well as the larger amount of chin secretions obtained from field-caught rabbits clearly reflects their higher level of territoriality. Earlier it has been demonstrated that the size and secretory activities of odour-producing skin glands is directly related to the territorial activities of an individual (Mykytowcyz and Dudzinski, 1966).

The chin gland differs from the inguinal and anal glands in not having an odour detectable by the human nose. The "rabbity" odours of the inguinal and anal secretions are associated with the lipid components. The chin gland secretion consists largely of proteins and carbohydrate substances with very low volatility (Goodrich and Mykytowycz, 1972). However, the results presented here show conclusively that the pure secretion of the chin gland obtained directly from the secretory pores is detected olfactorily by the rabbits. There are also other reports of the likely implication of macro-molecules in chemosensory-based communication in mammals (Berüter, Beauchamp & Muetterties, 1973). In addition to the evidence given above we are in possession of data showing changes in the heart-rate rhythm of rabbits exposed to olfactory stimulation by chin gland secretion (Hesterman *et al* in prep.).

Further studies aimed at elucidating the odoriferous properties of the chin gland secretion are in progress. Chemical and ethological techniques are being combined to explain the full role of chin gland secretion in rabbit behaviour.

Summary

Experimental proof is given that in the European wild rabbit, *Oryctolagus cuniculus,* the secretion from its own or sexual partner's chin gland reinforces territorial behaviour.

When two rabbits, unknown to one another were introduced into a neutral pen the one whose chin gland secretion was also experimentally placed there dominated the other one (P O.001). The confidence of an individual and the outcome of the confrontation reversed on reversal of the source of the gland secretion.

Acknowledgements

Thanks are due to my colleagues M.L. Dudzinski, B.V. Fennessy, S. Gambale, E.R. Hesterman and D. Wood for their assistance in the preparation of this paper.

References

Berüter, J., G.K. Beauchamp, and E.L. Muetterties (1973). Biochem. Biophys. Res. Comm. 53: 264-271.
Goodrich, B.S. and R. Mykytowycz (1972). J. Mammal. 53: 540-548.
Johnson, R.P. (1973). − Anim. Behav. 21: 521-535.
Mykytowycz, R. (1968) − Sci. Amer. 218: 116-126.
Mykytowycz, R. (1973) − J. Reprod. Fert., Suppl. 19: 433-446.
Mykytowycz, R. (1974) − In "Pheromones". Ed. M.C. Birch. Pp 327-343. North-Holland Publ. Co.
Mykytowycz, R. and M.L. Dudzinski (1966) − CSIRO Wildl. Res. 11: 31-47.
Mykytowcz, R. and E.R. Hesterman (in press) − Behaviour.

Olfaction and Pheromones in Animal Behaviour

Chairman's Summary

John E. Amoore, Western Regional Research Laboratory, U.S.D.A., Berkeley, CA 94710, U.S.A.

Discussion of Dr. Michael's paper

Henderson: I am checking human male cyclic activity at the moment. I have found so far that women on the pill change the male parameters that I am looking at. Have you got data on the male monkey behaviour towards females on the pill?

Michael: We have done work on a variety of oral contraceptive compounds, and they do depress the behaviour of the male. This is not immediately relevant to the problem of pheromone production, because if you treat female monkeys with these oral contraceptives, they become less receptive, so that they increase the number of refusals they make. A female can signal that she is not interested in other ways, than just by a pheromone mechanism.

Boeckh: Is anything known about the type of bacteria, because if it is known you could possibly guess at the substrate they are feeding on?

Michael: You can produce these substances from soya broth, beef broth, or from previously sterilized vaginal secretion, if you inoculate. There are about 8 or 9 species of *Lactobacilli* present, as well as *Staphylococcus albus.* We were disappointed because no one of these bacteria would produce this spectrum of acids on its own. A mixture of bacteria is acting cooperatively.

Amoore: The proportions of those fatty acids, that you report in the case of the human vaginal secretions, are such that isovaleric acid would make by far the largest contribution to the odor.

Schafer: Have you tried looking at under-arm secretions? It may be some sort of non-specific secretion that you are looking at.

Michael: We tried to collect underarm secretions in humans, by wearing a pad. However, we could not extract anything out of the pad.

Stuart: Were the male monkeys you were using naive? There may be a distinction between an olfactory cue that a monkey can learn to associate with sex, and a pheromone itself in a classical sense.

Michael: These monkeys were all caught in the jungle when mature.

Stuart: You would agree then that these monkeys might have associated the odors with females that they may have courted before, and been repulsed at certain non-receptive stages?

Michael: I think that learning must play a very big part.

Stuart: Then I would say that this is not so much a pheromone in the classical sense, but rather an olfactory cue associated with the mating opportunity.

Michael: Why would you say that it isn't a pheromone? In Karlson and Butenandt's definition, a pheromone is a substance, externally secreted by one individual, with an effect on the behaviour or physiology of a conspecific. They never said anything about an immediate response, a specific molecule, or a specific receptor.

Stuart: Acetic acid is produced by the bacteria, not by the monkey. So if it were a pheromone, it would have to produce an effect on another bacterium. Chemical releasers were described by Tinbergen and Lorentz, who definitely implied that there is no gross learning effect in the reactions; that they are, in the old terminology, innate.

Michael: Taking the original definition, it seems to me that this category of substances quite suitably and conveniently falls under the rubric of pheromone. I think that since the meaning of the word was established in the '50s it is quite legitimate to expand the concept in the '70s.

Discussion of Dr. Mykytowycz's paper

Michell: When the male rabbit mounts the female, presumably the chin gland leaves her with his scent. Does that then inhibit other males from approaching her?

Mykytowycz: I don't think so. We saw one male interfering with another one of the course of copulation. In a population of rabbits, each one knows where it fits. There is very little hostility between the 1st. ranker and 3rd. or 4th. rankers, because the lower ranking animal knows that it won't win. But the 2nd. ranker is always in competition with the dominant individual.

Bardach: What is the spacing of marked sites in a field territory?

Mykytowycz: Actually you can say it is saturated with the marking. In a pen like this, it would be a matter of a few centimeters.

Bardach: Would you think that the proteinaceous nature of the material in the secretion and its low volatility has something to do with the persistence of the marked sites?

Mykytowycz: Certainly. It is quite likely that slow bacterial decomposition may have something to do with the persistence of the odor within their territory. This way of marking may be an advantage. By the way, this is not the only means of territorial marking. The anal gland also plays a very important role.

Tucker: Can humans smell the chin gland secretion?

Mykytowycz: No. I cannot, and other people I know cannot smell it.

Tucker: Have you tried diluting it to see how sensitive the rabbits are to it?

Mykytowycz: We used in the course of this experiment just a few drops at a time of a 1 in 1000 dilution. So it must be a very potent signal for the rabbit.

Tucker: In your observations of their behaviour, do they seem to be just sniffing, or do they perhaps touch the marked site?

Mykytowycz: You think perhaps that licking may be involved? Well, it seems to be just straight olfaction.

SUMMARY OF GENERAL DISCUSSION OF OLFACTION

General Discussion of Olfaction

Discussion centered on pheromones in both insects and mammals and on olfactory coding. An influential concept in interpretations of the properties of insect odor receptors has been the "specialist" "generalist" dichotomy as defined by Schneider and his co-workers. In summarizing the development of this concept Dr. Schneider recalled that when they first started recording from single cells in receptors of male silkmoths they found that about 50 per cent of the receptors were tuned to the female sex attractant while the remaining receptors responded to a variety of different compounds. The first type they called "specialists" and the second "generalists". But at this time they did not have a full picture of the response characteristics of the receptors studied.

Recent evidence, from other species of insects, suggests a more complex situation. Dr. den Otter questioned the validity of the "generalist" — "specialist" distinction citing his own work on a fruit tortrix moth. In the male, the same receptor cells respond to the "cis" and "trans" isomers of tetradecenyl acetate. The "cis" isomer is a component of the sex attractant and excites the cells while the "trans" isomers are strongly inhibitory and elicit a completely different behavioural response. Dr. Halpern then suggested that the choice of the silkmoth for the initial experiments was a historical accident since this species has been cultured for several thousand years. Perhaps domestication had led to the loss not only of flight but also of all pheromones except one. Responding to these comments Dr. Schneider emphasized that every odor receptor is, in a sense, a "specialist" having its own response spectrum. While he felt that the term had served a useful purpose he recommended that we do not use it in its earlier sense any more but rather define in every case what excites a cell. To reach a full definition we may need to specify the intensity curve for each compound tested.

Evidence of the complexity of response to odors is also emerging from behavioral studies. Dr. Schneider cited the example of the gypsy moth. The male is attracted to the female by the pheromone "disparlure". But a precursor also exists which in field trials is about 1,000-10,000 times less effective than disparlure. If combined with disparlure, however, it repels the males from the females. In the case of the silkmoth, bombykol — the main sex attractant — is not the only compound which can attract the male. As Dr. Pfaffmann pointed out, this suggests that once the neural channel associated with bombykol receptor has been excited the same behavior results regardless of the stimulus. The alternative assumption — that a non-bombykol stimulus acts through a separate channel — involves a more complex situation for the brain.

Further discussion emphasized the importance of context in determining which response an insect will make to the same odor. How much more important is this likely to be when we turn to mammalian behavior? Experience is also a significant factor in considering experiments with mammalian pheromones. As Dr. Michael pointed out in responding to comments on his work by Dr. Stuart, it is much more difficult to find a naive monkey than a naive moth. But could one, in any case, work with a naive monkey? Harlow found that a monkey requires much experience to develop and maintain its behavior and animals raised in isolation cannot distinguish between themselves and other members of the same species. For example, they are as likely to bite themselves when angry as others. Developmental experience and learning are also essential for mating. In this context a pheromone may act as a conditional stimulus. For example, the male rhesus monkey may learn to associate higher fatty acids present in the vagina with the attractiveness of the female. This, however, does not necessarily invalidate the concept that it is a pheromone.

The remainder of the discussion centered on olfactory coding in vertebrates. According to work outlined by Dr. MacLeod the frog olfactory receptors are broadly tuned. Maximum selectivity in a single cell is about 10% (i.e., the cell responded to only 10% of the many stimuli that were presented). The counterpart of this poor selectivity is the large number of receptors and the dependence on spatial coding. He believes that the efficiency of the topographic coding is such that there is no need for temporal patterning to give a more precise definition of either intensity or quality.

The discussion then turned to the relation between the time required for an animal to identify a stimulus and the information generated by the receptors in that time. It was generally agreed that for a mammal this time is considerably less than a second (Dr. Laing cited his own frame-by-frame analysis of rats responding to odors to support this point) and may well be in the order of 100-200 m.sec. For frogs, no data is available but the resting firing frequency of the receptors is low and even maximum frequencies during odor stimulation are no more than 30 spikes/sec. Thus hardly any change in frequency could occur if reaction time is assumed to be in the order of 100 m. sec. According to Dr. MacLeod this again supports the view that temporal patterning would not be an effective coding mechanism. However, it is conceivable that an amphibian takes significantly longer than a mammal to respond to an odor.

Dr. Miller questioned whether there is any real difference between intensity and quality coding. In reply, Dr. Schneider stressed that each

440

sensory channel has a certain destination in the brain. The meaning of the message lies in this connection. When activated it elicits a certain behavior, etc. This is quality coding. For Dr. MacLeod the distinction lay in the comparison of the single unit, which when taken alone codes only intensity, and all units which — when taken simultaneously — convey two kinds of message: the overall increase in activity related to intensity coding and the spatial patterning that codes quality.

At the close of the discussion Dr. Garcia inserted a recommendation that workers in olfaction consider the use of radiation in probing the properties of the olfactory receptors. It produces a punctuate stimulus with a very brief half life lacking the lag in delivery time associated with odor stimulation. Using this tool it should be possible to demonstrate more clearly that the olfactory receptors can follow a flickering stimulus. Rates of at least 10/sec are attainable. Dr. Moulton cautioned against any assumption that odors cannot be punctately delivered. Using a vacuum surround and concentric pipettes one can restrict an odor pulse to a circular area of 450μ in diameter.

D.G. Moulton, Monell Chemical Senses Center and Department of Physiology, University of Pennsylvania, and Veterans Administration Hospital, Philadelphia, Pennsylvania, 19104, U.S.A.

SYMPOSIUM BANQUET

Held in the Great Hall, National Gallery of Victoria
October 31st, 1974

The Art of the Australian Winemaker

Great Western Imperial Brut Champagne
Wynns Romalo Vintage Champagne — Brut de Brut 1966
Leasingham 1974 Bin 7 Rhine Riesling
Lake's Folly 1974 Chardonnay
Penfolds 1955 Grange Hermitage
Wynns Coonawarra Estate 1966 Cabernet Sauvignon
Seppeltsfield Mt. Rufus Tawny Port

Acknowledgements

Lake's Folly
Penfolds Wines
B. Seppelt & Sons Ltd.
Stanley Wine Company
Wynn Winegrowers Ltd.

Carlton & United Breweries
The Australian Dairy Produce Board

SUBJECT INDEX

Note: Numbers in italics refer to pages on which tables or figures appear

A

Acanthopleura granulata, amino acid release, *127*
Acetazolamide
  CO_2 receptor blockade in honeybee, 207-208
  effect on insect CO_2 response, 196
Acetic acid
  carp, response to, 3
  chorda tympani responses in mammals, 6
  nerve fiber response in rat, *63*
  neural response in fetal and adult sheep, 92-95
  response
    in the chicken, 4, 5
    in pigeons, 4
    in the puffer, 148
  taste cell response in frog, 83
  in vaginal secretions, 418, 422
Acids
  bullhead and catfish response to, 3
  carp, response to, 3
  detection in hummingbirds, 5
  depolarizing responses in frog taste cells, 83
  effect on membrane potentials, 77, *78*
  frog taste sensitivity to, 4
  pontine taste area response in rats, 203, *204-205*
  response
    in fish, *71-72*, *72*
    in insects, *72-73*
    in invertebrates, 71
    in man, *74-75*
    in rat, *73-74*
Acid response
  in hamsters, 9
  lack of specificity, 76
  stimulating effects in the puffer, 148
  taste responsiveness in regenerating taste buds, 99-102
Acid mucopolysaccharide, in pores of Locusta palps, 228
Acropora muricata, amino acid release, *127*
Adrenal glomerulosa, changes in Na deficiency, 253
Adrenalectomy, and Na intake in sheep and rat, 257
Adrenocorticotrophic hormone, neuroendocrine release, 273

Aiptasia anulata, see Anemone
Alanine, range of taste response, 158, *159*
Alcohols, *see also* Aliphatic alcohols
  chemoreception by the fly, 195
  organoleptic testing in man, 313, *315*
Alcohol-terpene, receptor spectra in cockroach, 239, 241-244
Aldehydes
  chemoreception by the fly, 195
  organoleptic testing in man, 314, *315*
Aldosterone
  action of Na-K ATP'ase, 261
  changes in Na deficiency, 253, 259
  neuroendocrine release, 273
  presence in vertebrates, 256
Algae, complex chemical signals, 128
Aliphatic alcohols, receptor spectra in cockroach, *241*
Alliesthesia, 56
Alosa pseudoharengus, chemical signals, *125*
Amines
  organoleptic testing in man, 313, *315*
  receptor spectra in cockroach, 125, *241-244*
Amino acids
  bullhead response to, 4
  carp response to, 4
  catfish response to, 4
  chemical attraction
    in eels, 156
    in flounder, hagfish, silversides, 154
  effect in *Aplysia* chemoreception, 135-136
  feeding responses in invertebrates and fish, 155
  gustatory response in catfish, 157
  olfactory stimuli in fish, 157
  pontine taste area response in rats, *204*
  relative prevalence in aquatic animals, 125-126
  release by aquatic invertebrates, 126, *127*
  responses in hamsters, 9
  sensitivities in aquatic animals, 128
  single fibre response in the puffer, *150*, 151
  specificity of functional groups in catfish, 158, *159-160*
  stimulating effect in the puffer, 149
  stimuli in *Homarus*, 145

445

447

A 5
B 6
C 7
D 8
E 9
F 0
G 1
H 2
I 3
J 4